AF618716

# Lattice Boltzmann Methodology for Single-Phase and Multiphase Nanoparticle Modeling

Hee Min Lee • Joon Sang Lee

# Lattice Boltzmann Methodology for Single-Phase and Multiphase Nanoparticle Modeling

Hee Min Lee
Department of Mechanical Engineering
Yonsei University
Seoul, Korea (Republic of)

Joon Sang Lee
Department of Mechanical Engineering
Yonsei University
Seoul, Korea (Republic of)

ISBN 978-981-95-9116-9 ISBN 978-981-95-9117-6 (eBook)
https://doi.org/10.1007/978-981-95-9117-6



This Springer imprint is published by the registered company Springer Nature Singapore Pte Ltd.
The registered company address is: 152 Beach Road, #21-01/04 Gateway East, Singapore 189721, Singapore

*This book is dedicated to all members of the Multi-scale Fluid Dynamics Lab.*

# Preface

Advances in modern manufacturing, energy systems, and biomedical technologies increasingly rely on precise control of multiphase flows and nanoparticle transport. Processes such as inkjet printing, droplet evaporation, porous-media infiltration, and colloidal self-assembly are governed by complex interactions among liquid, gas, and solid phases across multiple spatial and temporal scales. Accurately modeling these phenomena remains a major challenge, particularly when interfacial dynamics, particle–particle interactions, and phase change processes are strongly coupled.

The lattice Boltzmann method (LBM) has gained considerable attention as an effective numerical framework for addressing such challenges. Unlike conventional computational fluid dynamics approaches based on the direct solution of the Navier–Stokes equations, LBM adopts a mesoscopic kinetic perspective that enables natural treatment of complex interfaces, evolving contact lines, and multiphysics coupling. Its local and explicit algorithmic structure further makes LBM highly amenable to large-scale parallelization, providing a strong foundation for high-performance simulations on modern GPU architectures.

This book aims to provide a comprehensive and application-driven introduction to the LBM for simulating single-phase and multiphase flows, particle dynamics, and evaporation-induced transport phenomena, with particular emphasis on inkjet printing and nanofluid systems. Chapters 1–3 establish the theoretical and numerical foundations of LBM, covering single-phase formulations and widely used multiphase models, including the color gradient, and pseudopotential approaches. Key modeling considerations such as numerical stability, density and viscosity contrasts, and wetting behavior are discussed in detail.

Chapters 4 and 5 extend the framework from fluid-only systems to particle-laden processes and their technological context. Chapter 4 introduces particle dynamics models, including the smoothed profile method, immersed boundary techniques, and nanoparticle interaction models based on the Derjaguin, Landau, Verwey, and Overbeek (DLVO) theory and Langevin dynamics, enabling systematic investigation of particle transport, aggregation, and deposition. Chapter 5 serves as a conceptual and technological bridge between numerical methodology and industrial application by providing a comprehensive overview of inkjet printing processes. It discusses the physical principles of drop-on-demand inkjet printing and the fluid dynamic constraints governing stable droplet generation, jetting regimes, and deposition accuracy, thereby motivating the detailed numerical studies presented in the later chapters.

Chapters 6–10 are devoted to application-driven studies that demonstrate how the developed LB frameworks can be applied to realistic engineering problems dominated by interfacial physics and particle transport. These chapters cover surfactant-covered interfacial dynamics in complex multiphase systems, gas–liquid transport in porous media relevant to energy storage technologies, particle-laden drop-on-demand inkjet printing, and evaporation-driven particle deposition in sessile droplets.

Throughout the text, emphasis is placed on the physical interpretation of numerical models and the connection between simulation parameters and real-world processes. Exercises are provided at the end of each chapter to reinforce theoretical understanding and support independent study. Implementation aspects relevant to efficient computation, including GPU-oriented considerations, are also addressed where appropriate.

This book is intended for graduate students, researchers, and engineers working in computational fluid dynamics, multiphase flows, and advanced manufacturing technologies. It is hoped that the material presented will serve both as an educational resource for readers new to the lattice Boltzmann method and as a practical reference for those seeking to apply LBM to complex multiphase and particle-laden systems.

Seoul, South Korea Hee Min Lee
Seoul, South Korea Joon Sang Lee

# Acknowledgements

This work was supported by the National Research Foundation of Korea (NRF) grant funded by the Korea Government (MSIT) (RS-2022-NR070832).

# Contents

# Chapter 1
# Introduction

## 1.1 Background and Motivation

Multiphase flows, nanoparticle transport, and inkjet-based droplet manipulation constitute fundamental processes across a wide range of scientific and industrial applications. The interaction among liquid, gas, and solid phases governs phenomena such as droplet formation, spreading, evaporation, and particle deposition, all of which are central to advanced manufacturing and material-processing technologies. In particular, the ability to accurately model the dynamics of fluid interfaces and suspended nanoparticles has become increasingly important as product scales shrink and functional requirements grow more demanding.

Nanofluids and inkjet printing technologies exemplify this trend. In high-resolution OLED and QD-OLED fabrication, inkjet printing enables precise material placement, high pattern fidelity, and efficient material utilization, forming a core component of next-generation display manufacturing. Beyond display technologies, nanoparticle-laden droplets are widely employed in biotechnology, additive manufacturing, and energy-related fields, including biosensor fabrication, solar cells, porous-media transport, and the development of advanced functional materials (Fig. 1.1). In these applications, nanofluids offer tunable thermal, electrical, and rheological properties, while inkjet processes provide scalable and contact-free deposition capabilities. As device architectures become more heterogeneous and microstructured, the underlying physics governing droplet–substrate interaction, nanoparticle assembly, and interfacial behavior requires increasingly refined numerical analysis.

Despite substantial progress in computational fluid dynamics (CFD), conventional Navier–Stokes-based methods face intrinsic limitations when applied to such complex multiphase and particle-laden systems. Explicit interface tracking or reconstruction often leads to numerical diffusion, geometric complexity, and stability issues, especially when interfaces undergo topological changes such as breakup,

 
H. M. Lee, J. S. Lee, *Lattice Boltzmann Methodology for Single-Phase and Multiphase Nanoparticle Modeling*,
https://doi.org/10.1007/978-981-95-9117-6_1

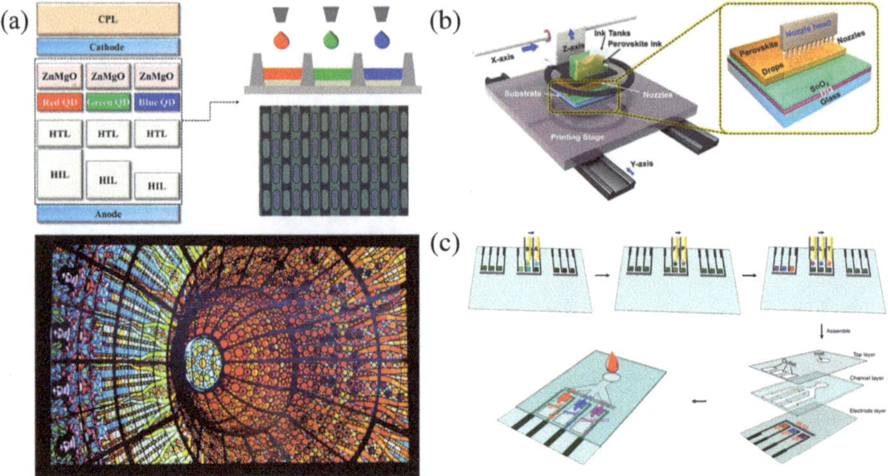


**Fig. 1.1** Applications of inkjet printing technology. Fabrication of (**a**) AM QD-LED display panels [1], (**b**) solar cell panels [2], and (**c**) multiplexed biosensors [3] using inkjet printing processes

coalescence, or wetting transitions. Moreover, capturing interfacial forces, colloidal interactions, evaporation-driven flows, and dynamic contact-line behavior demands finely resolved grids and sophisticated coupling strategies that impose significant computational overhead. These challenges become more pronounced when simulating multiscale particle dynamics or irregular microstructured surfaces. Additionally, traditional CFD approaches are not naturally aligned with large-scale parallelization, making it difficult to achieve the computational efficiency required for industrial-scale simulations.

These considerations highlight the need for a numerical framework that can robustly and efficiently handle complex interfacial physics, nanoparticle dynamics, and high-throughput simulation demands. The lattice Boltzmann method (LBM), rooted in a mesoscopic description of fluid flow, offers a flexible and computationally efficient alternative particularly suited for multiphase systems and large-scale parallel computation. Its intuitive treatment of interfaces, compatibility with particle-level models, and inherent alignment with GPU-based high-performance computing create a compelling foundation for next-generation simulation tools. This motivation forms the basis of the present textbook, which aims to provide a comprehensive and practical understanding of LBM for modeling multiphase flows, nanoparticle transport, and inkjet printing phenomena.

**Exercise 1.1 Nanoparticles in Inkjet Printing Systems**
Discuss how these phenomena become more complex when they are coupled with nanoparticle transport, and explain the mechanisms through which droplets and suspended particles interact.

**Solution**

Inkjet droplets containing nanoparticles exhibit significantly more complex behavior than pure-fluid droplets due to the following reasons:

1. Internal flow and particle coupling
   - Evaporation enhances capillary-driven internal circulation toward the contact line.
   - Suspended nanoparticles are transported along this flow, leading to edge accumulation (coffee-ring effect).
   - In functional inks, this directly affects film uniformity, electrical connectivity, and optical performance.
2. Interfacial and colloidal interactions
   - Nanoparticle–substrate and particle–particle forces (DLVO, van der Waals, electrostatic repulsion, steric effects) influence adhesion and aggregation.
   - These interactions introduce microscale heterogeneity that cannot be captured by single-phase flow descriptions.
3. Dynamic contact-line behavior
   - Particles can pin or unpin the contact line, altering spreading, evaporation rate distribution, and final deposition patterns.
   - Contact-line pinning strongly influences droplet-shape evolution and drying dynamics.
4. Multiscale coupling
   - The droplet exhibits macroscopic deformation (μm–mm), while nanoparticles evolve on nm–μm scales.
   - This requires a modeling framework capable of capturing macroscale hydrodynamics and microscale particulate interactions simultaneously.

## 1.2 Introduction of the Lattice Boltzmann method

The LBM is a mesoscopic numerical approach that simulates fluid flow by evolving particle distribution functions on a discrete lattice over time, as illustrated in Fig. 1.2. Rather than directly solving the Navier–Stokes equations, LBM employs a simplified form of the Boltzmann equation and recovers macroscopic fluid behavior through statistical averaging. Owing to this kinetic foundation, LBM provides an intuitive and flexible framework for handling various fluid phenomena, including interfacial dynamics, phase separation, and complex geometrical boundary conditions.

Compared with traditional CFD methods based on the finite volume method (FVM) or finite element method (FEM), the lattice Boltzmann method offers several distinct advantages (Fig. 1.3). First, because the algorithm is entirely local on the

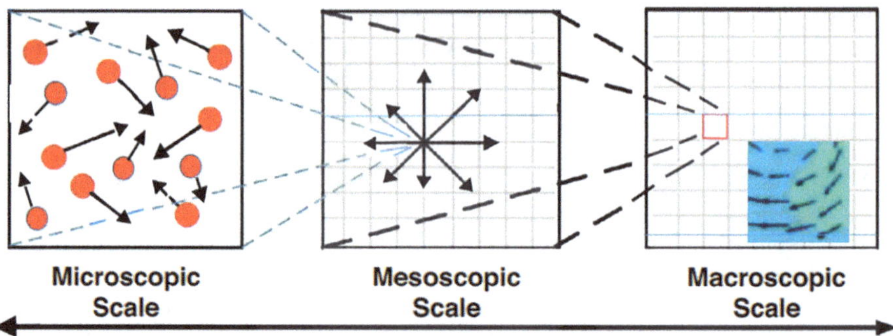


Fig. 1.2 Classification of CFD methodologies according to scale. The LBM corresponds to the mesoscopic scale

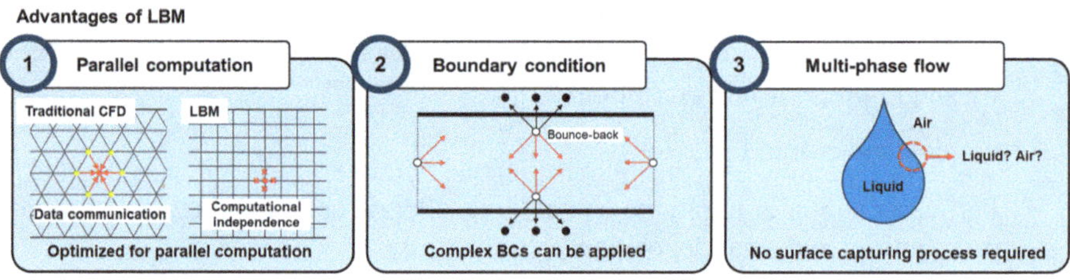


Fig. 1.3 Advantages of LBM

lattice, it achieves exceptionally high parallel efficiency and delivers ideal performance on GPU architectures, where memory-access patterns are critical. Second, LBM enables efficient computation on boundaries with complex geometries—such as porous media—through the straightforward application of bounce-back boundary conditions. Third, LBM naturally incorporates interfacial physics such as surface tension, phase separation, and wetting behavior through mesoscopic interaction forces or free-energy-based formulations, thereby avoiding the geometric complexity inherent in explicit interface-tracking schemes used in conventional CFD. These characteristics make LBM particularly suitable for challenging problems involving multiphase and multicomponent flows, complex surfaces, and dynamic contact-line behavior.

**Exercise 1.2 Simulation Scale**

Figure 1.2 illustrates the microscopic, mesoscopic, and macroscopic scales of CFD methodologies. Using this figure as a reference, answer the following:

(a) Explain, from a physical perspective, what it means for the LBM to operate at the mesoscopic scale.
(b) In multiphysics problems such as inkjet droplet dynamics—where multiple scales coexist—discuss the advantages of adopting a mesoscopic approach.

**Solution**

(a) Meaning of the mesoscopic scale

The microscopic scale resolves individual molecular or atomic motion (e.g., molecular dynamics). The macroscopic scale describes flow using continuum-level averaged quantities governed by PDEs such as the Navier–Stokes equations.

The mesoscopic scale, where LBM operates, represents an intermediate statistical level:

- It does not track individual molecules.
- Instead, it evolves discrete particle distribution functions associated with a finite set of velocities.
- Macroscopic flow variables (density, velocity, and stress) are recovered through statistical moments of these distributions.

Thus, LBM captures essential interfacial and kinetic effects while avoiding the prohibitive cost of full microscopic simulations.

(b) Advantages of the mesoscopic approach in inkjet dynamics

Inkjet flows involve multiscale coupling: microscopic (nanoparticle interactions, DLVO forces), mesoscopic (surface tension, wetting, evap-oration-induced flows), and macroscopic (droplet shape, spreading radius).

A mesoscopic method benefits these systems by:

1. More flexible representation of interfacial physics (surface tension, phase separation, Marangoni effects) without explicit geometric reconstruction.
2. Natural incorporation of micro-level effects (e.g., nanoparticle interactions) via mesoscopic force models without relying on empirical macroscopic closure laws.
3. Lower computational cost than microscopic MD simulations while remaining more physically expressive than macroscopic PDE solvers.

As a result, LBM is well suited for problems where droplet deformation, contact-line motion, internal circulation, and nanoparticle redistribution occur simultaneously across different spatial and temporal scales.

Multiphase LBM models can be broadly classified into three main categories: the color-gradient model, the pseudopotential (Shan–Chen) model, and the free-energy model. The color-gradient approach distinguishes phases using separate distribution functions and employs color-gradient operators to reconstruct interfaces, providing independent control of surface tension and viscosity ratios. The pseudopotential model introduces interparticle forces based on a density-dependent potential, allowing phase separation to emerge naturally from attractive interactions with simple implementation and computational efficiency. The free-energy model constructs interfaces by minimizing a thermodynamic free-energy functional, offering improved thermodynamic consistency and accurate control of interfacial properties, though it may require solving additional differential equations such as the Poisson equation. Table 1.1 presents a comparison of the multiphase LBM models.

**Table 1.1** Comparison of multiphase LBM models

| Model | Advantages | Limitations | Representative applications |
|---|---|---|---|
| Color-gradient model | – Independent control of surface tension and viscosity ratios<br>– Sharp, well-defined interfaces<br>– Good accuracy for strong interfacial tension-dominated flows | – Requires two distribution functions → higher memory and computational cost.<br>– More complex implementation than pseudopotential | Surfactant-laden flows, droplet breakup, coalescence |
| Pseudopotential model | – Very simple implementation<br>– High computational efficiency<br>– No explicit interface tracking required | – Surface tension, density ratio, and EOS parameters are coupled (less independent control).<br>– Thermodynamic inconsistency | Boiling and condensation, evaporation, porous-media flow |
| Free-energy model | – Thermodynamically consistent<br>– Accurate dynamic contact angles and wetting behavior<br>– Independent control over interfacial properties | – Requires solving additional PDEs (e.g., Poisson equation) → higher computational cost | Complex wetting, film formation, high density-ratio flows |

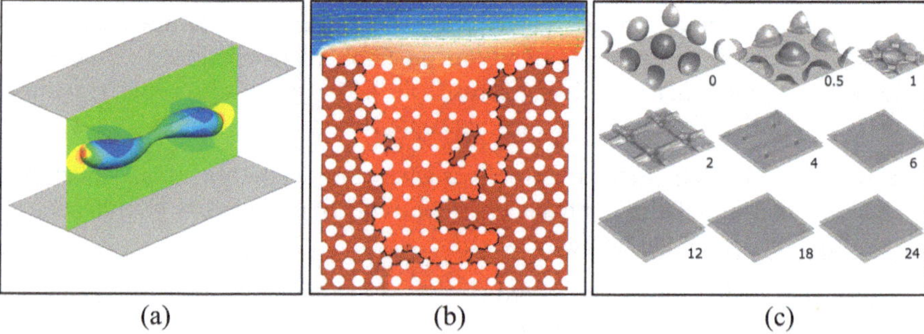


**Fig. 1.4** Previous studies employing multiphase LBM models. (**a**) Observation of surfactant-laden droplet breakup in shear flow using the color-gradient model [4]. (**b**) Analysis of evaporation phenomena in porous media using the multicomponent pseudopotential model [5]. (**c**) Simulation of film formation during multiple droplet collisions using the free-energy model [6]

These multiphase LBM models have been widely employed across numerous prior studies (Fig. 1.4). The color-gradient model has been primarily used in problems where interfacial tension plays a dominant role—such as droplet breakup, coalescence, and surfactant-laden flows—demonstrating high accuracy. The pseudopotential model has been extensively applied to various phenomena, including droplet impact on solid surfaces, phase change, capillary flow in porous media, and wetting transitions influenced by substrate anisotropy. The free-energy model has shown strong capability in accurately simulating equilibrium interface shapes, dynamic contact angles, and complex wetting behavior on anisotropic surfaces, particularly under high density-ratio conditions. Collectively, these studies clearly demonstrate the flexibility and potential of LBM in comprehensively capturing complex multiphase flow physics.

## 1.3 Conclusions

This chapter explained why a new numerical approach is required to understand the diverse physical phenomena that arise in multiphase flows, nanoparticle transport, and inkjet printing processes. It was shown that conventional CFD alone is insufficient in modern processing environments where complex interfacial behavior, microscale interactions, and large-scale parallel computation requirements coexist. Within this context, the fundamental principles of the LBM and the structural advantages it offers were introduced. Furthermore, by describing how multiphase LBM models—such as the color-gradient, pseudopotential, and free-energy approaches—implement interfacial physics and phase separation, and by illustrating how these models have been applied in previous studies, the suitability of LBM for handling complex flow problems was clarified. Based on this discussion, the reader can understand how the subsequent chapters will develop the mathematical formulation of LBM, its algorithmic implementation, and its various application-level simulations.

**Summary**

1. **Multiphase nanoparticle-laden droplets**
   - Inkjet droplets are governed by coupled liquid–gas–solid interactions that determine formation, spreading, evaporation, and final deposition.
   - Nanoparticle transport modifies internal capillary flow, contact-line dynamics, and pattern formation.
   - Microscale DLVO interactions and Brownian fluctuations strongly influence macroscale drying outcomes.
2. **Limitations of conventional CFD**
   - Navier–Stokes solvers rely on VOF/level-set interface tracking, leading to numerical diffusion and geometric reconstruction challenges.
   - Topological changes such as breakup, coalescence, and wetting transitions are difficult to capture robustly.
   - High computational cost arises when modeling microstructured substrates, colloidal interactions, or multiscale particle dynamics.

3. **LBM as a mesoscopic alternative**
   - LBM evolves particle distribution functions on discrete lattice directions and extracts macroscopic fields through statistical moments.
   - Local collision-streaming operations enable excellent GPU scalability and simple parallelization.
   - LBM handles complex boundaries, interfacial forces, and wetting phenomena naturally and couples efficiently with particle-level models.
4. **Multiphase LBM categories**
   - Color-gradient models offer sharp interfaces and independent control of $\sigma$ and viscosity but require two distribution sets.
   - Pseudopotential models provide simple and efficient phase separation with EOS-governed surface tension.
   - Free-energy models ensure thermodynamic consistency and accurate wetting but require solving additional Poisson-type equations.
5. **Application scope**
   - Droplet breakup, coalescence, and jetting dynamics.
   - Surfactant-driven interface behavior and Marangoni stresses.
   - Porous-media infiltration, evaporation-driven flows, and nanoparticle assembly.
   - Inkjet printing, functional film formation, and pattern control in additive manufacturing.

## References

1. M. Park, Y. I. Kim, Y. K. Jung, J. G. Kang, S. Kim, J. Ha, Y. G. Yoon, and C. Lee, "All inkjet-printed 6.95 ″217 ppi active matrix qd-led display with rgb cd-free qds in the top-emission device structure," Journal of the Society for Information Display **30**, 433-440 (2022).
2. L. Li, L. Pan, Z. Ma, K. Yan, W. Cheng, Y. Shi, and G. Yu, "All inkjet-printed amperometric multiplexed biosensors based on nanostructured conductive hydrogel electrodes," Nano letters **18**, 3322-3327 (2018).
3. V. V. Satale, H. B. Lee, B. Tyagi, M. M. Ovhal, S. Chowdhury, A. Mohamed, D.-H. Kim, and J.-W. Kang, "Ink engineering using 1, 3-dimethyl-2-imidazolidinone solvent for efficient inkjet-printed triple-cationic perovskite solar cells," Chemical Engineering Journal **493**, 152541 (2024).
4. Y. Ba, H. Liu, W. Li, and W. Yang, "Modelling the effect of soluble surfactants on droplet deformation and breakup in simple shear flow," Journal of Fluid Mechanics **988**, A41 (2024).
5. L. Fei, F. Qin, J. Zhao, D. Derome, and J. Carmeliet, "Lattice boltzmann modelling of isothermal two-component evaporation in porous media," Journal of Fluid Mechanics **955**, A18 (2023).
6. W. Zhou, D. Loney, A. G. Fedorov, F. L. Degertekin, and D. W. Rosen, "Lattice boltzmann simulations of multiple-droplet interaction dynamics," Physical Review E **89**, 033311 (2014).

# Chapter 2
# Lattice Boltzmann Method: Single Phase

The LBM is a computational technique for simulating fluid dynamics, which has gained popularity due to its unique approach to solving the Navier–Stokes equations. Unlike traditional methods that solve macroscopic equations directly, LBM is based on mesoscopic kinetic equations that describe the evolution of particle distribution functions on a discrete lattice grid. This framework offers a simple and intuitive way to model fluid flow by simulating the microscopic interactions of particles, leading to emergent macroscopic fluid behavior.

LBM's primary strength lies in its ability to handle complex boundary conditions and geometries with relative ease. This is largely due to the local nature of the particle collisions and streaming processes, which simplifies the implementation of boundaries compared to conventional numerical methods. Additionally, LBM is highly efficient for parallel computing, as its algorithmic structure lends itself well to parallelization, making it suitable for simulations requiring substantial computational resources. Another advantage of LBM is its flexibility in handling multiphase and multicomponent flows. The method is inherently capable of simulating complex fluid behaviors such as phase separation, mixing, and droplet dynamics. This flexibility has led to its widespread use in fields ranging from microfluidics and porous media to biomedical engineering and inkjet printing.

This chapter covers algorithms for single-phase LBM and methodologies for implementing LBM.

## 2.1 LBM Algorithm

Compared to traditional computational fluid dynamics (CFD), LBM has a simple algorithmic structure because it does not solve nonlinear equations and can compute fluid dynamics through "collision" and "propagation" processes. The LBM algorithm consists of the following steps.



H. M. Lee, J. S. Lee, *Lattice Boltzmann Methodology for Single-Phase and Multiphase Nanoparticle Modeling*,
https://doi.org/10.1007/978-981-95-9117-6_2

1. **Initialization:** This is the process of distributing the particle distribution functions, $f_i(\mathbf{x}, t)$, across all nodes. The particle distribution function, $f_i$, is set based on the equilibrium function, $f_i^{eq}$, which is determined by the initial density, $\rho_0$, and velocity, $\mathbf{u}_0$, values, $f_i(\mathbf{x}, t=0) = f_i^{eq}(\rho_0, \mathbf{u}_0)$.
2. **Moment update**: This is the process of calculating the macroscopic density and velocity. The calculated macroscopic properties are used in subsequent steps.
3. **Collision**: The post-collision particle distribution function, $f_i^*$, is calculated using the equilibrium distribution function, $f_i^{eq}$. $f_i^*(\mathbf{x}, t) = f_i(\mathbf{x}, t) - \frac{1}{\tau}[f_i(\mathbf{x}, t) - f_i^{eq}(\mathbf{x}, t)]$.
4. **Propagation**: The post-collision particle distribution function is propagated in the $i^{th}$ direction. $f_i(\mathbf{x} + \mathbf{e}_i\Delta t, t + \Delta t) = f_i^*(\mathbf{x}, t)$.
5. **Boundary conditions**: Apply no-slip, periodic, and open boundary conditions.
6. **Timestep update**: Update to a new timestep and return to step 2.

In the subsequent subchapters, detailed explanations of each process will be provided.

## 2.2 BGK Collision Operator

LBM can be derived as the Bhatnagar–Gross–Krook (BGK) approximation of the Boltzmann kinetic equation [1], as shown in (2.1).

$$\frac{\partial f}{\partial t} + \boldsymbol{\xi} \cdot \nabla f = -\frac{f - f^{eq}}{\tau}, \tag{2.1}$$

where $f$ is the particle distribution function, $\boldsymbol{\xi}$ is the macroscopic velocity, $\tau$ is the relaxation time due to collision, and $f^{eq}$ is the Boltzmann–Maxwellian distribution function. In the BGK method, the Boltzmann equation is represented using the discretized particle distribution function, $f_i$.

$$f_i(\mathbf{x} + \mathbf{e}_i\Delta t, t + \Delta t) = f_i(\mathbf{x}, t) - \frac{1}{\tau}[f_i(\mathbf{x}, t) - f_i^{eq}(\mathbf{x}, t)], \tag{2.2}$$

where $i$ is the direction of discrete velocity, $\mathbf{e}_i$ is the discrete velocity, $t$ is the time step, and $\Delta t$ is the lattice time step. $\tau$ is a variable related to kinematic viscosity, $\nu$, which is calculated as follows.

$$\nu = c_s^2(\tau - 0.5)\Delta t, \tag{2.3}$$

where $c_s$ is the speed of sound, and it is generally given as $c_s = 1/\sqrt{3}$. The pressure is given by $p = \rho c_s^2$. As shown in (2.3), $\tau$ must be greater than 0.5. $f_i^{eq}$ is calculated as follows:

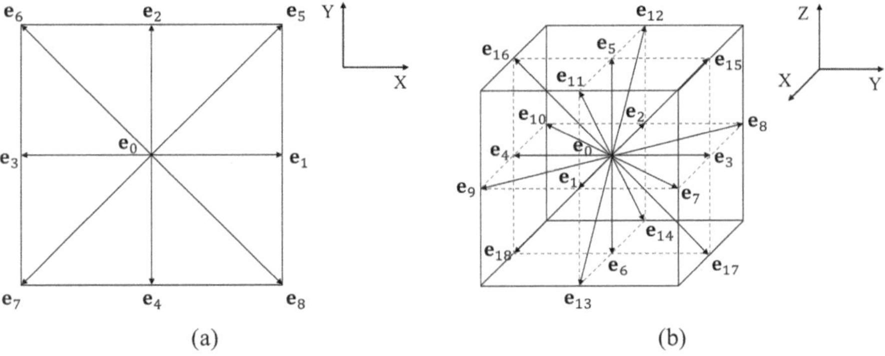


**Fig. 2.1** Set of velocity vectors for (**a**) D2Q9 and (**b**) D3Q19

$$f_i^{eq}(\mathbf{x}, t) = \rho\omega_i\left[1 + \frac{\mathbf{e}_i \cdot \mathbf{u}}{c_s^2} + \frac{(\mathbf{e}_i \cdot \mathbf{u})^2}{2c_s^4} - \frac{|\mathbf{u}|^2}{2c_s^2}\right], \tag{2.4}$$

where $\omega_i$ is the weighting factor.

In LBM, the model is determined by the dimensions and the degree of discretization of the space. Generally, a model that discretizes $d$ dimensions into $q$ discrete directions is referred to as D$d$Q$q$. For example, a model consisting of nine velocity vectors in two dimensions is called D2Q9. Figure 2.1 illustrates the set of velocity vectors for D2Q9 and D3Q19. The discrete velocity set of the D2Q9 model is as follows.

$$[\mathbf{e}_0, \mathbf{e}_1, \mathbf{e}_2, \mathbf{e}_3, \mathbf{e}_4, \mathbf{e}_5, \mathbf{e}_6, \mathbf{e}_7, \mathbf{e}_8] = c\begin{bmatrix} 0 & 1 & 0 & -1 & 0 & 1 & -1 & -1 & 1 \\ 0 & 0 & 1 & 0 & -1 & 1 & 1 & -1 & -1 \end{bmatrix}. \tag{2.5}$$

And the discrete velocity set of the D3Q19 model is as follows.

$$\begin{aligned} &[\mathbf{e}_0, \mathbf{e}_1, \mathbf{e}_2, \mathbf{e}_3, \mathbf{e}_4, \mathbf{e}_5, \mathbf{e}_6, \mathbf{e}_7, \mathbf{e}_8, \mathbf{e}_9, \mathbf{e}_{10}, \mathbf{e}_{11}, \mathbf{e}_{12}, \mathbf{e}_{13}, \mathbf{e}_{14}, \mathbf{e}_{15}, \mathbf{e}_{16}, \mathbf{e}_{17}, \mathbf{e}_{18}] \\ &= c\begin{bmatrix} 0 & 1 & -1 & 0 & 0 & 0 & 0 & 1 & -1 & 1 & -1 & 1 & -1 & 1 & -1 & 0 & 0 & 0 & 0 \\ 0 & 0 & 0 & 1 & -1 & 0 & 0 & 1 & 1 & -1 & -1 & 0 & 0 & 0 & 0 & 1 & -1 & 1 & -1 \\ 0 & 0 & 0 & 0 & 0 & 1 & -1 & 0 & 0 & 0 & 0 & 1 & 1 & -1 & -1 & 1 & 1 & -1 & -1 \end{bmatrix} \end{aligned} \tag{2.6}$$

where $c$ is the lattice velocity which is defined as $c = \Delta x/\Delta t = 1$. $\omega_i$ varies depending on the lattice link, and the values of $\omega_i$ for each model are summarized in Table 2.1 [2]. The macroscopic property is calculated as follows.

**Table 2.1** The weighting factors for the lattice models

| Model | $\omega_i$ |
|---|---|
| D2Q9 | $\frac{4}{9}\left\{i:\|\mathbf{e}_i\|^2=0\right\}, \frac{1}{9}\left\{i:\|\mathbf{e}_i\|^2=1\right\}, \frac{1}{36}\left\{i:\|\mathbf{e}_i\|^2=2\right\}$ |
| D3Q15 | $\frac{2}{9}\left\{i:\|\mathbf{e}_i\|^2=0\right\}, \frac{1}{9}\left\{i:\|\mathbf{e}_i\|^2=1\right\}, \frac{1}{72}\left\{i:\|\mathbf{e}_i\|^2=3\right\}$ |
| D3Q19 | $\frac{1}{3}\left\{i:\|\mathbf{e}_i\|^2=0\right\}, \frac{1}{18}\left\{i:\|\mathbf{e}_i\|^2=1\right\}, \frac{1}{36}\left\{i:\|\mathbf{e}_i\|^2=2\right\}$ |
| D3Q27 | $\frac{8}{27}\left\{i:\|\mathbf{e}_i\|^2=0\right\}, \frac{2}{27}\left\{i:\|\mathbf{e}_i\|^2=1\right\}, \frac{1}{54}\left\{i:\|\mathbf{e}_i\|^2=2\right\}, \frac{1}{216}\left\{i:\|\mathbf{e}_i\|^2=3\right\}$ |

$$\rho=\sum_i f_i=\sum_i f_i^{eq}, \tag{2.7}$$

$$\rho\mathbf{u}=\sum_i f_i\mathbf{e}_i=\sum_i f_i^{eq}\mathbf{e}_i. \tag{2.8}$$

So far, the processes of collision, propagation, and moment update in the LBM have been covered. The following presents the equations for each process in the form of pseudocode. First, we start with the moment update process described by (2.7) and (2.8), following the LBM algorithm.

```
//Moment update
for all (x,y) do
  if (x,y) is fluid node then
    rho[x][y] = ux[x][y] = uy[x][y] = 0.0
    for all i do
      rho[x][y] += f[x][y][i]
      ux[x][y] += f[x][y][i] * ex[i]
      uy[x][y] += f[x][y][i] * ey[i]
    end for
    ux[x][y] /= rho[x][y]
    uy[x][y] /= rho[x][y]
  end if
end for
```

Next, the collision process, represented by the right side of (2.2), can be expressed as follows.

```
//Collision
for all (x,y) do
  if (x,y) is fluid node then
    udotu = ux[x][y] * ux[x][y] + uy[x][y] * uy[x][y]
  for all i do
      udote = ux[x][y] * ex[i] + uy[x][y] * ey[i]
      feq = omega[i] * rho[x][y] * (1.0 + 3.0 * udote + 4.5 * udote * udote
        - 1.5 * udotu)
```

```
      f[x][y][i] += - (f[x][y][i] - feq) / tau
    end for
  end if
end for
```

After the collision process, the calculation of the propagation process is performed. In the propagation process, $f_i$ spreads in the direction of each $i$. Therefore, a function is needed to store the nodes that are moved by $\mathbf{e}_i\Delta t$ from a specific node. This process takes place during the initialization step.

```
//Setup for propagation
//XL and YL are the maximum values of the x and y domains, respectively
for all (x,y) do
  for all i do
    nx[x][y][i] = x + ex[i]
    ny[x][y][i] = y + ey[i]
    if nx[x][y][i] > XL - 1 then
      nx[x][y][i] = 0
    else if nx[x][y][i] < 0 then
      nx[x][y][i] = XL - 1
    end if
    if ny[x][y][i] > YL - 1 then
      ny[x][y][i] = 0
    else if ny[x][y][i] < 0 then
      ny[x][y][i] = YL - 1
    end if
  end for
end for
```

The $n_x(\mathbf{x}, i)$ and $n_y(\mathbf{x}, i)$ functions in the above pseudocode can be used to obtain the nodes corresponding to the direction $\mathbf{e}_i\Delta t$ as they spread from the $\mathbf{x}$. However, if the values of $n_x$ and $n_y$ exceed the domain limits, it will result in an error. Therefore, the code is designed to reset the propagated node to the minimum value if it exceeds the maximum domain limit, and to the maximum value if it falls below the minimum limit. This method is known as the periodic boundary condition, and related topics will be discussed in detail in the"2.3 Boundary condition." The propagation step using the $n_x$ and $n_y$ is as follows.

```
//Propagation
for all (x,y) do
  for all i do
    xtemp = nx[x][y][i]
    ytemp = ny[x][y][i]
```

```
    ftemp[xtemp][ytemp][i] = f[x][y][i]
  end for
end for
for all (x,y) do
  for all i do
    f[x][y][i] = ftemp[x][y][i]
  end for
end for
```

**Exercise 2.1 Basic Lattice Boltzmann Equation**
Consider the lattice Boltzmann equation in (2.2). Derive the continuity equation from the lattice Boltzmann equation. The continuity equation is as follows.

$$\frac{\partial \rho}{\partial t} + \nabla \cdot (\rho \mathbf{u}) = 0$$

**Solution**
To derive the continuity equation, we sum the lattice Boltzmann equation over all lattice directions

$$\sum_i f_i(\mathbf{x} + \mathbf{e}_i \Delta t, t + \Delta t) = \sum_i f_i(\mathbf{x}, t) - \sum_i \frac{1}{\tau}[f_i(\mathbf{x}, t) - f_i^{eq}(\mathbf{x}, t)]$$

When the advection term is expanded using a Taylor series, we get:

$$f_i(\mathbf{x} + \mathbf{e}_i \Delta t, t + \Delta t) \approx f_i(\mathbf{x}, t) + \Delta t \left( \frac{\partial f_i}{\partial t} + \nabla f_i \cdot \mathbf{e}_i \right)$$

Summing this over all lattice directions, we get:

$$\sum_i f_i(\mathbf{x} + \mathbf{e}_i \Delta t, t + \Delta t) \approx \sum_i f_i(\mathbf{x}, t) + \Delta t \sum_i \left[ \frac{\partial f_i}{\partial t} + \nabla f_i \cdot \mathbf{e}_i \right]$$

From (2.7), we can see that the sum of the collision terms cancels out. Therefore, the right-hand side can be simplified as follows.

$$\sum_i f_i(\mathbf{x}, t) - \sum_i \frac{1}{\tau}[f_i(\mathbf{x}, t) - f_i^{eq}(\mathbf{x}, t)] = \sum_i f_i(\mathbf{x}, t)$$

The simplified equation is as follows.

$$\Delta t \sum_i \left[ \frac{\partial f_i}{\partial t} + \mathbf{e}_i \cdot \nabla f_i \right] = 0$$

Finally, using Equations (2.7) and (2.8), the continuity equation can be derived.

$$\frac{\partial \rho}{\partial t} + \nabla \cdot (\rho \mathbf{u}) = 0$$

## 2.3 Boundary Conditions

Boundary conditions are a crucial aspect of the LBM, as they define how the fluid interacts with the boundaries of the computational domain. Proper implementation of boundary conditions ensures the physical accuracy of simulations, especially in complex fluid flows involving walls, periodic systems, and open domains. In LBM, different types of boundary conditions are applied depending on the nature of the problem being modeled. Each of these conditions handles the distribution functions of the fluid particles in distinct ways to reflect realistic interactions at the boundaries. The boundary conditions covered in this chapter are as follows.

- Periodic boundaries: These are used to model systems where the fluid properties at one boundary seamlessly continue to the opposite boundary. They are particularly useful for simulating flow in infinite or cyclic domains.
- Bounce-back boundaries: These conditions are employed to model solid boundaries. When a fluid particle strikes a boundary, it reflects back, simulating no-slip conditions, which is essential for accurately modeling walls or obstacles within the flow domain.
- Open boundaries: In cases where the domain is open, such as an inlet or outlet, open boundary conditions are applied. These allow fluid to flow into or out of the domain, maintaining the continuity of mass and momentum without artificial reflection or accumulation at the boundaries.

Each of these boundary conditions plays a vital role in ensuring the fidelity of LBM simulations, and the choice of boundary condition depends heavily on the physical scenario being modeled.

### *2.3.1 Periodic Boundaries*

The periodic boundary condition is one of the easiest boundary conditions to implement in LBM from a technical standpoint. For periodic boundary conditions, the fluid particles leaving one side of the domain should re-enter the domain from the opposite side. This means that the distribution functions on one side of the boundary are directly copied to the opposite boundary and vice versa. Figure 2.2 illustrates this

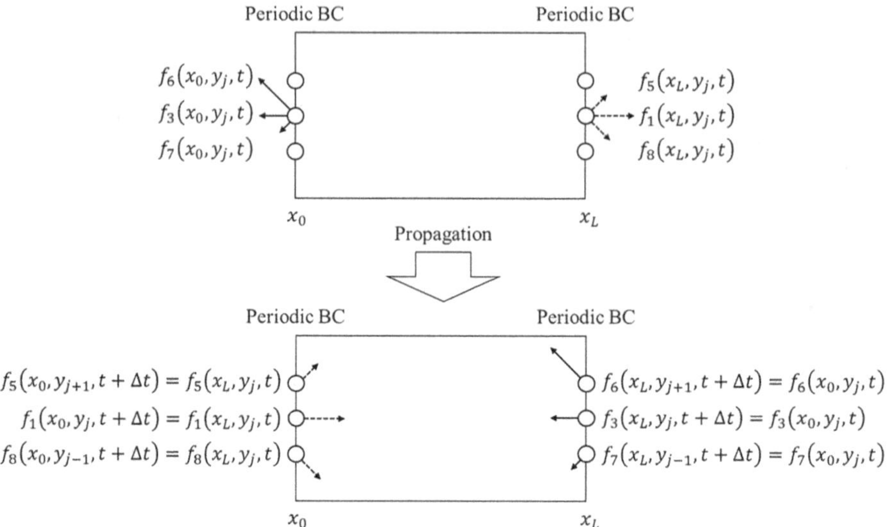


**Fig. 2.2** The schematic for applying the periodic boundary condition

process in schematic. Before propagation, the distribution functions at the boundaries beyond the domain in the direction of the periodic boundaries are propagated and move in the opposite direction. Before propagation, the distribution functions located at the periodic boundaries ($\mathbf{x} = \mathbf{x}_0$, $\mathbf{x} = \mathbf{x}_L$) that point outside the domain are propagated and move toward their respective opposite boundaries. This can be represented in the equations as follows.

$$
\begin{aligned}
f_1\left(x_0, y + e_{1,y}\Delta t, t + \Delta t\right) &= f_1(x_L, y, t),\\
f_5\left(x_0, y + e_{5,y}\Delta t, t + \Delta t\right) &= f_5(x_L, y, t),\\
f_8\left(x_0, y + e_{8,y}\Delta t, t + \Delta t\right) &= f_8(x_L, y, t),
\end{aligned}
\tag{2.9}
$$

and

$$
\begin{aligned}
f_3\left(x_L, y + e_{3,y}\Delta t, t + \Delta t\right) &= f_3(x_0, y, t),\\
f_6\left(x_L, y + e_{6,y}\Delta t, t + \Delta t\right) &= f_6(x_0, y, t),\\
f_7\left(x_L, y + e_{7,y}\Delta t, t + \Delta t\right) &= f_7(x_0, y, t).
\end{aligned}
\tag{2.10}
$$

As can be seen in (2.9) and (2.10), the periodic boundaries do not require special methods and can be implemented within the propagation process. The pseudocode for periodic boundaries using $n_x(\mathbf{x}, i)$ and $n_y(\mathbf{x}, i)$ can be found in the previous chapter.

### 2.3.2 Bounce-Back Boundaries

The bounce-back method is a widely used boundary condition in LBM for simulating fluid flow near solid boundaries. It effectively models no-slip boundary conditions, where the fluid velocity at the boundary matches that of the solid wall, leading to realistic flow behavior. The bounce-back method assumes that when a fluid particle encounters a solid boundary, it reflects back in the opposite direction. This approach for solid boundaries can be consistently and easily applied to complex geometries such as porous media, giving LBM an advantage over other CFD methods.

The bounce-back process is illustrated in Fig. 2.3. $x_B$ and $x_S$ represent the fluid node in contact with the solid and the solid node, respectively. The dotted line at the boundary between the white and gray areas represents the solid interface, and in the bounce-back method, the solid interface is positioned midway between the fluid and solid nodes. The post-collision distribution function of the $x_B$ node is moved by the standard propagation process. The distribution function moved to the $x_S$ node reverses its direction and is propagated again. This process can be represented mathematically as follows.

$$f_{\bar{i}}(\mathbf{x}_B, t + \Delta t) = f_i(\mathbf{x}_B, t), \tag{2.11}$$

where $\bar{i}$ is the direction of discrete velocity that satisfies $\mathbf{e}_{\bar{i}} = -\mathbf{e}_i$. (2.11) is applicable only when $\mathbf{x}_B + \mathbf{e}_i \Delta t = \mathbf{x}_S$. The pseudocode for the bounce-back process is presented below.

```
//Bounce-back
//opp[i] is the index representing the opposite direction of i
```

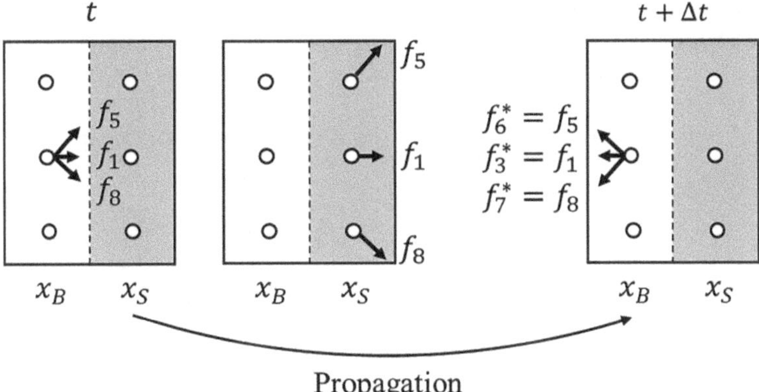


**Fig. 2.3** The process of bounce-back method. The white area ($x_B$) represents the fluid nodes adjacent to the solid, while the gray area ($x_S$) represents the solid nodes. $f^*$ is the post-propagation distribution function

```
for all (x,y) do
  if (x,y) is solid node then
    for all i do
      itemp = opp[i]
      xtemp = nx[x][y][itemp]
      ytemp = ny[x][y][itemp]
      if (xtemp,ytemp) is fluid node then
        f[xtemp][ytemp][itemp] = f[x][y][i]
      end if
    end for
  end if
end for
```

Here, opp[i] represents $\bar{i}$, and when it has a discrete velocity as in (2.5), it can be expressed as follows.

$$opp(i) = [0\ 3\ 4\ 1\ 2\ 7\ 8\ 5\ 6]. \tag{2.12}$$

### 2.3.3 Open Boundaries

Open boundary conditions are used to model flow at the boundaries of a domain where fluid can freely enter or exit, such as inlets, outlets, or open systems. These conditions are crucial for ensuring that the fluid behaves realistically at the domain boundaries without artificial reflection of fluid particles, which can disturb the simulation. In this chapter, we introduce the Zou and He [3] method, which is widely used for boundary conditions in LBM.

Figure 2.4 shows the distribution function at the open south boundary. On the south side, after the propagation process, the values of $f_2, f_5$, and $f_6$ remain unknown. To set the boundary condition, it is necessary to determine the unknown distribution functions. Using (2.7) and (2.8), the following three equations can be derived.

$$f_2 + f_5 + f_6 = \rho - (f_0 + f_1 + f_3 + f_4 + f_7 + f_8), \tag{2.13}$$

$$f_5 - f_6 = \rho u_x - (f_1 - f_3 - f_7 + f_8), \tag{2.14}$$

$$f_2 + f_5 + f_6 = \rho u_y + (f_4 + f_7 + f_8). \tag{2.15}$$

Through Eqs. (2.13) and (2.15), the following equations are derived.

$$\rho = \frac{1}{1 - u_y}[f_0 + f_1 + f_3 + 2(f_4 + f_7 + f_8)], \tag{2.16}$$

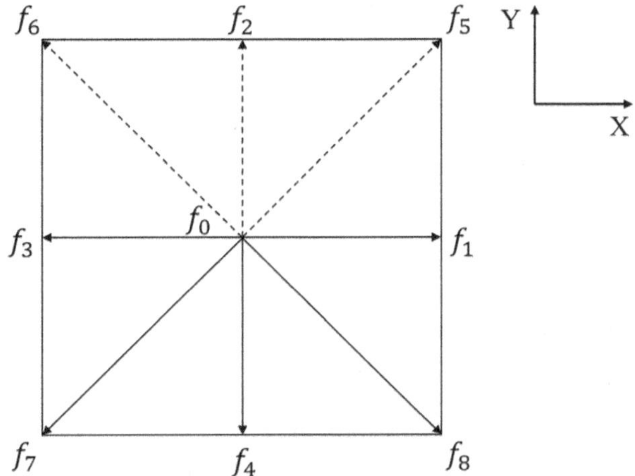


**Fig. 2.4** The distribution function on the south side. The unknown values ($f_2, f_5$, and $f_6$) are represented by dashed lines

$$u_y = 1 - \frac{1}{\rho}[f_0 + f_1 + f_3 + 2(f_4 + f_7 + f_8)]. \tag{2.17}$$

If the boundary condition assumes constant pressure and $u_x$ is fixed, there are four unknowns—$f_2, f_5, f_6$, and $u_y$—but only three equations, making it impossible to solve the system of equations. To address this issue, we assume that the bounce-back rule is applicable to the non-equilibrium part in the direction perpendicular to the boundary ($f_2^{neq} = f_2 - f_2^{eq} = f_4 - f_4^{eq} = f_4^{neq}$). The final calculated unknown distribution functions are as follows.

$$f_2 = f_4 + \frac{2}{3}\rho u_y, \tag{2.18}$$

$$f_5 = f_7 - \frac{1}{2}(f_1 - f_3) + \frac{1}{2}\rho u_x + \frac{1}{6}\rho u_y, \tag{2.19}$$

$$f_6 = f_8 + \frac{1}{2}(f_1 - f_3) - \frac{1}{2}\rho u_x + \frac{1}{6}\rho u_y. \tag{2.20}$$

Under constant pressure conditions, use $\rho = \rho_{BC}$ and (2.17), while for constant velocity conditions, use $u_y = u_{y,\ BC}$ and (2.16). The same methodology can be applied to other sides to set the boundary conditions. The pseudocode for the Zou and He pressure boundary condition is provided below.

```
//Zou and He pressure boundary condition (south side)
for all (x,y = 0) do
  uy_temp = 1.0 - (f[x][y][0] + f[x][y][1] + f[x][y][3] + 2.0 * (f[x][y]
[4] +
     f[x][y][7] + f[x][y][8])) / rho_BC
  f[x][y][2] = f[x][y][4] + (2.0 / 3.0) * rho_BC * uy_temp
```

```
 f[x][y][5] = f[x][y][7] - (1.0 / 2.0) * (f[x][y][1] - f[x][y][3]) + (1.0
/ 6.0)
        * rho_BC * uy_temp
 f[x][y][6] = f[x][y][8] + (1.0 / 2.0) * (f[x][y][1] - f[x][y][3]) + (1.0
/ 6.0)
        * rho_BC * uy_temp
end for
```

**Exercise 2.2 Boundary Condition Example**

We have discussed applying the boundary condition on the south side in the D2Q9 LBM using the Zou and He method. Apply the velocity boundary condition ($u_y = 0$) on the west side in the form of (2.18) to (2.20) by using this approach.

**Solution**

On the west side, the unknown distribution functions are $f_1, f_5$, and $f_8$. Organize the equations around the unknown distribution functions in the form of (2.13) to (2.15).

$$f_1 + f_5 + f_8 = \rho - (f_0 + f_2 + f_3 + f_4 + f_6 + f_7)$$
$$f_1 + f_5 + f_8 = \rho u_x + (f_3 + f_6 + f_7)$$
$$f_5 - f_8 = -f_2 + f_4 - f_6 + f_7$$

The density at the boundary can be calculated using the following equation.

$$\rho = \frac{1}{1 - u_x}[f_0 + f_2 + f_4 + 2(f_3 + f_6 + f_7)]$$

By applying the non-equilibrium bounce-back, the velocity boundary condition can be implemented on the west side.

$$f_1 = f_3 + \frac{2}{3}\rho u_x$$

$$f_5 = f_7 - \frac{1}{2}(f_2 - f_4) + \frac{1}{6}\rho u_x$$

$$f_8 = f_6 + \frac{1}{2}(f_2 - f_4) + \frac{1}{6}\rho u_x$$

## 2.4 Forcing Schemes

Forcing schemes are employed to introduce external forces (such as gravity, body forces, or pressure gradients) into the fluid dynamics simulation. These schemes modify the basic LBM formulation to account for forces acting on the fluid, making the method more versatile and capable of handling a wider range of flow problems.

Especially in particle-laden flows, the forcing scheme plays a critical role in accurately capturing the interactions between fluid and particles. Properly accounting for forces in flows is essential for predicting fluid behavior, particle dynamics, and system efficiency. In particle-laden flows, the motion of particles is influenced by the surrounding fluid, and vice versa. Forcing schemes allow for the incorporation of forces such as drag, lift, and buoyancy, which dictate how particles move through the fluid. Without a proper forcing scheme, these interactions cannot be accurately represented, leading to incorrect predictions of particle motion and fluid dynamics.

In LBM, various forcing schemes can be used depending on the simulation environment, but in this chapter, we introduce the Guo et al.'s method [4] and exact difference method (EDM) [5] which are widely used for particle-laden flow.

### 2.4.1 *Guo et al.'s Method*

Guo et al.'s forcing scheme is one of the most widely used methods for introducing external forces into LBM. It is known for its accuracy and simplicity in implementation. In this scheme, the external force $\mathbf{F}$ is directly incorporated into collision step via a forcing term, $S_i$.

$$f_i(\mathbf{x}+\mathbf{e}_i\Delta t, t+\Delta t) = f_i(\mathbf{x},t) - \frac{1}{\tau}[f_i(\mathbf{x},t) - f_i^{eq}(\mathbf{x},t)] + S_i. \tag{2.21}$$

With the introduction of the external force term, the velocity calculated by (2.8) is modified as follows.

$$\mathbf{u}^{eq} = \frac{1}{\rho}\sum_i f_i\mathbf{e}_i + \frac{\mathbf{F}\Delta t}{2\rho}. \tag{2.22}$$

The equilibrium velocity $\mathbf{u}^{eq}$ is also used to calculate $f_i^{eq}$. $S_i$ can be obtained as follows.

$$S_i = \left(1 - \frac{1}{2\tau}\right)\omega_i\left[\frac{\mathbf{e}_i - \mathbf{u}^{eq}}{c_s^2} + \frac{\mathbf{e}_i\cdot\mathbf{u}^{eq}}{c_s^4}\mathbf{e}_i\right]\cdot\mathbf{F}. \tag{2.23}$$

The Guo et al.'s method is presented in pseudocode form below.

```
//Guo's forcing scheme (Collision step)
for all (x,y) do
   if (x,y) is fluid node then
      ux[x][y] += 0.5 * Fx[x][y] / rho[x][y]
      uy[x][y] += 0.5 * Fy[x][y] / rho[x][y]
      udotu = ux[x][y] * ux[x][y] + uy[x][y] * uy[x][y]
```

```
      for all i do
        udote = ux[x][y] * ex[i] + uy[x][y] * ey[i]
      S[i] = (1.0 - 0.5 / tau) * omega[i] * ((3.0 * (ex[i] - ux[x][y]) + 9.0
          * udote * ex[i]) * Fx[x][y] + (3.0 * (ey[i] - uy[x][y]) + 9.0
          * udote * ey[i]) * Fy[x][y])
        feq = omega[i] * rho[x][y] * (1.0 + 3.0 * udote + 4.5 * udote * udote
         - 1.5 * udotu)
        f[x][y][i] += - (f[x][y][i] - feq) / tau + S[i]
      end for
    end if
end for
```

## 2.4.2 *Exact Difference Method*

The exact difference method (EDM) is another forcing scheme that adds an external force term directly to the particle distribution functions. Unlike Guo et al.'s method, in the EDM, $f_i^{eq}$ remains the same as before, and only $S_i$ takes a different form. The forcing term for the EDM is as follows.

$$S_i = f_i^{eq}(\rho, \mathbf{u} + \Delta\mathbf{u}) - f_i^{eq}(\rho, \mathbf{u}), \tag{2.24}$$

where $\Delta\mathbf{u} = \mathbf{F}\Delta t/\rho$. As can be seen in (2.24), this method is independent of the relaxation time. With the addition of the force, the actual velocity also changes, as given by (2.22). The pseudocode for EDM is provided below.

```
//Exact difference method (Collision step)
for all (x,y) do
  if (x,y) is fluid node then
    delux = Fx[x][y] / rho[x][y]
    deluy = Fy[x][y] / rho[x][y]
    udotu = ux[x][y] * ux[x][y] + uy[x][y] * uy[x][y]
   udotu2 = (ux[x][y] + delux) * (ux[x][y] + delux) + (uy[x][y] + deluy) *
       (uy[x][y] + deluy)
    for all i do
      udote = ux[x][y] * ex[i] + uy[x][y] * ey[i]
      udote2 = (ux[x][y] + delux) * ex[i] + (uy[x][y] + deluy) * ey[i]
      feq = omega[i] * rho[x][y] * (1.0 + 3.0 * udote + 4.5 * udote * udote
       - 1.5 * udotu)
      feq2 = omega[i] * rho[x][y] * (1.0 + 3.0 * udote2 + 4.5 * udote2 *
        udote2 - 1.5 * udotu2)
    S[i] = feq2 - feq
      f[x][y][i] += - (f[x][y][i] - feq) / tau + S[i]
```

```
    end for
  end if
end for
```

## 2.5 MRT Collision Operator

The multi-relaxation-time (MRT) model is an important extension of LBM that improves stability and accuracy, especially in simulations involving complex flows such as those with high Reynolds numbers, turbulent flows, or non-equilibrium effects. The MRT model generalizes the simpler single-relaxation-time (SRT) model, also known as the BGK model, by allowing different relaxation times for various moments of the particle distribution function. This makes the MRT model more robust and versatile for a wider range of fluid flow problems. The advantages of the MRT model are as follows.

- Improved stability: In the MRT model, different physical processes (such as viscous stress and heat flux) relax at different rates, which significantly improves numerical stability. This is especially important in simulations with high Reynolds numbers or near solid boundaries, where the SRT model may become unstable.
- Better accuracy in non-equilibrium flows: Non-equilibrium flows, such as turbulent flows or flows with large velocity gradients, are better captured by the MRT model because it allows higher-order moments (related to stress and heat flux) to relax at appropriate rates. This results in more accurate simulation of non-equilibrium effects.
- Reduced viscosity dependence: In the SRT model, the viscosity of the fluid is directly related to the relaxation time, meaning that changing the relaxation time also affects the stability of the simulation. In the MRT model, the relaxation times of higher-order moments can be independently tuned, reducing the dependency of the model's stability on the fluid viscosity.

In the MRT model, the distribution function is first transformed into a set of moments that represent macroscopic quantities like density, momentum, and higher-order moments (e.g., stress, energy). Each of these moments can relax at a different rate, allowing the simulation to more accurately capture both equilibrium and non-equilibrium dynamics. The evolution equation in MRT is:

$$f_i(\mathbf{x} + \mathbf{e}_i\Delta t, t + \Delta t) = f_i(\mathbf{x}, t) - \mathbf{M}^{-1}\mathbf{S}\mathbf{M}[f_i(\mathbf{x}, t) - f_i^{eq}(\mathbf{x}, t)], \tag{2.25}$$

where $\mathbf{M}$ is the transformation matrix and $\mathbf{S}$ is the relaxation matrix. When $\mathbf{m} = \mathbf{M}\mathbf{f}$, the set of moments $\mathbf{m}$ for the D2Q9 model with the discrete velocity set from (2.5) are as follows [6].

$$\mathbf{m} = \left(\rho, e, \varepsilon, j_x, q_x, j_y, q_y, p_{xx}, p_{xy}\right)^T, \tag{2.26}$$

where $\rho$ is the density, $e$ is the energy, $\varepsilon$ is the energy square, $j_x$ and $j_y$ are the momentum flux, $q_x$ and $q_y$ are the energy flux, and $p_{xx}$ and $p_{xy}$ are the diagonal and off-diagonal stress tensor. The transformation matrix $\mathbf{M}$ is structured as follows.

$$\mathbf{M} = \begin{bmatrix} 1 & 1 & 1 & 1 & 1 & 1 & 1 & 1 & 1 \\ -4 & -1 & -1 & -1 & -1 & 2 & 2 & 2 & 2 \\ 4 & -2 & -2 & -2 & -2 & 1 & 1 & 1 & 1 \\ 0 & 1 & 0 & -1 & 0 & 1 & -1 & -1 & 1 \\ 0 & -2 & 0 & 2 & 0 & 1 & -1 & -1 & 1 \\ 0 & 0 & 1 & 0 & -1 & 1 & 1 & -1 & -1 \\ 0 & 0 & -2 & 0 & 2 & 1 & 1 & -1 & -1 \\ 0 & 1 & -1 & 1 & -1 & 0 & 0 & 0 & 0 \\ 0 & 0 & 0 & 0 & 0 & 1 & -1 & 1 & -1 \end{bmatrix}, \tag{2.27}$$

The inverse matrix $\mathbf{M}^{-1}$ can also be obtained from the transformation matrix.

$$\mathbf{M}^{-1} = \frac{1}{36} \begin{bmatrix} 4 & -4 & 4 & 0 & 0 & 0 & 0 & 0 & 0 \\ 4 & -1 & -2 & 6 & -6 & 0 & 0 & 9 & 0 \\ 4 & -1 & -2 & 0 & 0 & 6 & -6 & -9 & 0 \\ 4 & -1 & -2 & -6 & 6 & 0 & 0 & 9 & 0 \\ 4 & -1 & -2 & 0 & 0 & -6 & 6 & -9 & 0 \\ 4 & 2 & 1 & 6 & 3 & 6 & 3 & 0 & 9 \\ 4 & 2 & 1 & -6 & -3 & 6 & 3 & 0 & -9 \\ 4 & 2 & 1 & -6 & -3 & -6 & -3 & 0 & 9 \\ 4 & 2 & 1 & 6 & 3 & -6 & -3 & 0 & -9 \end{bmatrix}. \tag{2.28}$$

Next, when $\mathbf{m}^{eq} = \mathbf{M}\mathbf{f}^{eq}$, the equilibrium moments $\mathbf{m}^{eq}$ is

$$\mathbf{m}^{eq} = \left(\rho^{eq}, e^{eq}, \varepsilon^{eq}, j_x^{eq}, q_x^{eq}, j_y^{eq}, q_y^{eq}, p_{xx}^{eq}, p_{xy}^{eq}\right)^T. \tag{2.29}$$

The non-conserved moments of equilibrium values are given as

$$\begin{aligned} e^{eq} &= -2\rho + 3\left(j_x^2 + j_y^2\right)/\rho, \\ \varepsilon^{eq} &= \rho - 3\left(j_x^2 + j_y^2\right)/\rho, \\ q_x^{eq} &= -j_x, \\ q_y^{eq} &= -j_y, \end{aligned} \tag{2.30}$$

$$p_{xx}^{eq} = \left(j_x^2 - j_y^2\right)/\rho,$$

$$p_{xy}^{eq} = j_x j_y/\rho.$$

(2.29) can be simplified in terms of density and velocity.

$$\mathbf{m}^{eq} = \rho\left(1, \ -2 + 3\mathbf{u}^2, 1 - 3\mathbf{u}^2, u_x, \ -u_x, u_y, \ -u_y, u_x^2 - u_y^2, u_x u_y\right)^T. \tag{2.31}$$

So far, we have obtained $\mathbf{m}$ and $\mathbf{m}^{eq}$. If we denote the collision process as $\overline{\mathbf{m}} = -\mathbf{S}(\mathbf{m} - \mathbf{m}^{eq})$, (2.25) can be simplified as follows.

$$f_i(\mathbf{x} + \mathbf{e}_i\Delta t, t + \Delta t) = f_i(\mathbf{x}, t) + \sum_j M_{ij}^{-1}\overline{m}_j. \tag{2.32}$$

The relaxation matrix $\mathbf{S}$ is a diagonal matrix composed of nine relaxation rates.

$$\mathbf{S} = \mathrm{diag}(s_0, s_1, s_2, s_3, s_4, s_5, s_6, s_7, s_8) = \mathrm{diag}\left(0, s_e, s_\zeta, 0, s_q, 0, s_q, s_\nu, s_\nu\right), \tag{2.33}$$

where $s_0$, $s_3$, and $s_5$ are set to 0 to conserve mass and momentum. $s_\nu$ and $s_e$ are the parameters related to shear viscosity $\mu$ and bulk viscosity $\mu_b$, respectively.

$$\mu = \rho c_s^2\left(\frac{1}{s_\nu} - \frac{1}{2}\right), \mu_b = \rho c_s^2\left(\frac{1}{s_e} - \frac{1}{2}\right). \tag{2.34}$$

$s_\zeta$ and $s_q$ are the free parameters. In previous study [6], $s_e = 1.63$, $s_\zeta = 1.14$, and $s_q = 1.92$ were set. If $s_\nu = s_e = s_\zeta = s_q = 1/\tau$, the MRT model can be recovered into the SRT model.

**Exercise 2.3 Orthogonality of the Transformation Matrix**

(a) Show that $\mathbf{M}$ is orthogonal.

(b) Show the differences between the MRT and SRT models due to the orthogonality of $\mathbf{M}$.

**Solution**

(a) $\mathbf{M}$ possesses orthogonality when it satisfies the following equation.

$$\sum_i M_{mi}M_{ni} = 0 \ (m \neq n)$$

(b) If two vectors are orthogonal, their inner product is zero, meaning that one vector component is not projected onto the direction of the other vector. When the matrix $\mathbf{M}$ possesses orthogonality, each axis in the moment space becomes a completely separate and independent axis. Consequently, the relaxation applied to one moment during the collision process acts only on that axis and does not interfere with other moment axes. This is analogous to how, in a coordinate system where the $x$, $y$, and $z$ axes are orthogonal, movement in the $x$-direction does not affect the y or $z$ coordinates. Therefore, if the matrix $\mathbf{M}$ is orthogonal, it ensures that relaxation operations in the moment space occur independently without interference. By analogy, while SRT attenuates motion in all directions simultaneously using a single relaxation parameter, MRT applies different relaxation parameters to each orthogonal coordinate axis, allowing the relaxation of each moment to be individually controlled as needed.

Next, the MRT model with the external force term introduced using the method by Guo et al. will be explained. Earlier, we defined $\overline{\mathbf{m}} = -\mathbf{S}(\mathbf{m} - \mathbf{m}^{eq})$. When the external force term is introduced, $\overline{\mathbf{m}}$ is redefined as follows.

$$\overline{\mathbf{m}} = -\mathbf{S}(\mathbf{m} - \mathbf{m}^{eq}) + (\mathbf{I} - 0.5\mathbf{S})\mathbf{MF}, \tag{2.35}$$

where $\mathbf{F}$ is the source term defined in (2.23). $\mathbf{MF}$ is given by the following.

$$\mathbf{MF} = \Big(0, 6\big(F_x u_x + F_y u_y\big), -6\big(F_x u_x + F_y u_y\big), F_x, -F_x, F_y, -F_y, \\ 2\big(F_x u_x - F_y u_y\big), F_y u_x + F_x u_y\Big)^T. \tag{2.36}$$

As with the SRT method, the equilibrium velocity is used in $\mathbf{m}^{eq}$ and $\mathbf{MF}$. The MRT model discussed so far is presented in pseudocode as follows.

```
//MRT operator with Guo's forcing scheme
for all (x,y) do
  if (x,y) is fluid node then
    ux[x][y] += 0.5 * Fx[x][y] / rho[x][y]
    uy[x][y] += 0.5 * Fy[x][y] / rho[x][y]
    //Calculate m
    for all i do
      sum = 0.0
      for all j do
        sum += TransM[i][j] * f[x][y][j]
      end for
      m[i] = sum
    end for
    //Calculate meq
    meq[0] = rho[x][y]
    meq[1] = rho[x][y] * (-2.0 + 3.0 * (ux[x][y] * ux[x][y] + uy[x][y] *
      uy[x][y]))
    meq[2] = rho[x][y] * (1.0 - 3.0 * (ux[x][y] * ux[x][y] + uy[x][y] *
      uy[x][y]))
    meq[3] = rho[x][y] * ux[x][y]
```

```
    meq[4] = - rho[x][y] * ux[x][y]
    meq[5] = rho[x][y] * uy[x][y]
    meq[6] = - rho[x][y] * uy[x][y]
    meq[7] = rho[x][y] * (ux[x][y] * ux[x][y] - uy[x][y] * uy[x][y])
    meq[8] = rho[x][y] * ux[x][y] * uy[x][y]
    //Calculate MF
    MF[0] = 0.0
    MF[1] = 6.0 * (Fx[x][y] * ux[x][y] + Fy[x][y] * uy[x][y])
    MF[2] = - 6.0 * (Fx[x][y] * ux[x][y] + Fy[x][y] * uy[x][y])
    MF[3] = Fx[x][y]
    MF[4] = - Fx[x][y]
    MF[5] = Fy[x][y]
    MF[6] = - Fy[x][y]
    MF[7] = 2.0 * (Fx[x][y] * ux[x][y] - Fy[x][y] * uy[x][y])
    MF[8] = Fy[x][y] * ux[x][y] + Fx[x][y] * uy[x][y]
    //Calculate m_bar
    for all i do
      m_bar[i] = - S[i] * (m[i] - meq[i]) + (1.0 - 0.5 * S[i]) * MF[i]
    end for
    for all i do
      sum = 0.0
      for all j do
        sum += TransM_Inv[i][j] * m_bar[j]
      end for
      f[x][y][i] += sum
    end for
  end if
end for
```

## 2.6 Nondimensionalization

In the study of fluid dynamics, the process of nondimensionalization plays a crucial role in simplifying complex equations and revealing the fundamental relationships between physical parameters. By transforming the governing equations into dimensionless form, one can eliminate redundant physical scales and focus on the key factors that govern the behavior of a system. Such nondimensionalization is also important in the LBM. All the LBM models explained so far are in a nondimensionalized form. Since all variables are nondimensionalized, numerical analysis becomes more convenient. However, for simulations with specified physical dimensions, the process of nondimensionalizing physical dimensions is essential. In addition, selecting an appropriate nondimensionalization strategy can enhance the stability, accuracy, and efficiency of numerical analysis.

The process of nondimensionalizing physical dimensions is as follows. First, a conversion factor $C$ is needed to nondimensionalize the physical dimensions. For nondimensionalization, three conversion factors corresponding to mass, length, and time must be determined, each possessing physical dimensions. The relationship between the conversion factors and LBM variables is as follows.

$$\delta_m^{LB} C_m = \delta_m, \delta_l^{LB} C_l = \delta_L, \delta_t^{LB} C_t = \delta_t, \tag{2.37}$$

where $\delta^{LB}$ is a nondimensional LBM variables. Through (2.37), dimensionalization of LBM variables or nondimensionalization of physical variables can be achieved. For instance, if a nozzle with a diameter of 20 μm is represented by 50 lu (lattice units) in the LBM simulation, the length conversion factor can be calculated as follows.

$$C_l = \frac{D}{D^{LB}} = \frac{20 \times 10^{-6}\text{m}}{50} = 4.0 \times 10^{-7}\text{m} \tag{2.38}$$

Using a similar approach, the conversion factors for mass and time can also be calculated. If the fluid density is set to 1000 kg/m$^3$ in physical units and 1 in the LBM, the mass conversion factor is as follows.

$$C_m = \frac{\rho}{\rho^{LB}} C_l^3 = \frac{1000\text{kg}/\text{m}^3}{1} \left(4.0 \times 10^{-7}\text{m}\right)^3 = 6.4 \times 10^{-17}\text{kg} \tag{2.39}$$

If the kinematic viscosity of this fluid is $1.0 \times 10^{-6}$ m$^2$/s and the relaxation time is set to 1, then the time conversion factor is as follows.

$$C_t = \frac{\nu^{LB}}{\nu} C_l^2 = \frac{(1.0 - 0.5)/3}{1.0 \times 10^{-6}\text{m}^2/\text{s}} \left(4.0 \times 10^{-7}\text{m}\right)^2 = 2.7 \times 10^{-8}\text{s} \tag{2.40}$$

where $\nu^{LB}$ can be determined using (2.3).

Through the above process, the nondimensionalization procedure has been examined. This allows problems with physical dimensions to be transferred onto the LBM simulation domain. However, selecting appropriate nondimensional variables in the LBM is also essential. The following points should be considered when setting LBM nondimensional variables.

1. As the relaxation time approaches 0.5, the simulation can become unstable, and as it exceeds 1, it may become less accurate.
2. The reference velocity should not be greater than $c_s = 1/\sqrt{3} \approx 0.577$.
3. Pressure is defined as a relative value.
4. As the lattice size decreases, the resolution of the simulation improves, but computational efficiency may decline, and memory requirements can increase.

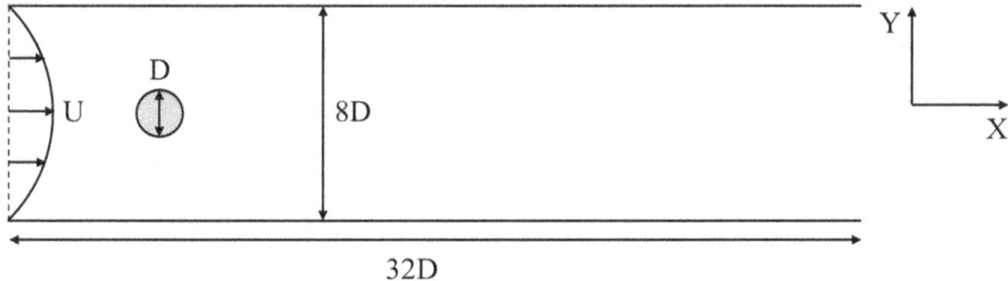


**Fig. 2.5** Example of the nondimensionalization process. 2D channel flow over the cylinder

Let's apply nondimensionalization to a 2D channel flow with a cylinder, as shown in Fig. 2.5, based on the topics covered in this chapter. A cylinder with a diameter $D = 5$ cm is positioned within a 2D channel. The channel has a width of $8D$ and a length of $32D$. At the inlet, a parabolic velocity profile is applied as the boundary condition, with the centerline velocity set to $U = 0.045$ m/s. The fluid has a density of 1kg/m$^3$ and a kinematic viscosity of $1.5 \times 10^{-5}$m$^2$/s. The Reynolds number, *Re* can be calculated using the information provided.

$$Re = \frac{\overline{U}D}{\nu} = \frac{(0.03\ \mathrm{m/s})(5\ \mathrm{cm})}{1.5 \times 10^{-5}\ \mathrm{m^2/s}} = 100, \tag{2.41}$$

where $\overline{U}$ is the average velocity at the inlet, given by $\overline{U} = \frac{2}{3}U$. If $D^{LB} = 40$, $\tau^{LB} = 1$, and $\rho^{LB} = 1$ are set, the conversion factors are as follows.

$$C_l = 0.00125\ \mathrm{m}, \quad C_m = 1.95 \times 10^{-9}\ \mathrm{kg}, \quad C_t = 0.0174\ \mathrm{s}, \tag{2.42}$$

The velocity conversion factor can be calculated as $C_v = C_l/C_t = 0.072$ m/s using (2.42). Therefore, the centerline velocity in the LB simulation is $U^{LB} = 0.625$. However, since this reference velocity $U^{LB} > c_s$, it is not suitable for the simulation. Then, let only $D^{LB}$ be changed to 200. In this case, $U^{LB} = 0.125$, satisfying the condition $U^{LB} < c_s$. However, as $D^{LB}$ increases by a factor of 5, the domain size grows by 25 times, and the time conversion factor decreases by 25 times, resulting in approximately 625 times the computational time required to reach the same physical time. As the number of lattice increases, the simulation may become more stable and accurate, but it can also consume excessive computational resources.

Then, let's set $D^{LB} = 60$, $\tau^{LB} = 0.6$, and $\rho^{LB} = 1$. In this case, $U = 0.0833 < c_s$, and computational efficiency can also be maintained. The relaxation time is also not excessively close to 0.5, ensuring a stable simulation. Through this example, it can be seen that the nondimensionalization process has a significant impact on the computational stability, accuracy, and efficiency of the LB simulation.

**Exercise 2.4 Pressure in LBM**

(a) Explain the meaning of "3. Pressure is defined as a relative value."
(b) In the example from Fig. 2.5, where the outlet is at atmospheric pressure and the inlet has a pressure boundary condition of 0.1 Pa, calculate the constant density at the inlet. (In this case, $D^{LB} = 60$, $\tau^{LB} = 0.6$, and $\rho^{LB} = 1$.)

**Solution**

(a) In the Navier–Stokes equation, the absolute pressure term does not appear; only the pressure gradient term, $\nabla p$, is present. This means that the change in pressure influences the motion of the fluid.
(b) The calculated conversion factors based on the given conditions are as follows.

$$C_l = 0.000833 \text{ m}, C_m = 5.79 \times 10^{-10} \text{ kg}, C_t = 0.00154 \text{ s}.$$

The pressure conversion factor calculated through this is $C_p = C_m / C_l C_t^2 = 0.2916$ Pa. When $\Delta p = p_{in} - p_{out}$, then $\Delta p = \Delta \rho^{LB} \left(c_s^{LB}\right)^2$.

$$\Delta \rho^{LB} = \frac{\Delta p}{\left(c_s^{LB}\right)^2} \frac{1}{C_p} = \frac{3(0.1\text{Pa})}{0.2916\text{Pa}} = 1.03$$

As a result, the density at the inlet is $\rho_{inlet}^{LB} = \rho^{LB} + \Delta \rho^{LB} = 2.03$.

## 2.7 Parallel Computing

A graphics processing unit (GPU) is a hardware device initially designed to handle the intensive parallel computations required for rendering graphics. While early GPUs were developed mainly for 3D rendering, rapid advancements in computational capabilities and parallel processing power have extended their use to general-purpose computing (GPGPU). Consequently, GPUs are now widely utilized across diverse fields, including scientific computing, artificial intelligence, data analysis, and engineering disciplines like fluid dynamics.

The primary advantage of GPUs lies in their ability to perform multiple operations concurrently through thousands of cores. Unlike central processing units (CPUs), which typically consist of a few high-performance cores suited for sequential processing of complex tasks, GPUs are optimized for parallel processing of simpler computations across many cores. Figure 2.6 represents the allocation ratio of transistors integrated into each processor, illustrated by the area in the diagram. A core is a computational unit that performs actual data calculations, such as addition, subtraction, and multiplication. The control unit manages tasks like the execution order of instructions, conditional branching, and scheduling. Cache and DRAM are used to retrieve and output data to be processed. The CPU allocates a significant portion of its resources to control orders and data caching. In contrast, the GPU

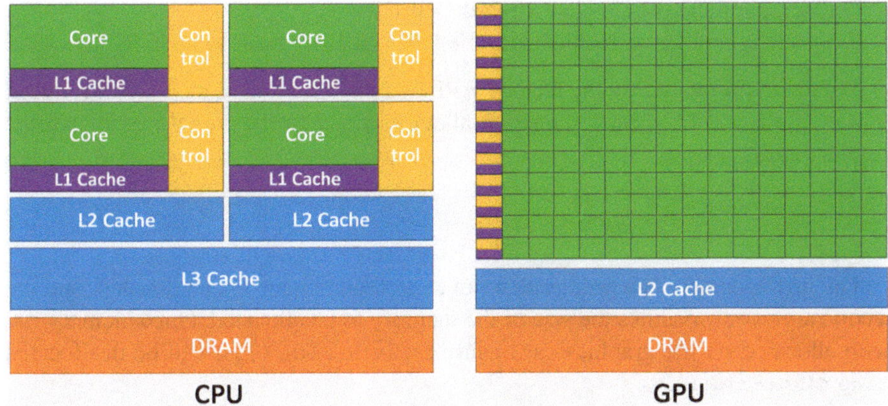


**Fig. 2.6** The ratio of transistor resource allocation between the CPU and GPU. (CUDA C++ Programming Guide)

dedicates much of its resources to cores to handle massive parallel computations, such as 3D graphics processing. This design makes GPUs particularly effective for tasks involving data parallelism, such as matrix or vector operations.

Many CFD methodologies have attempted to enhance computational efficiency by using GPGPU. Among them, LBM is particularly suited for GPU parallel computation due to the following characteristics [7].

- LBM works on a Cartesian grid, with each cell functioning independently.
- Calculations at each cell are simple, but there are usually a large number of cells.
- Transfer of data between the cell centers is ordered and can be utilized to make data access patterns conductive for implementing on a GPU.

Due to these characteristics of LBM, many researchers have achieved computational acceleration using GPU-based LBM [8–11]. This chapter introduces GPU-based LBM, and its optimization methodologies based on the compute unified device architecture (CUDA), an integrated development platform for GPGPU developed by NVIDIA.

### *2.7.1 Basics of CUDA*

To utilize the parallel computing capabilities of a graphics card, CUDA requires transferring input data stored in the CPU memory to the GPU memory, and then returning the computed results back to the CPU memory. This process follows the steps outlined below.

1. Allocate memory space on the graphics card.
2. Copy input data from the CPU memory to the graphics card's memory.

3. Perform parallel computation using the GPU.
4. Copy the computation results from the GPU memory back to the CPU memory.

The functions provided by the CUDA API for performing each step are introduced below. The `cudaMalloc()` function allocates space in the global memory of the graphics card.

```
cudaMalloc (void** devPtr, size_t size);
```

The first argument is a pointer that points to the memory to be allocated, and the second argument specifies the size of the memory to be allocated. Once memory has been allocated on the graphics card using `cudaMalloc()`, it can be deallocated using `cudaFree()`.

```
cudaFree (void* devPtr);
```

The argument is a pointer that points to the memory to be deallocated. The `cudaMemcpy()` function is used for data transfer between the CPU and GPU.

```
cudaMemcpy (void* dst, const void* src, size_t count, cudaMemcpyKind
kind);
```

The first argument is the pointer to the destination, the second argument is the pointer to the source, and the third argument specifies the size of the data to be copied. The fourth argument specifies the direction of data transfer. The types of `cudaMemcpyKind` for data transfer direction are as follows.

`cudaMemcpyHostToHost`: Copies data from the CPU (host) to the CPU (host).
`cudaMemcpyHostToDevice`: Copies data from the CPU to the GPU.
`cudaMemcpyDeviceToHost`: Copies data from the GPU to the CPU.
`cudaMemcpyDeviceToDevice`: Copies data from the GPU to the GPU.

The memory in CUDA is organized in a hierarchical structure according to the GPU's architecture, as shown in Fig. 2.7. Understanding and efficiently utilizing the CUDA memory hierarchy is crucial for enhancing computational efficiency in LBM simulations. The CUDA memory hierarchy consists of several levels, each with distinct characteristics in terms of size, speed, and scope of access:

**Registers:** Registers are the fastest type of memory available to each thread. They store variables that are frequently used by a thread, providing extremely low latency. However, the number of registers is limited, and efficient utilization of registers is key to achieving high performance in CUDA programs.

**Local Memory:** Despite its name, local memory is not physically local to a thread but rather resides in the global memory. It is used to store data that cannot fit into registers, typically arrays or dynamically allocated variables. Access to local

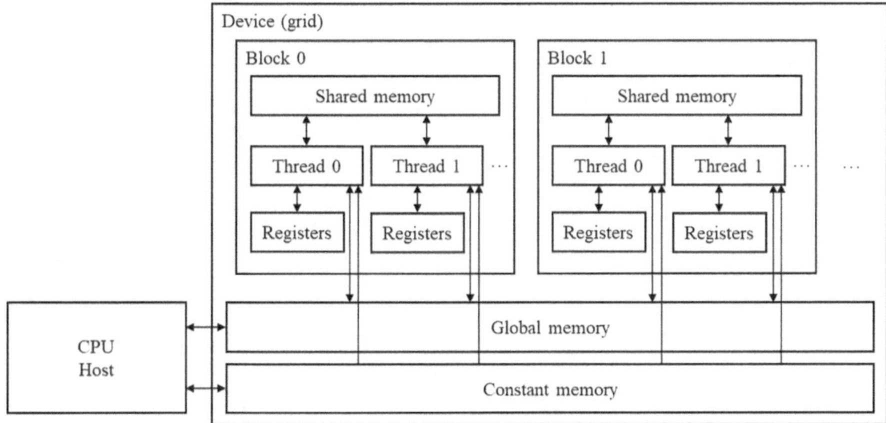


**Fig. 2.7** CUDA memory hierarchy

memory is slower compared to shared memory or registers, and minimizing its use can help improve performance.

**Shared Memory:** Shared memory is a smaller, faster memory that is shared among threads within the same block. It allows for efficient data exchange between threads, reducing the need for multiple global memory accesses. Shared memory is particularly useful for optimizing algorithms that require data reuse, as it can significantly improve memory access times compared to global memory.

**Global Memory:** This is the largest memory space available on the GPU, and it can be accessed by all threads across different blocks. However, global memory has relatively high latency compared to other types of memory, which means accessing it can be slow if not used carefully.

**Constant Memory:** Constant memory is specialized read-only memory spaces optimized for specific use cases. Constant memory is used to store data that remains constant across kernel launches and is cached for efficient access by all threads.

CUDA thread block architecture is a fundamental concept for understanding how to leverage the parallel processing capabilities of GPUs. CUDA threads are organized in a hierarchical structure of thread-block-grid, as shown in Fig. 2.8. A block is a collection of threads. Depending on the GPU device, a single block can contain up to 512 or 1024 threads. Threads within a block can be arranged in 1D, 2D, or 3D configurations, and in Fig. 2.8, the threads are arranged in a 2D layout. Each thread is assigned an index, and the `threadIdx` variable is used to distribute tasks to specific threads. Similarly, multiple thread blocks form a grid, and each block is scheduled independently by the GPU. Blocks within a grid can also be organized in a 3D configuration. `blockIdx` is a struct that contains the index of the block to which the current thread belongs, and `blockDim` is a struct that holds the dimensional information of the block.

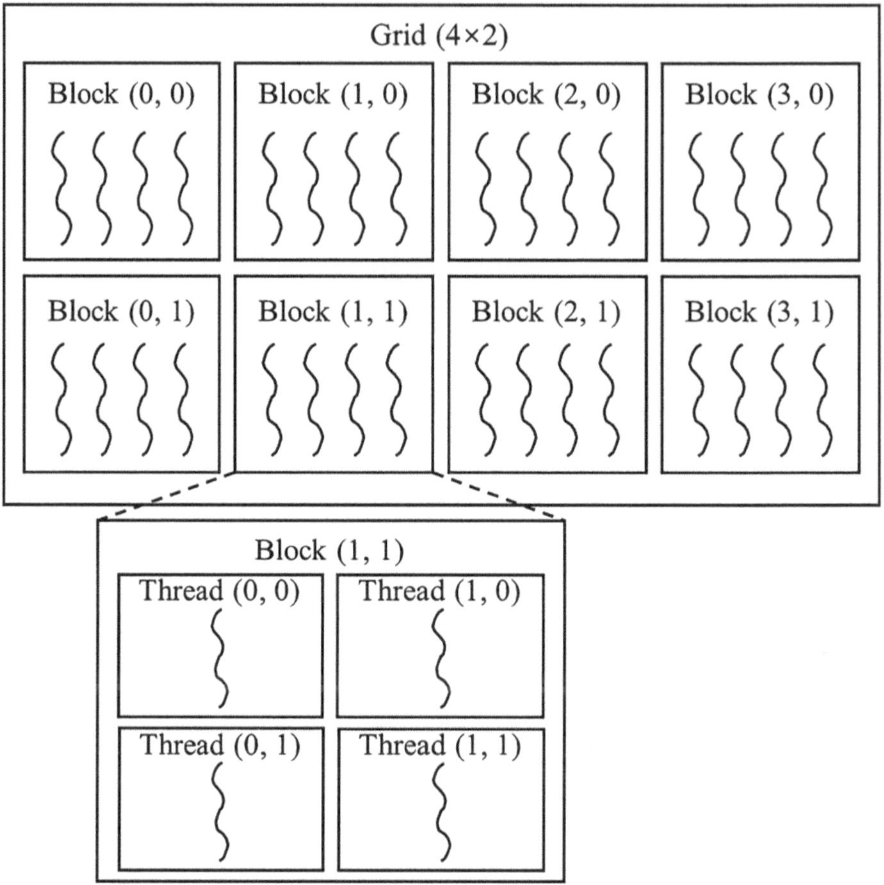


**Fig. 2.8** CUDA thread hierarchy

A kernel function is a function that runs on the GPU and is executed in parallel by multiple threads. Threads and blocks can be configured when the kernel function is executed. The syntax for calling a kernel function is as follows.

```
SomeKernel<<<StructureOfGrid, StructureOfBlock>>>(...);
```

Within <<< >>>, the structure of the grid and the structure of the block are configured. The grid and block structure are specified using `dim3`. For example, to set up a thread-block structure like in Fig. 2.8 and call the kernel function, the code would be as follows:

```
dim3 DimGrid(4, 2, 1);
dim3 DimBlock(2, 2, 1);
SomeKernel<<<DimGrid, DimBlock>>> (...);
```

This is an example of creating a 2D thread-block structure. To configure it in 3D, modify the third argument in `dim3`.

Having examined kernel function calls, we will now implement a kernel function through a matrix multiplication example and compare implementations in C and CUDA. Let **A** and **B** be square matrices of size 64×64. We aim to calculate their matrix product **C** = **AB**. The value at the *i*-th row and *j*-th column of matrix **C** can be calculated as follows:

$$C_{ij} = \sum_k A_{ik} B_{kj}. \tag{2.43}$$

(2.43) can be implemented in C as follows

```
void MatrixMultiplyC (int* A, int* B, int* C)
{
  int i, j, k, index, sum;
  for (i = 0; i < WIDTH; i++) {
    for (j = 0; j < HEIGHT; j++) {
      index = i * WIDTH + j;
      sum = 0;
      for (k = 0; k < WIDTH; k++) {
        sum += *(A + i * WIDTH + k) * *(B + k * WIDTH + j);
      }
      *(C + index) = sum;
    }
  }
}
```

As seen in the code, matrix multiplication requires a triple nested loop. In other words, a total of 262,144 operations are required to compute the multiplication of two 64×64 matrices. Let's implement the same matrix multiplication using a CUDA kernel function:

```
__global__ void MatrixMultiplyCUDA(int* A, int* B, int* C)
{
  int i = threadIdx.x + blockIdx.x * blockDim.x;
  int j = threadIdx.y + blockIdx.y * blockDim.y;
  int k, sum;
  sum = 0;
  for (k = 0; k < WIDTH; k++) {
    sum += A[i * WIDTH + k] * B[k * WIDTH + j];
  }
  C[i * WIDTH + j] = sum;
}
```

To call the `MatrixMultiplyCUDA` kernel, use the following code:

```
dim3 DimGrid(8, 8, 1);
dim3 DimBlock(8, 8, 1);
MatrixMultiplyCUDA<<<DimGrid, DimBlock>>> (dev_A,
dev_B, dev_C);
```

In the CUDA code, 4,096 threads are assigned to the task, with each thread performing only 64 operations. Through this parallel computation technique, the efficiency of calculations increases significantly as the number of operations grows.

## *2.7.2 Optimization Strategies for GPU-Based LBM*

Optimization is necessary to achieve maximum computational efficiency in GPU-based LBM. There are various optimization methodologies for GPU-based LBM, but this section introduces two simple yet highly efficient optimization strategies. The first is increasing data coalescence, and the second is modifying the propagation method.

Data coalescence refers to arranging data so that when threads within a warp access global memory, each thread reads or writes data located in contiguous memory addresses. This allows the GPU to handle the data of multiple threads in a single memory transaction. As shown in Fig. 2.9, data coalescence can be achieved through proper data storage methods. Figure 2.9a illustrates the Array of Structures (AoS) pattern, which is commonly used in CPU-based LBM. In the AoS pattern, the distribution functions at the same node are arranged in a one-dimensional array in the order of lattice velocities. As observed in the data access patterns of individual threads, the AoS pattern does not ensure coalescence because threads do not access

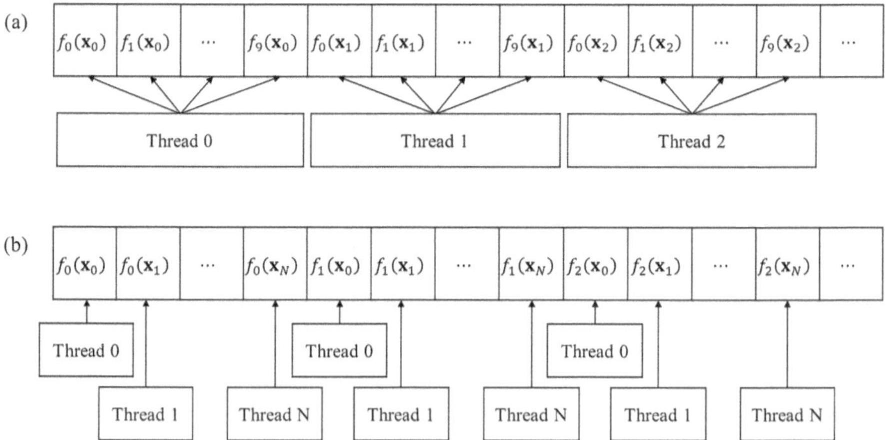


**Fig. 2.9** Data storage methods. (**a**) AoS scheme and (**b**) SoA scheme

contiguous memory locations. Figure 2.9b shows the Structure of Arrays (SoA) pattern, where distribution functions are grouped and arranged in memory based on their lattice velocities. The SoA pattern maximizes memory bandwidth utilization, enabling efficient parallel computation. Below is an example of how the indices of the distribution functions are organized in AoS and SoA formats:

```
index_AoS = i + x * Q + y * Q * XL + z * Q * XL * YL;
index_SoA = x + y * XL + z * XL * YL + i * XL * YL * ZL;
```

where `XL`, `YL`, and `ZL` represent the domain sizes along the x-, y-, and z-axes, respectively, and `Q` denotes the total number of discrete directions.

Data coalescence can also be achieved by modifying the propagation method. As discussed in the previous section, LBM consists of calculating the collision process of distribution functions at each node, followed by propagating the distribution functions to adjacent nodes in sequence. This method is called the push scheme, as illustrated in Fig. 2.10a. However, the pull scheme shown in Fig. 2.10b first gathers distribution functions from adjacent nodes and then calculates the collision process. In the push scheme, coalesced reads occur during the collision process, while uncoalesced writes occur during the propagation process. Conversely, in the pull scheme, uncoalesced reads occur during the collision process, and coalesced writes occur during the propagation process. According to previous research, uncoalesced reads are known to be faster than uncoalesced writes, making the pull scheme more efficient for achieving higher computational performance.

In addition to the pull scheme, minimizing access to global memory can enhance computational efficiency. This can be achieved by merging the propagation kernel and the collision kernel. If the algorithm is composed of two kernels, one for collision and the other for propagation, it requires two global memory read and write operations. However, by merging the propagation and collision into a single kernel, these operations can be reduced to a single read and write process.

## 2.8 Project: Vortex Shedding

Kármán vortex refers to the alternating vortex street that forms behind an object when fluid flows past it at a constant speed. It is most commonly observed behind cylindrical obstacles, where a sequence of vortices develops periodically downstream of the object (Fig. 2.11). Kármán vortex shedding occurs at moderate Reynolds numbers, typically in the range of 40 to 200. At these Reynolds numbers, vortices are shed alternately from either side of the obstacle, creating an oscillating wake pattern that can cause instabilities and induce vibrations in the object. The Kármán vortex is an important phenomenon in various practical applications. For instance, it is used to explain vibration issues in structures such as chimneys, bridges,

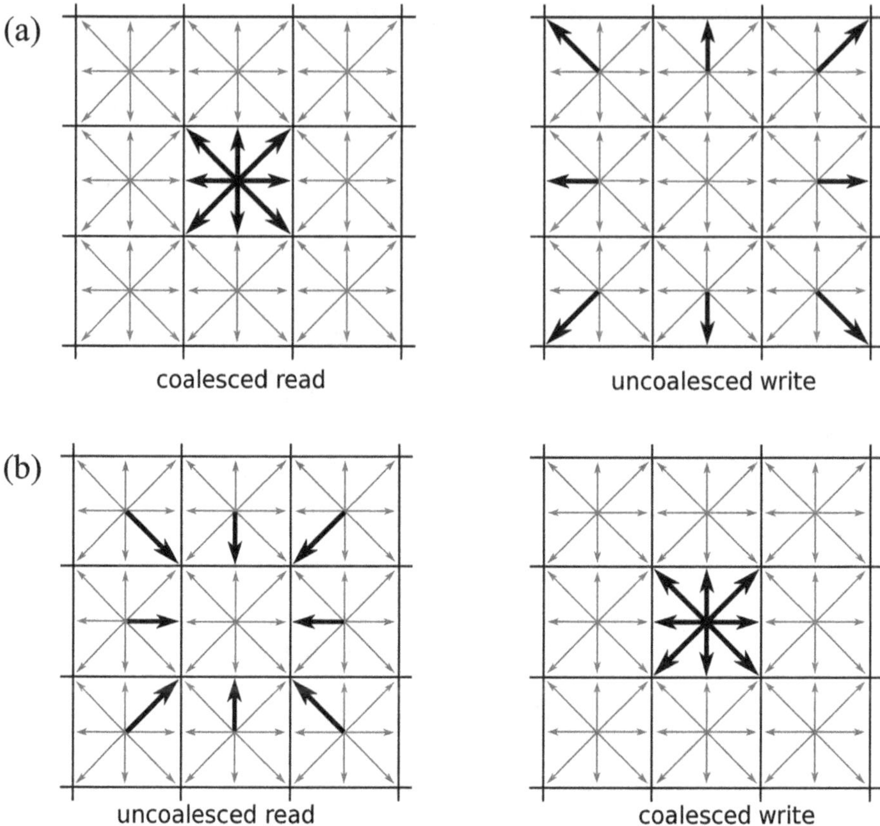


**Fig. 2.10** (**a**) Push scheme and (**b**) pull scheme [12]

and pipelines. As shown in Fig. 2.12, it can also be observed in meteorological phenomena.

In this project, the Kármán vortex phenomenon is simulated based on the fundamentals of LBM discussed in this chapter. Through this example simulation, readers will gain an overall understanding of the LBM algorithm, visually comprehend the vortex formation process, and analyze the flow characteristics.

Figure 2.13 shows a schematic diagram of the simulation domain for observing the vortex shedding phenomenon. A cylinder with a diameter $D$ is placed along the centerline of the domain. A parabolic velocity profile is applied at the inlet. The simulation conditions are as follows.

- The height of the domain is $H = 8D$, the length is $L = 32D$, and the center of the cylinder is located $6D$ from the inlet.

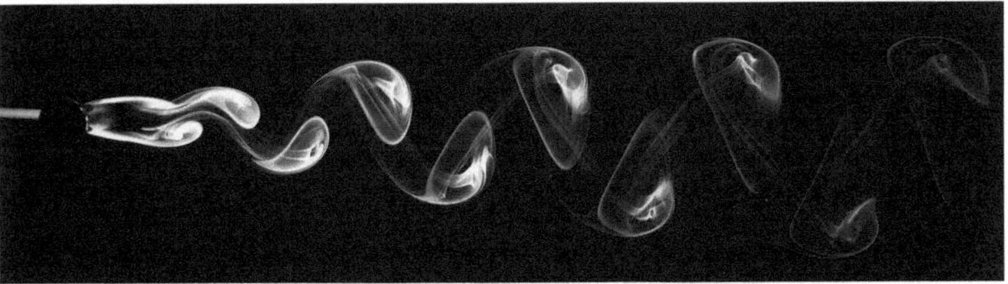

**Fig. 2.11** Visualization of the vortex street behind a circular cylinder

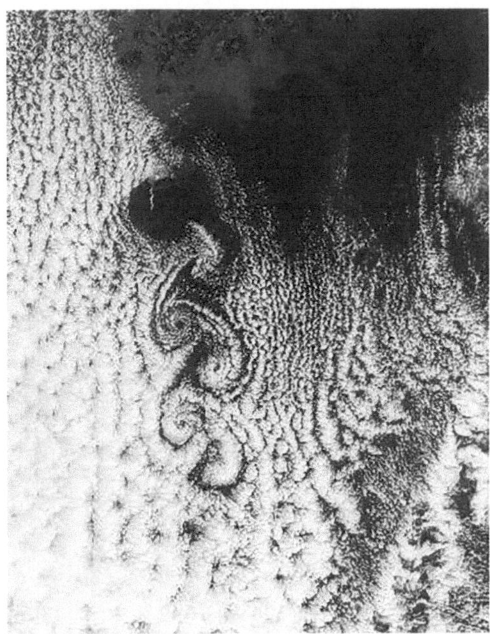

**Fig. 2.12** Kármán vortex caused by wind flow around Jeju Island

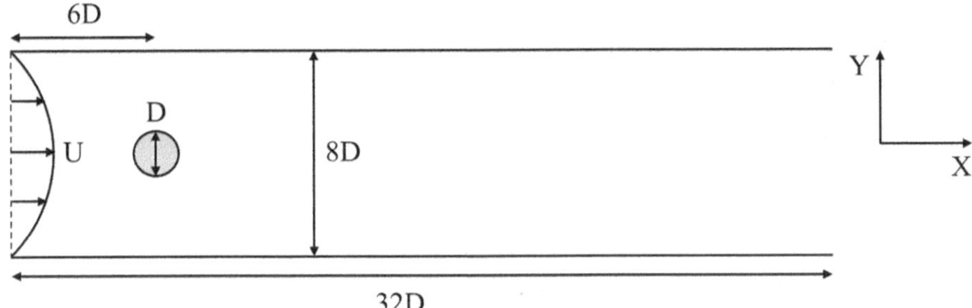


**Fig. 2.13** Schematic diagram of simulation domain for vortex shedding

- The inlet (left side) applies a velocity boundary condition, the outlet (right side) uses a constant pressure boundary condition, and no-slip boundary conditions are applied to the top and bottom surfaces as well as the cylinder.
- The velocity profile at the inlet is applied as follows:

$$U_{inlet}(y) = \frac{4Uy(H-y)}{H^2}. \tag{2.44}$$

Simulate for $Re = 5$, $Re = 60$, and $Re = 150$, and analyze the differences in flow characteristics for each case. $Re$ is defined in (2.41).

**Summary**

1. **Lattice Boltzmann background**

   - LBM is derived from Boltzmann's equation using velocity discretization.
   - Macroscopic behavior arises through Chapman–Enskog expansion.
   - Collision–streaming structure enables locality and parallelism.

2. **Single-Relaxation-Time (SRT or BGK) model**

   - Collision operator uses a single relaxation time $\tau$.
   - Viscosity relates directly to $\tau$:

   $$\nu = c_s^2\left(\tau - \frac{1}{2}\right)\Delta t$$

   - Advantages: simplicity, computational efficiency.
   - Limitations: stability issues for low viscosity; limited control over higher-order moments.

3. **Streaming step**

   - Distributions move along discrete lattice directions $f_i(\mathbf{x} + \mathbf{e}_i\Delta t, t + 1)$.

4. **Boundary conditions**

   - Bounce-back: simple, accurate for no-slip at walls.
   - Halfway bounce-back improves second-order accuracy.
   - Zou–He and Guo schemes handle pressure/velocity boundaries.

5. **Forcing schemes**

   - Important for modeling: body forces, immersed particles, gravity, magnetic/thermal effects.
   - Guo forcing term ensures second-order accuracy.
   - EDM computes forcing by exact integration of the Boltzmann equation, improving stability for strong forces.

6. **Multiple-Relaxation-Time (MRT) model**
   - Uses moment space $\mathbf{m} = \mathbf{Mf}$.
   - Each moment relaxes independently via a diagonal matrix $\mathbf{S}$.
   - Improves stability, reduces non-physical (ghost) modes, enhances accuracy.
7. **GPU computation**
   - Streaming–collision steps consist of uniform, local operations $\rightarrow$ ideal for GPU parallelism.
   - Structure-of-arrays (SoA) memory layout ensures coalesced memory access.
   - Pull-gather scheme reduces memory divergence and cache thrashing.
   - Achieves significant speedups compared to CPU-based Navier–Stokes solvers.

## References


1. X. He, and L.-S. Luo, "Theory of the lattice boltzmann method: From the boltzmann equation to the lattice boltzmann equation," Physical review E **56**, 6811 (1997).
2. S. Leclaire, A. Parmigiani, O. Malaspinas, B. Chopard, and J. Latt, "Generalized three-dimensional lattice boltzmann color-gradient method for immiscible two-phase pore-scale imbibition and drainage in porous media," Phys. Rev. E **95**, 033306 (2017).
3. Q. Zou, and X. He, "On pressure and velocity boundary conditions for the lattice boltzmann bgk model," Phys. Fluids **9**, 1591-1598 (1997).
4. Z. Guo, C. Zheng, and B. Shi, "Discrete lattice effects on the forcing term in the lattice boltzmann method," Physical review E **65**, 046308 (2002).
5. A. L. Kupershtokh, D. A. Medvedev, and D. I. Karpov, "On equations of state in a lattice boltzmann method," Comput. Math. with Appl. **58**, 965-974 (2009).
6. P. Lallemand, and L.-S. Luo, "Theory of the lattice boltzmann method: Dispersion, dissipation, isotropy, galilean invariance, and stability," Physical review E **61**, 6546 (2000).
7. O. Navarro-Hinojosa, S. Ruiz-Loza, and M. Alencastre-Miranda, "Physically based visual simulation of the lattice boltzmann method on the gpu: A survey," J. Supercomput. **74**, 3441-3467 (2018).
8. M. Wen, S. Shen, and W. Li, "Gpu parallel implementation of a finite volume lattice boltzmann method for incompressible flows," Comput. Fluids 106460 (2024).
9. S. Fu, and L. Wang, "Gpu-based unresolved lbm-dem for fast simulation of gas–solid flows," Chemical Engineering Journal **465**, 142898 (2023).
10. H. M. Lee, and J. S. Lee, "Effects of heat transfer on particle suspended drop-on-demand inkjet printing using lattice boltzmann method," Appl. Therm. Eng. **213**, (2022).
11. H. M. Lee, and J. S. Lee, "Particle deposition dynamics in evaporating droplets using lattice boltzmann and magnetic particle simulation," Phys. Fluids **35**, (2023).
12. N. Delbosc, J. L. Summers, A. Khan, N. Kapur, and C. J. Noakes, "Optimized implementation of the lattice boltzmann method on a graphics processing unit towards real-time fluid simulation," Comput. Math. with Appl. **67**, 462-475 (2014).

# Chapter 3
# Lattice Boltzmann Method: Multiphase

## 3.1 Overview of Multiphase LBM

Multiphase flow refers to the simultaneous flow of different phases, such as liquid, gas, or solid, within a system. Multiphase flows are prevalent in both natural and industrial settings, making their understanding crucial for a wide range of applications. In particular, inkjet printing technology requires precise generation, control, and deposition of micron-sized droplets, which are ejected from a nozzle and travel through a gaseous environment before impacting a substrate. The multiphase nature of this process involves complex interactions between the liquid phase of the ink and the surrounding gas phase, including effects such as droplet breakup, evaporation, and spreading upon impact. Accurate modeling of these interactions is critical to optimize print quality, prevent clogging of the nozzle, and ensure consistent droplet formation.

The LBM has emerged as a powerful tool for simulating multiphase flows. Unlike traditional CFD methods that solve the Navier–Stokes equations directly, LBM is based on the mesoscopic Boltzmann equation, allowing for a more intuitive and flexible treatment of interface dynamics and phase interactions. This mesoscopic approach bridges the microscopic particle-level interactions and the macroscopic continuum-scale flow properties, making it well-suited for simulating flows with complex interface dynamics. LBM's ability to naturally incorporate interface dynamics without the need for explicitly tracking the interface position is particularly advantageous. This makes it especially effective for handling complex boundary conditions, intricate interfacial phenomena such as bubble formation, coalescence, and droplet breakup, and situations involving dynamic interfaces that change shape or topology.

LBM for multiphase flow typically employs the color-gradient model, pseudopotential model, or the free-energy model. The color-gradient model, proposed by Rothman and Keller [1], is the first multiphase LBM model. The initial

 
H. M. Lee, J. S. Lee, *Lattice Boltzmann Methodology for Single-Phase and Multiphase Nanoparticle Modeling*,
https://doi.org/10.1007/978-981-95-9117-6_3

color-gradient model was numerically unstable at different density ratios, but it was later improved through extensive research. This model can apply surface tension independently of density and viscosity ratios, making it suitable for simulating phenomena where capillary flow plays a significant role. However, since two distribution functions are used, computational efficiency may decrease. The pseudopotential model is a multiphase model proposed by Shan and Chen [2] in 1993. It introduces a pseudopotential function that induces interparticle forces, allowing for the phase separation. The pseudopotential model is widely used due to its simplicity in simulating attractive and repulsive interactions between phases. However, this model has limitations, as the magnitude of surface tension varies with density and viscosity ratios, and it suffers from thermodynamic inconsistency. The free-energy model [3], on the other hand, is well-suited for applications requiring a thermodynamically consistent description of interfacial properties, such as surface tension and phase coexistence. Subsequently, Inamuro et al. [4] developed an improved free-energy model that is stable even at high density and viscosity ratios. However, this model has the limitation of requiring the computationally intensive solution of the Poisson equation. Each of these models has its strengths and limitations, making them suitable for different multiphase flow problems depending on the required level of accuracy, computational resources, and specific physical phenomena being studied.

This chapter introduces the fundamentals of multiphase LBM and methodologies for implementing its models. Among the various multiphase models, the color-gradient model and the pseudopotential model, which are widely used for inkjet printing simulations, are primarily discussed. The color-gradient model has been utilized for observing droplet shapes influenced by capillary effects during ejection, such as in DoD inkjet printing [5] and direct writing [6]. On the other hand, the pseudopotential model has been employed to study phase change phenomena induced by heat transfer, such as in thermal inkjet printing [7] and droplet evaporation [8, 9]. The methodologies of the multiphase LBM models presented in this chapter enable the simulation of high-density and high-viscosity ratio DoD inkjet printing processes as well as thermodynamically coupled droplet evaporation phenomena.

**Exercise 3.1 Fundamentals of Multiphase LBM**
Explain why LBM is advantageous for simulating multiphase flows compared to traditional CFD methods.

**Solution**
LBM is advantageous for multiphase flows because it is based on mesoscopic distribution functions rather than directly solving the Navier–Stokes equations. This kinetic formulation offers:

- A simple streaming–collision algorithm on uniform lattices, which is highly suitable for parallel computation (e.g., GPU).

- Natural handling of complex interfaces and topological changes (breakup, coalescence, wetting/dewetting) without explicit interface reconstruction or front tracking.
- Straightforward implementation of complex boundaries and wetting conditions through local rules (e.g., bounce-back, virtual density, contact angle control).
- Easy extension to multi-physics (thermal effects, phase change, external forces) by adding or modifying distribution functions and source terms.

## 3.2 D3Q19 Color-Gradient LBM

The color-gradient model uses two distribution functions to model different phases. The two distribution functions are distinguished by the phase index $k$ and are commonly referred to as red and blue fluid. The collision and streaming processes of the red and blue fluid are performed independently of each other. The collision function in the color-gradient model consists of three parts.

$$f_i^k(\mathbf{x}+\mathbf{e}_i\Delta t, t+\Delta t) - f_i^k(\mathbf{x},t) = \left(\Omega_i^k\right)^{(3)}\left[\left(\Omega_i^k\right)^{(1)} + \left(\Omega_i^k\right)^{(2)}\right], \tag{3.1}$$

where $\left(\Omega_i^k\right)^{(1)}$ is the single-phase collision operator, $\left(\Omega_i^k\right)^{(2)}$ is the perturbation operator, and $\left(\Omega_i^k\right)^{(3)}$ is the recoloring operator. The density and velocity of each phase can be calculated as follows:

$$\rho_k = \sum_i f_i^k, \rho = \sum_k \rho_k, \rho\mathbf{u} = \sum_i\sum_k \mathbf{e}_i f_i^k, \tag{3.2}$$

While Chap. 2 used D2Q9 LBM as an example, this chapter extends the explanation to D3Q19. The discrete velocity set of the D3Q19 model is

$$\mathbf{e}_i = \begin{bmatrix} 0 & 1 & -1 & 0 & 0 & 0 & 0 & 1 & -1 & 1 & -1 & 1 & -1 & 1 & -1 & 0 & 0 & 0 & 0 \\ 0 & 0 & 0 & 1 & -1 & 0 & 0 & 1 & -1 & -1 & 1 & 0 & 0 & 0 & 0 & 1 & -1 & 1 & -1 \\ 0 & 0 & 0 & 0 & 0 & 1 & -1 & 0 & 0 & 0 & 0 & 1 & -1 & -1 & 1 & 1 & -1 & -1 & 1 \end{bmatrix} c. \tag{3.3}$$

The single-phase collision operator is identical to the BGK model.

$$\left(\Omega_i^k\right)^{(1)} = -\frac{1}{\tau}\left[f_i^k(\mathbf{x},t) - f_i^{k,eq}(\mathbf{x},t)\right]. \tag{3.4}$$

However, the equilibrium function is modified as follows to maintain stability under different density ratios:

$$f_i^{k,eq} = \rho_k \left\{ \phi_i^k + \omega_i \left[ \frac{3}{c^2} (\mathbf{e}_i \cdot \mathbf{u}) + \frac{9}{2c^2} (\mathbf{e}_i \cdot \mathbf{u})^2 - \frac{3}{2c^2} |\mathbf{u}|^2 \right] \right\}. \tag{3.5}$$

In the D3Q19 model, $\phi_i^k$ is defined as follows:

$$\phi_i^k = \begin{cases} \alpha_k, & i = 0 \\ (1 - \alpha_k)/12, & i = 1, \cdots, 6 \\ (1 - \alpha_k)/24, & i = 7, \cdots 18 \end{cases}, \tag{3.6}$$

where $\alpha_k$ is a parameter that varies depending on the density ratio.

$$\frac{\rho_R}{\rho_B} = \frac{1 - \alpha_B}{1 - \alpha_R}. \tag{3.7}$$

In the color-gradient model, the speed of sound is determined by $\alpha_k$, and the pressure follows the relationship $p_k = \rho_k \left(c_s^k\right)^2 = 0.5\rho_k c^2 (1 - \alpha_k)$.

In single-phase flow of Newtonian fluids, only one viscosity value is required, and thus the relaxation time is also constant. However, in multiphase flows, the viscosities of each phase need to be specified, and the relaxation times of the red and blue fluids are defined as $\tau^R$ and $\tau^B$, respectively. At the interface where the two phases are mixed, the $\tau$ should take a value between $\tau^R$ and $\tau^B$. To achieve this, the $\tau$ is formulated as follows [10]:

$$\tau = \begin{cases} \tau^R, & \rho^N > \delta \\ s_1 + s_2 \rho^N + s_3 \left(\rho^N\right)^2, & \delta \geq \rho^N > 0 \\ t_1 + t_2 \rho^N + t_3 \left(\rho^N\right)^2, & 0 \geq \rho^N \geq -\delta \\ \tau^B, & \rho^N < -\delta \end{cases}, \tag{3.8}$$

where $\delta$ is a free parameter related to the interface thickness, and $s$ and $t$ are the coefficients of the parabolic functions, defined as follows:

$$s_1 = \frac{2\tau^R \tau^B}{\tau^R + \tau^B}, s_2 = \frac{2(\tau^R - s_1)}{\delta}, s_3 = -\frac{s_2}{2\delta},$$
$$t_1 = \frac{2\tau^R \tau^B}{\tau^R + \tau^B}, t_2 = \frac{2(t_1 - \tau^B)}{\delta}, t_3 = \frac{t_2}{2\delta}. \tag{3.9}$$

$\rho^N$ is a phase parameter that represents the state of the phases.

$$\rho^N = \frac{\rho_R/\rho_R^{in} - \rho_B/\rho_B^{in}}{\rho_R/\rho_R^{in} + \rho_B/\rho_B^{in}}, \tag{3.10}$$

where $\rho^{in}$ is the initial density. $\rho^N$ ranges between -1 and 1, where values closer to 1 represent the red fluid, values closer to -1 represent the blue fluid, and a value of 0 indicates the fluid interface.

Below, (3.2) and (3.8)–(3.10) are implemented in pseudocode:

```
//Moment update
for all (x,y,z) do
  if (x,y,z) is fluid node then
   sum_R = sum_B = sum_ux = sum_uy = sum_uz = 0.0
    for all i do
     sum_R += f[x][y][z][i][0]
     sum_B += f[x][y][z][i][1]
     sum_ux += (f[x][y][z][i][0] + f[x][y][z][i][1]) * ex[i]
     sum_uy += (f[x][y][z][i][0] + f[x][y][z][i][1]) * ey[i]
     sum_uz += (f[x][y][z][i][0] + f[x][y][z][i][1]) * ez[i]
    end for
    rho[x][y][z][0] = sum_R
    rho[x][y][z][1] = sum_B
    rho[x][y][z] = sum_R + sum_B
    ux[x][y][z] = sum_ux / (sum_R + sum_B)
    uy[x][y][z] = sum_uy / (sum_R + sum_B)
    uz[x][y][z] = sum_uz / (sum_R + sum_B)
    rhoN[x][y][z] = (rho[x][y][z][0] / rho_R_in - rho[x][y][z][1] /
rho_B_in)
          / (rho[x][y][z][0] / rho_R_in + rho[x][y][z][1] / rho_B_in)
    pressure[x][y][z][0] = 0.5 * rho[x][y][z][0] * (1.0 - ALPHA[0])
    pressure[x][y][z][1] = 0.5 * rho[x][y][z][1] * (1.0 - ALPHA[1])
    if rhoN[x][y][z] > delta then
      tau[x][y][z] = tau_R
    else if rhoN[x][y][z] > 0.0 then
      tau[x][y][z] = s1 + s2 * rhoN[x][y][z] + s3 * rhoN[x][y][z] *
           rhoN[x][y][z]
    else if rhoN[x][y][z] > - delta then
      tau[x][y][z] = t1 + t2 * rhoN[x][y][z] + t3 * rhoN[x][y][z] *
           rhoN[x][y][z]
    else
      tau[x][y][z] = tau_B
    end if
  end if
end for
```

The pseudocode for the single-phase collision operator is as follows:

```
//Collision function 1
```

```
for all (x,y,z) do
  if (x,y,z) is fluid node then
   udotu = ux[x][y][z] * ux[x][y][z] + uy[x][y][z] * uy[x][y][z] + uz[x]
[y][z]
        * uz[x][y][z]
     for all i do
      udote = ux[x][y][z] * ex[i] + uy[x][y][z] * ey[i] + uz[x][y][z] * ez
[i]
      for all k do
        feq = rho[x][y][z][k] * (phi[i][k] + omega[i] * (3.0 * udote + 4.5
          * udote * udote - 1.5 * udotu))
        f[x][y][z][i][k] += - (f[x][y][z][i][k] - feq) / tau[x][y][z]
      end for
     end for
  end if
end for
```

To generate surface tension, the second collision function, known as the perturbation operator, $\left(\Omega_i^k\right)^{(2)}$ is required [11].

$$\left(\Omega_i^k\right)^{(2)} = \frac{A_k}{2}\left|\nabla\rho^N\right|\left[\omega_i\frac{(\mathbf{e}_i\cdot\nabla\rho^N)^2}{|\nabla\rho^N|^2} - B_i\right], \tag{3.11}$$

where $B_i$ is

$$B_i = \begin{cases} -1/3, & i=0, \\ 1/18, & 1\leq i\leq 6, \\ 1/36, & 7\leq i\leq 18. \end{cases} \tag{3.12}$$

The perturbation operator includes the gradient of the phase parameter, $\nabla\rho^N$. As depicted schematically in Fig. 3.1, there is little variation in $\rho^N$ within the red and blue fluids, while the greatest change occurs at the interface. In other words, surface tension is automatically applied at the fluid interface, where $\nabla\rho^N$ is at its maximum. This is why the model is called the color-gradient model. In LBM, the gradient is calculated by discretization as follows:

$$\nabla\phi \approx \frac{3}{c^2\Delta t}\sum_i \omega_i\phi(\mathbf{x} + \mathbf{e}_i\Delta t)\mathbf{e}_i, \tag{3.13}$$

where $\phi$ is an arbitrary scalar field. The magnitude of the surface tension derived using (3.11) is expressed as

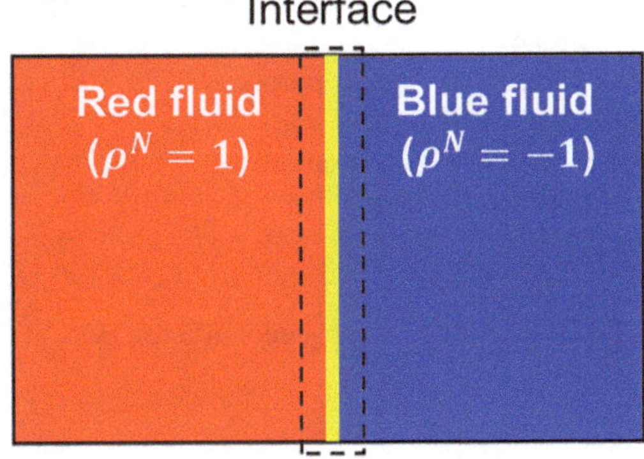


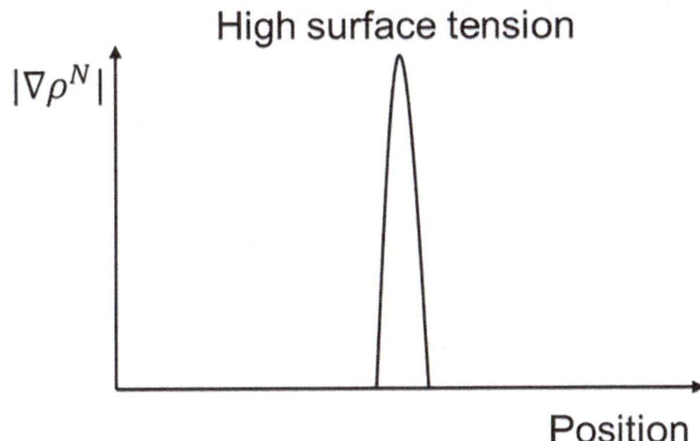


**Fig. 3.1** Schematic representing the magnitude of surface tension at the interface between red and blue fluids

$$\sigma = \frac{2}{9}(A_R + A_B)\tau c^4 \Delta t, \tag{3.14}$$

As seen in (3.11), the magnitude of the surface tension, $\sigma$ can be adjusted independently of the relaxation time through the tuning of $A_k$.

To calculate $\left(\Omega_i^k\right)^{(2)}$, $\nabla\rho^N$ needs to be determined first. The following is the pseudocode for calculating $\nabla\rho^N$:

```
//Gradient
for all (x,y,z) do
  if (x,y,z) is fluid node then
   sumx = sumy = sumz = 0.0
     for all i do
       xtemp = nx[x][y][z][i]
       ytemp = ny[x][y][z][i]
       ztemp = nz[x][y][z][i]
     sumx += 3.0 * omega[i] * rhoN[xtemp][ytemp][ztemp] * ex[i]
     sumy += 3.0 * omega[i] * rhoN[xtemp][ytemp][ztemp] * ey[i]
```

```
      sumz += 3.0 * omega[i] * rhoN[xtemp][ytemp][ztemp] * ez[i]
     end for
     gradx[x][y][z] = sumx
     grady[x][y][z] = sumy
    gradz[x][y][z] = sumz
  end if
end for
```

The following is the pseudocode for $\left(\Omega_i^k\right)^{(2)}$:

```
//Collision function 2
for all (x,y,z) do
  if (x,y,z) is fluid node then
   grad = sqrt(gradx[x][y][z] * gradx[x][y][z] + grady[x][y][z] * grady
[x][y][z]
       + gradz[x][y][z] * gradz[x][y][z])
    if grad > 0.0 then
      for all i do
        for all k do
          f[x][y][z][i][k] += 0.5 * A[k] * grad * (omega[i] * pow((ex[i]
                   * gradx + ey[i] * grady + ez[i] * gradz) /
                  grad, 2) - B[i])
        end for
      end for
    end if
  end if
end for
```

Finally, the recoloring operator $\left(\Omega_i^k\right)^{(3)}$ is applied to thin the thickness of the diffuse interface (Fig. 3.2). To address the lattice pinning problem, the recoloring operator proposed by Latva-Kokko and Rothman [12] is as follows:

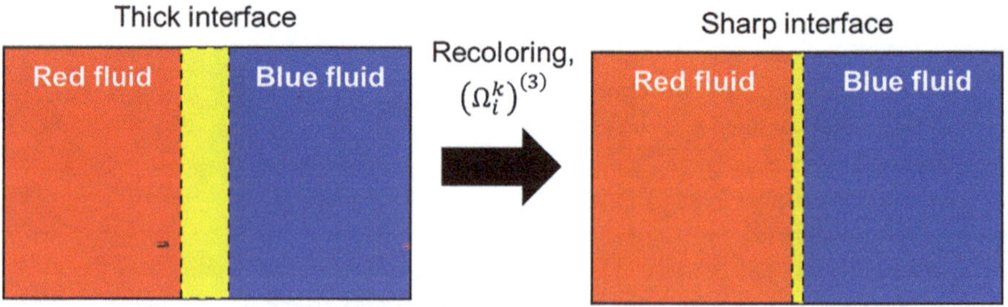


**Fig. 3.2** Recoloring operator introduced to ensure a sufficiently thin interface thickness.

$$\left(\Omega_i^R\right)^{(3)} = \frac{\rho_R}{\rho}\left(f_i^R + f_i^B\right) + \beta\frac{\rho_R\rho_B}{\rho^2}\cos(\varphi_i)\sum_k f_i^{k,eq}(\rho_k, \alpha_k, \mathbf{u}=0), \tag{3.15}$$

$$\left(\Omega_i^B\right)^{(3)} = \frac{\rho_B}{\rho}\left(f_i^R + f_i^B\right) - \beta\frac{\rho_R\rho_B}{\rho^2}\cos(\varphi_i)\sum_k f_i^{k,eq}(\rho_k, \alpha_k, \mathbf{u}=0), \tag{3.16}$$

where $\beta$ is a parameter between 0 and 1 that controls the thickness of the interface. The closer $\beta$ is to 1, the thinner the interface becomes, but numerical instability may increase. $\varphi_i$ is the angle between $\nabla\rho^N$ and the direction $\mathbf{e}_i$. Thus $\cos(\varphi_i)$ can be expressed as $\cos(\varphi_i) = (\mathbf{e}_i \cdot \nabla \rho^N)/(|\mathbf{e}_i| \nabla \rho^N)$. The pseudocode for the recoloring operator can be represented as follows:

```
//Collision function 3
for all (x,y,z) do
  if (x,y,z) is fluid node then
   grad = sqrt(gradx[x][y][z] * gradx[x][y][z] + grady[x][y][z] * grady
[x][y][z]
      + gradz[x][y][z] * gradz[x][y][z])
    if grad > 0.0 then
      for all i do
        edote = ex[i] * ex[i] + ey[i] * ey[i] + ez[i] * ez[i]
        if i is 0 then
          cosRB[i] = 0
        else
          cosRB[i] = (ex[i] * gradx[x][y][z] + ey[i] * grady[x][y][z] +
            ez[i] * gradz[x][y][z]) / (sqrt(edote) * grad)
        end if
        f[x][y][z][i][0] = (f[x][y][z][i][0] + f[x][y][z][i][1]) *
          rho[x][y][z][0] / rho[x][y][z] + BETA * (rho[x][y][z][0] * rho
[x][y][z][1] / (rho[x][y][z] * rho[x][y][z])) * cosRB * (rho[x][y][z]
[0] * phi[i][0] + rho[x][y][z][1] * phi[i][1])
        f[x][y][z][i][1] = (f[x][y][z][i][0] + f[x][y][z][i][1]) *
          rho[x][y][z][1] / rho[x][y][z] - BETA * (rho[x][y][z][0] * rho
[x][y][z][1] / (rho[x][y][z] * rho[x][y][z])) * cosRB * (rho[x][y][z]
[0] * phi[i][0] + rho[x][y][z][1] * phi[i][1])
      end for
    end if
  end if
end for
```

**Exercise 3.2 D3Q19 Color-Gradient Model**

(a) Write the form of the BGK collision operator in the D3Q19 color-gradient framework and explain the meaning of the modified equilibrium function.
(b) Discuss why two distribution functions (red/blue) are needed and how the density and velocity fields are reconstructed.

**Solution**

(a) In the D3Q19 color-gradient framework, each phase (e.g., red $r$ and blue $b$) has its own distribution function $f_i^k(k=r,b)$. The BGK collision operator is

$$\Omega_i^k = \frac{1}{\tau_k}\left(f_i^k - f_{ik}^{k,eq}\right)$$

where $f_{ik}^{k,eq}$ is a modified equilibrium distribution that includes not only the usual hydrodynamic terms (density, velocity) but also additional corrections to account for:

- local mixture velocity (common velocity for both phases),
- the effect of interfacial tension through perturbation terms,
- possibly higher-order moments needed for stability.

Thus, the modified equilibrium ensures that the recovered macroscopic equations include appropriate pressure and surface tension contributions at the interface.

(b) Two distribution functions are needed because the color-gradient model treats each phase as a separate "fluid component":

- For each phase $k$, $f_i^k$ carries the partial density and momentum of that phase.
- The phase densities are obtained by summation:

$$\rho_k = \sum_i f_i^k, \rho = \rho_r + \rho_b$$

- The mixture velocity is reconstructed from the total momentum:

$$\rho\mathbf{u} = \sum_k \sum_i f_i^k \mathbf{c}_i + \frac{\Delta t}{2}\mathbf{F}$$

The recoloring and perturbation operators then redistribute the post-collision distributions between red and blue to maintain phase segregation and impose surface tension while evolving a single hydrodynamic velocity field.

## 3.3 Improved Color-Gradient LBM

So far, the BGK color-gradient LBM has been explained. The initial color-gradient model proposed by Rothman and Keller [1] had the limitation of being able to simulate only cases with a density ratio of 1. However, the introduction of a modified equilibrium function enabled simulations with density ratios close to 10. Despite these advancements, the color-gradient model still faced challenges in simulating dynamic flows with density ratios exceeding 100 and suffered from a lack of Galilean invariance. To address these issues, in 2012, Li et al. [13] proposed an improved equilibrium function and an error correction term.

$$f_i^{k,eq} = \rho_k\left(\phi_i^k + \omega_i\left\{\frac{3}{c^2}(\mathbf{e}_i \cdot \mathbf{u}) + \frac{9}{2c^2}(\mathbf{e}_i \cdot \mathbf{u})^2 - \frac{3}{2c^2}|\mathbf{u}|^2\right.\right. \tag{3.17}$$

$$\left.\left. + \frac{3(\mathbf{e}_i \cdot \mathbf{u})}{2c^2}\left[\frac{3\left(c_s^k\right)^2}{c^2} - 1\right]\left(\frac{3|\mathbf{e}_i|^2}{c^2} - 5\right)\right\}\right).$$

To eliminate errors in the diagonal elements of the third-order moment of $f_i^{k,eq}$, an error correction term $G_i^k$ is introduced into $\left(\Omega_i^k\right)^{(1)}$.

$$\left(\Omega_i^k\right)^{(1)} = -\frac{1}{\tau}\left[f_i^k(\mathbf{x},t) - f_i^{k,eq}(\mathbf{x},t)\right] + \Delta t\left(1 - \frac{1}{2\tau}\right)G_i^k(\mathbf{x},t). \tag{3.18}$$

Subsequently, Ba et al. [14] proposed a D2Q9 MRT color-gradient model incorporating the improved equilibrium function and error correction term. Wen et al. [15] extended this to a three-dimensional model, demonstrating stable simulations in environments with density ratios as high as 1000. Representing (3.18) in the MRT form yields the following [15]:

$$\left(\Omega_i^k\right)^{(1)} = \mathbf{M}^{-1}\left[-\mathbf{S}\left(\mathbf{m}^k - \mathbf{m}^{k,eq}\right) + \Delta t\left(\mathbf{I} - \frac{\mathbf{S}}{2}\right)\mathbf{C}^k\right], \tag{3.19}$$

where $\mathbf{M}$ is the transformation matrix, $\mathbf{I}$ is the unit matrix, and $\mathbf{S}$ is the relaxation rate matrix. The set of moments, $\mathbf{m}^k$ and set of equilibrium moments, $\mathbf{m}^{k,\,eq}$ are defined as $\mathbf{m}^k = \mathbf{M}\mathbf{f}^k$, $\mathbf{m}^{k,\,eq} = \mathbf{M}\mathbf{f}^{k,\,eq}$. $\mathbf{C}^k = \mathbf{M}\mathbf{G}^k$ is the error correction term in the moment space. The non-orthogonal transformation matrix and its inverse are as follows:

$$
\mathbf{M}=\begin{bmatrix}
1&1&1&1&1&1&1&1&1&1&1&1&1&1&1&1&1&1&1\\
0&1&-1&0&0&0&0&1&-1&1&-1&1&-1&1&-1&0&0&0&0\\
0&0&0&1&-1&0&0&1&-1&-1&1&0&0&0&0&1&-1&1&-1\\
0&0&0&0&0&1&-1&0&0&0&0&1&-1&-1&1&1&-1&-1&1\\
0&1&1&1&1&1&1&2&2&2&2&2&2&2&2&2&2&2&2\\
0&2&2&-1&-1&-1&-1&1&1&1&1&1&1&1&1&-2&-2&-2&-2\\
0&0&0&1&1&-1&-1&1&1&1&1&-1&-1&-1&-1&0&0&0&0\\
0&0&0&0&0&0&0&1&1&-1&-1&0&0&0&0&0&0&0&0\\
0&0&0&0&0&0&0&0&0&0&0&1&1&-1&-1&0&0&0&0\\
0&0&0&0&0&0&0&0&0&0&0&0&0&0&0&1&1&-1&-1\\
0&0&0&0&0&0&0&1&-1&-1&1&0&0&0&0&0&0&0&0\\
0&0&0&0&0&0&0&1&-1&1&-1&0&0&0&0&0&0&0&0\\
0&0&0&0&0&0&0&0&0&0&0&1&-1&-1&1&0&0&0&0\\
0&0&0&0&0&0&0&0&0&0&0&1&-1&1&-1&0&0&0&0\\
0&0&0&0&0&0&0&0&0&0&0&0&0&0&0&1&-1&-1&1\\
0&0&0&0&0&0&0&0&0&0&0&0&0&0&0&1&-1&1&-1\\
0&0&0&0&0&0&0&1&1&1&1&0&0&0&0&0&0&0&0\\
0&0&0&0&0&0&0&0&0&0&0&1&1&1&1&0&0&0&0\\
0&0&0&0&0&0&0&0&0&0&0&0&0&0&0&1&1&1&1
\end{bmatrix},
\tag{3.20}
$$

$$\mathbf{M}^{-1}=\frac{1}{12}\begin{bmatrix}
12 & 0 & 0 & 0 & -12 & 0 & 0 & 0 & 0 & 0 & 0 & 0 & 0 & 0 & 0 & 0 & 12 & 12 & 12 \\
0 & 6 & 0 & 0 & 2 & 2 & 0 & 0 & 0 & 0 & 0 & -6 & 0 & -6 & 0 & 0 & -6 & -6 & 0 \\
0 & -6 & 0 & 0 & 2 & 2 & 0 & 0 & 0 & 0 & 0 & 6 & 0 & 6 & 0 & 0 & -6 & -6 & 0 \\
0 & 0 & 6 & 0 & 2 & -1 & 3 & 0 & 0 & 0 & -6 & 0 & 0 & 0 & 0 & -6 & -6 & 0 & -6 \\
0 & 0 & -6 & 0 & 2 & -1 & 3 & 0 & 0 & 0 & 6 & 0 & 0 & 0 & 0 & 6 & -6 & 0 & -6 \\
0 & 0 & 0 & 6 & 2 & -1 & -3 & 0 & 0 & 0 & 0 & 0 & -6 & 0 & -6 & 0 & 0 & -6 & -6 \\
0 & 0 & 0 & -6 & 2 & -1 & -3 & 0 & 0 & 0 & 0 & 0 & 6 & 0 & 6 & 0 & 0 & -6 & -6 \\
0 & 0 & 0 & 0 & 0 & 0 & 0 & 3 & 0 & 0 & 3 & 3 & 0 & 0 & 0 & 0 & 3 & 0 & 0 \\
0 & 0 & 0 & 0 & 0 & 0 & 0 & 3 & 0 & 0 & -3 & -3 & 0 & 0 & 0 & 0 & 3 & 0 & 0 \\
0 & 0 & 0 & 0 & 0 & 0 & 0 & -3 & 0 & 0 & -3 & 3 & 0 & 0 & 0 & 0 & 3 & 0 & 0 \\
0 & 0 & 0 & 0 & 0 & 0 & 0 & -3 & 0 & 0 & 3 & -3 & 0 & 0 & 0 & 0 & 3 & 0 & 0 \\
0 & 0 & 0 & 0 & 0 & 0 & 0 & 0 & 3 & 0 & 0 & 0 & 3 & 3 & 0 & 0 & 0 & 3 & 0 \\
0 & 0 & 0 & 0 & 0 & 0 & 0 & 0 & 3 & 0 & 0 & 0 & -3 & -3 & 0 & 0 & 0 & 3 & 0 \\
0 & 0 & 0 & 0 & 0 & 0 & 0 & 0 & -3 & 0 & 0 & 0 & -3 & 3 & 0 & 0 & 0 & 3 & 0 \\
0 & 0 & 0 & 0 & 0 & 0 & 0 & 0 & -3 & 0 & 0 & 0 & 3 & -3 & 0 & 0 & 0 & 3 & 0 \\
0 & 0 & 0 & 0 & 0 & 0 & 0 & 0 & 0 & 3 & 0 & 0 & 0 & 0 & 3 & 3 & 0 & 0 & 3 \\
0 & 0 & 0 & 0 & 0 & 0 & 0 & 0 & 0 & 3 & 0 & 0 & 0 & 0 & -3 & -3 & 0 & 0 & 3 \\
0 & 0 & 0 & 0 & 0 & 0 & 0 & 0 & 0 & -3 & 0 & 0 & 0 & 0 & -3 & 3 & 0 & 0 & 3 \\
0 & 0 & 0 & 0 & 0 & 0 & 0 & 0 & 0 & -3 & 0 & 0 & 0 & 0 & 3 & -3 & 0 & 0 & 3
\end{bmatrix}, \tag{3.21}$$

The relaxation matrix is defined as

$$\mathbf{S} = diag\Big(1, 1, 1, 1, \tau_e^{-1}, \tau_v^{-1}, \tau_v^{-1}, \tau_v^{-1}, \tau_v^{-1}, \tau_v^{-1}, \tau_q^{-1}, \tau_q^{-1}, \tau_q^{-1}, \tau_q^{-1}, \tau_q^{-1}, \tau_q^{-1}, \tau_\pi^{-1}, \tau_\pi^{-1}, \tau_\pi^{-1}\Big), \tag{3.22}$$

where $\tau_e$ and $\tau_v$ are the bulk and shear viscosities relaxation times, and $\tau_q$ and $\tau_\pi$ are related to nonhydrodynamic moments. The set of equilibrium moments is as follows

$$
\mathbf{m}^{k,eq} = \begin{bmatrix}
\rho_k \\
\rho_k u_x \\
\rho_k u_y \\
\rho_k u_z \\
3p_k + \rho_k |\mathbf{u}|^2 \\
\rho_k \left( 2u_x^2 - u_y^2 - u_z^2 \right) \\
\rho_k \left( u_y^2 - u_z^2 \right) \\
\rho_k u_x u_y \\
\rho_k u_x u_z \\
\rho_k u_y u_z \\
p_k u_y \\
p_k u_x \\
p_k u_z \\
p_k u_x \\
p_k u_z \\
p_k u_y \\
\left[ \rho_k \left( 1 - \alpha_k - |\mathbf{u}|^2 \right) + 2\rho_k c^2 \left( u_x^2 + u_y^2 \right) \right] /6 \\
\left[ \rho_k \left( 1 - \alpha_k - |\mathbf{u}|^2 \right) + 2\rho_k c^2 \left( u_x^2 + u_z^2 \right) \right] /6 \\
\left[ \rho_k \left( 1 - \alpha_k - |\mathbf{u}|^2 \right) + 2\rho_k c^2 \left( u_y^2 + u_z^2 \right) \right] /6
\end{bmatrix}. \tag{3.23}
$$

The error correction term, $\mathbf{C}^k$ is

$$\mathbf{C}^k = \begin{bmatrix} 0 \\ 0 \\ 0 \\ 0 \\ Q_x^k + Q_y^k + Q_z^k \\ 2Q_x^k - Q_y^k - Q_z^k \\ Q_y^k - Q_z^k \\ 0 \\ 0 \\ 0 \\ 0 \\ 0 \\ 0 \\ 0 \\ 0 \\ 0 \\ 0 \\ 0 \\ 0 \end{bmatrix}, \tag{3.24}$$

where $Q_x^k$, $Q_y^k$, and $Q_z^k$ are defined as follows.

$$Q_x^k = \partial_x\left\{\rho_k u_x\left[c^2 - 3\left(c_s^k\right)^2\right]\right\}, \tag{3.25}$$

$$Q_y^k = \partial_y\left\{\rho_k u_y\left[c^2 - 3\left(c_s^k\right)^2\right]\right\}, \tag{3.26}$$

$$Q_z^k = \partial_z\left\{\rho_k u_z\left[c^2 - 3\left(c_s^k\right)^2\right]\right\}. \tag{3.27}$$

The new $\left(\Omega_i^k\right)^{(2)}$, designed to ensure independence between surface tension and relaxation time while maintaining the MRT framework, is expressed as follows:

$$\left(\Omega_i^k\right)^{(2),MRT} = \mathbf{M}^{-1}\mathbf{S}\mathbf{M}\left(\Omega_i^k\right)^{(2)}, \tag{3.28}$$

With the removal of the relaxation time from the surface tension, the equation representing the magnitude of surface tension, (3.14), becomes

$$\sigma = \frac{2}{9}(A_R + A_B)c^4\Delta t. \tag{3.29}$$

The following is the pseudocode for the improved color-gradient model. Before calculating the MRT collision function, the error correction term needs to be determined first. At each node, compute $\overline{Q_\alpha^k} = \partial_\alpha\left\{\rho_k u_\alpha\left[c^2 - 3\left(c_s^k\right)^2\right]\right\}$, and then calculate the gradient of $\overline{Q_\alpha^k}$ as shown in (3.13).

```
//Error correction term
for all (x,y,z) do
  for all k do
    temp = rho[x][y][z][k] * (1.0 - 1.5 * (1.0 - ALPHA[k]))
   Qx_bar[x][y][z][k] = ux[x][y][z] * temp
    Qy_bar[x][y][z][k] = uy[x][y][z] * temp
   Qz_bar[x][y][z][k] = uz[x][y][z] * temp
  end for
end for
```

Then, calculate the MRT collision operator. Generally, to compute $\mathbf{m}^k = \mathbf{M}\mathbf{f}^k$, a triple `for` loop can be used. For example:

```
for all k do
  for all i do
    m[i][k] = 0.0
   for all j do
      m[i][k] += TransM[i][j] * f[x][y][z][j][k]
    end for
  end for
end for
```

However, as seen in (3.20) and (3.21), the non-orthogonal transformation matrix contains many zero entries, so it is unnecessary to perform calculations for all matrix elements. Therefore, to improve computational efficiency, the code is structured in an unrolled form of the `for` loops.

```
//Collision function 1 (improved color-gradient model)
for all (x,y,z) do
  if (x,y,z) is fluid node then
   u2 = ux[x][y][z] * ux[x][y][z] + uy[x][y][z] * uy[x][y][z] + uz[x][y]
[z] *
     uz[x][y][z]
    for all k do
      m[0] = rho[x][y][z][k]
      m[1] = rho[x][y][z][k] * ux[x][y][z]
```

```
        m[2] = rho[x][y][z][k] * uy[x][y][z]
        m[3] = rho[x][y][z][k] * uz[x][y][z]
        m[4] = f[x][y][z][1][k] + f[x][y][z][2][k] + f[x][y][z][3][k] +
        f[x][y][z][4][k] + f[x][y][z][5][k] + f[x][y][z][6][k] + 2.0 * (f
[x][y][z][7][k] + f[x][y][z][8][k] + f[x][y][z][9][k] + f[x][y][z]
[10][k] + f[x][y][z][11][k] + f[x][y][z][12][k] + f[x][y][z][13][k] + f
[x][y][z][14][k] + f[x][y][z][15][k] + f[x][y][z][16][k] + f[x][y][z]
[17][k] + f[x][y][z][18][k])
       m[5] = 2.0 * f[x][y][z][1][k] + 2.0 * f[x][y][z][2][k] - f[x][y][z]
[3][k]
         - f[x][y][z][4][k] - f[x][y][z][5][k] - f[x][y][z][6][k] + f[x]
[y][z][7][k] + f[x][y][z][8][k] + f[x][y][z][9][k] + f[x][y][z][10]
[k] + f[x][y][z][11][k] + f[x][y][z][12][k] + f[x][y][z][13][k] + f[x]
[y][z][14][k] - 2.0 * (f[x][y][z][15][k] + f[x][y][z][16][k] + f[x][y]
[z][17][k] + f[x][y][z][18][k])
        m[6] = f[x][y][z][3][k] + f[x][y][z][4][k] - f[x][y][z][5][k] -
        f[x][y][z][6][k] + f[x][y][z][7][k] + f[x][y][z][8][k] + f[x][y]
[z][9][k] + f[x][y][z][10][k] - f[x][y][z][11][k] - f[x][y][z][12]
[k] - f[x][y][z][13][k] - f[x][y][z][14][k]
        m[7] = f[x][y][z][7][k] + f[x][y][z][8][k] - f[x][y][z][9][k] -
          f[x][y][z][10][k]
      m[8] = f[x][y][z][11][k] + f[x][y][z][12][k] - f[x][y][z][13][k] -
        f[x][y][z][14][k]
      m[9] = f[x][y][z][15][k] + f[x][y][z][16][k] - f[x][y][z][17][k] -
        f[x][y][z][18][k]
       m[10] = f[x][y][z][7][k] - f[x][y][z][8][k] - f[x][y][z][9][k] +
        f[x][y][z][10][k]
       m[11] = f[x][y][z][7][k] - f[x][y][z][8][k] + f[x][y][z][9][k] -
        f[x][y][z][10][k]
      m[12] = f[x][y][z][11][k] - f[x][y][z][12][k] - f[x][y][z][13][k] +
        f[x][y][z][14][k]
      m[13] = f[x][y][z][11][k] - f[x][y][z][12][k] + f[x][y][z][13][k] -
        f[x][y][z][14][k]
      m[14] = f[x][y][z][15][k] - f[x][y][z][16][k] - f[x][y][z][17][k] +
        f[x][y][z][18][k]
      m[15] = f[x][y][z][15][k] - f[x][y][z][16][k] + f[x][y][z][17][k] -
        f[x][y][z][18][k]
       m[16] = f[x][y][z][7][k] + f[x][y][z][8][k] + f[x][y][z][9][k] +
        f[x][y][z][10][k]
      m[17] = f[x][y][z][11][k] + f[x][y][z][12][k] + f[x][y][z][13][k] +
        f[x][y][z][14][k]
      m[18] = f[x][y][z][15][k] + f[x][y][z][16][k] + f[x][y][z][17][k] +
        f[x][y][z][18][k]
```

```
meq[0] = rho[x][y][z][k]
meq[1] = rho[x][y][z][k] * ux[x][y][z]
meq[2] = rho[x][y][z][k] * uy[x][y][z]
meq[3] = rho[x][y][z][k] * uz[x][y][z]
meq[4] = 3.0 * pressure[x][y][z][k] + rho[x][y][z][k] * u2
meq[5] = rho[x][y][z][k] * (2.0 * ux[x][y][z] * ux[x][y][z] - uy[x]
[y][z]
    * uy[x][y][z] - uz[x][y][z] * uz[x][y][z])
meq[6] = rho[x][y][z][k] * (uy[x][y][z] * uy[x][y][z] - uz[x][y][z] *
    uz[x][y][z])
meq[7] = rho[x][y][z][k] * ux[x][y][z] * uy[x][y][z]
meq[8] = rho[x][y][z][k] * ux[x][y][z] * uz[x][y][z]
meq[9] = rho[x][y][z][k] * uy[x][y][z] * uz[x][y][z]
meq[10] = pressure[x][y][z][k] * uy[x][y][z]
meq[11] = pressure[x][y][z][k] * ux[x][y][z]
meq[12] = pressure[x][y][z][k] * uz[x][y][z]
meq[13] = pressure[x][y][z][k] * ux[x][y][z]
meq[14] = pressure[x][y][z][k] * uz[x][y][z]
meq[15] = pressure[x][y][z][k] * uy[x][y][z]
meq[16] = (rho[x][y][z][k] * (1.0 – ALPHA[k] – u2) + 2.0 * rho[x][y]
[z][k]
    * (ux[x][y][z] * ux[x][y][z] + uy[x][y][z] * uy[x][y][z])) / 6.0
meq[17] = (rho[x][y][z][k] * (1.0 – ALPHA[k] – u2) + 2.0 * rho[x][y]
[z][k]
    * (ux[x][y][z] * ux[x][y][z] + uz[x][y][z] * uz[x][y][z])) / 6.0
meq[18] = (rho[x][y][z][k] * (1.0 – ALPHA[k] – u2) + 2.0 * rho[x][y]
[z][k]
    * (uy[x][y][z] * uy[x][y][z] + uz[x][y][z] * uz[x][y][z])) / 6.0

for all i do
  m_bar[i] = - S[i] * (m[i] – meq[i]) + (1.0 – 0.5 * S[i]) * C[i][k]
end for

f[x][y][z][0][k] += m_bar[0] - m_bar[4] + m_bar[16] + m_bar[17] +
        m_bar[18]
f[x][y][z][1][k] += 0.5 * m_bar[1] + (m_bar[4] + m_bar[5]) / 6.0 - 0.5
        * (m_bar[11] + m_bar[13] + m_bar[16] + m_bar[17])
f[x][y][z][2][k] += - 0.5 * m_bar[1] + (m_bar[4] + m_bar[5]) / 6.0 +
        0.5 * (m_bar[11] + m_bar[13] - m_bar[16] - m_bar[17])
f[x][y][z][3][k] += 0.5 * m_bar[2] + m_bar[4] / 6.0 - m_bar[5] / 12.0
        + m_bar[6] / 4.0 - 0.5 * (m_bar[10] + m_bar[15] + m_bar[16] +
m_bar[18])
f[x][y][z][4][k] += - 0.5 * m_bar[2] + m_bar[4] / 6.0 - m_bar[5] / 12.0
        + m_bar[6] / 4.0 + 0.5 * (m_bar[10] + m_bar[15] - m_bar[16] -
m_bar[18])
```

```
        f[x][y][z][5][k] += 0.5 * m_bar[3] + m_bar[4] / 6.0 - m_bar[5] / 12.0
                  - m_bar[6] / 4.0 - 0.5 * (m_bar[12] + m_bar[14] + m_bar[17] +
m_bar[18])
        f[x][y][z][6][k] += - 0.5 * m_bar[3] + m_bar[4] / 6.0 - m_bar[5] / 12.0
                  - m_bar[6] / 4.0 + 0.5 * (m_bar[12] + m_bar[14] - m_bar[17] -
m_bar[18])
        f[x][y][z][7][k] += (m_bar[7] + m_bar[10] + m_bar[11] + m_bar[16]) /
                  4.0
        f[x][y][z][8][k] += (m_bar[7] - m_bar[10] - m_bar[11] + m_bar[16]) /
                  4.0
        f[x][y][z][9][k] += (- m_bar[7] - m_bar[10] + m_bar[11] + m_bar[16]) /
                  4.0
        f[x][y][z][10][k] += (- m_bar[7] + m_bar[10] - m_bar[11] + m_bar[16])
                  / 4.0
        f[x][y][z][11][k] += (m_bar[8] + m_bar[12] + m_bar[13] + m_bar[17]) /
                  4.0
        f[x][y][z][12][k] += (m_bar[8] - m_bar[12] - m_bar[13] + m_bar[17]) /
                  4.0
        f[x][y][z][13][k] += (- m_bar[8] - m_bar[12] + m_bar[13] + m_bar[17])
                  / 4.0
        f[x][y][z][14][k] += (- m_bar[8] + m_bar[12] - m_bar[13] + m_bar[17])
                  / 4.0
        f[x][y][z][15][k] += (m_bar[9] + m_bar[14] + m_bar[15] + m_bar[18]) /
                  4.0
        f[x][y][z][16][k] += (m_bar[9] - m_bar[14] - m_bar[15] + m_bar[18]) /
                  4.0
        f[x][y][z][17][k] += (- m_bar[9] - m_bar[14] + m_bar[15] + m_bar[18])
                  / 4.0
        f[x][y][z][18][k] += (- m_bar[9] + m_bar[14] - m_bar[15] + m_bar[18])
                  / 4.0
      end for
   end if
end for
```

## 3.4 Pseudopotential Model

The pseudopotential model introduces interparticle repulsive or attractive forces to achieve phase separation. Unlike the color-gradient model, which requires two distribution functions and three collision operators, the pseudopotential model can simulate multiphase flows solely by incorporating interparticle forces. This approach offers the advantages of high computational efficiency and methodological simplicity. In the pseudopotential model, the evolution equation for the distribution function is the same as that of the general LBM with external forces included.

$$f_i(\mathbf{x} + \mathbf{e}_i \Delta t, t + \Delta t) - f_i(\mathbf{x}, t) = -\frac{1}{\tau}[f_i(\mathbf{x}, t) - f_i^{eq}(\mathbf{x}, t)] + S_i, \quad (3.30)$$

where $S_i$ represents the external force source term when the EDM is employed. In Yuan and Schaefer model [16], the interparticle force is given by

$$\mathbf{F}_{int} = -c_0 \psi(\mathbf{x}) g \nabla \psi(\mathbf{x}), \quad (3.31)$$

where $g$ a parameter that controls the interaction force, $c_0$ is a constant determined by the lattice structure (In D2Q9 and D3Q19, $c_0 = 6.$), and $\psi$ is the effective mass. In 1994, Shan and Chen [2] proposed an effective mass.

$$\psi(\mathbf{x}) = \rho_0[1 - \exp(-\rho/\rho_0)], \quad (3.32)$$

where $\rho_0$ is an arbitrary density. When (3.31) is discretized, it can be expressed as

$$\mathbf{F}_{int} = -3c_0 g \psi(\mathbf{x}) \sum_i \omega_i \psi(\mathbf{x} + \mathbf{e}_i \Delta t) \mathbf{e}_i. \quad (3.33)$$

The EOS is an equation that describes the thermodynamic state of a substance through the interrelationship among pressure, volume, temperature, and other state variables. Depending on specific pressure, volume, and temperature conditions, a state where liquid and vapor coexist, that is, a multiphase system, can be formed. An example of a simple yet representative EOS is the van der Waals (vdW) EOS.

$$p = \frac{\rho RT}{1 - b\rho} - a\rho^2, \quad (3.34)$$

where $a$ is the attraction parameter, $b$ is the repulsion parameter, $R$ is the specific gas constant, and $T$ is the temperature.

Figure 3.3 shows a schematic diagram of the $p - v$ curve, illustrating the relationship between pressure and specific volume. If $T > T_c$, the pressure and specific volume have a one-to-one relationship. However, if $T < T_c$, three specific volume values ($v_1$, $v_3$, and $v_2$) can correspond to a single pressure value ($p_s$). Here, $v_1 = v_l$ and $v_2 = v_g$ represent the specific volumes of the liquid and gas, respectively. Gibbs free energy has equal values at liquid and vapor phases in equilibrium. This is the basic idea for the Maxwell construction.

$$\int_{v_l}^{v_g} p dv = p_s (v_g - v_l). \quad (3.35)$$

According to Maxwell construction, the coexistence of liquid and gas is maintained when the areas A and B in Fig. 3.3 are equal. This is why the Maxwell construction is also referred to as the "equal area rule." We have obtained three equations: $p_s = p($-$v_g, T)$, $p_s = p(v_l, T)$, and (3.35). Using these, we can solve for the three unknown

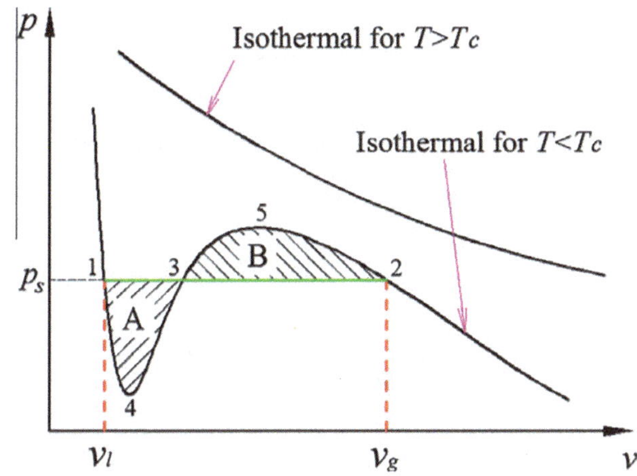


**Fig. 3.3** Schematic illustrations of the Maxwell construction. $v_l$ and $v_g$ denote the specific volume of the saturated liquid and gas, respectively. $p_s$ denote the saturated pressure corresponding to temperature $T$. $T_c$ is the critical temperature [17]

variables $v_g$, $v_l$, and $p_s$. If $v_g$ and $v_l$ plotted from a given EOS align with the theoretical values, the system can be considered thermodynamically consistent. However, the pseudopotential model has historically faced the limitation of being thermodynamically inconsistent. To address this issue, many researchers have proposed improved models.

The appropriate selection of the EOS is also crucial for achieving thermodynamic consistency. The Shan and Chen (S-C) EOS can be derived from (3.31) as follows

$$p = \rho c_s^2 + \frac{c_0}{2} g \psi^2(\mathbf{x}). \tag{3.36}$$

At the critical point, $(\partial p/\partial \rho)_T = (\partial^2 p/\partial \rho^2)_T = 0$ must be satisfied. Therefore, when the effective mass is as given in (3.32), $\rho_c$ and $g_c$ can be calculated as

$$\frac{\partial p}{\partial \rho} = c_s^2 + c_0 g_c \rho_0 [1 - \exp(-\rho_c/\rho_0)] \exp(-\rho_c/\rho_0) = 0, \tag{3.37}$$

$$\frac{\partial^2 p}{\partial \rho^2} = c_0 g_c \exp(-\rho_c/\rho_0)[2\exp(-\rho_c/\rho_0) - 1] = 0. \tag{3.38}$$

From the above equations, we can calculate $\rho_c = \rho_0 \ln 2$ and $g_c = -2/(9\rho_0)$. If $g < g_c$, a state where liquid and gas coexist is achieved. However, the original S-C EOS is thermodynamically inconsistent and has the limitation of being numerically unstable at high-density ratios. Yuan and Schaefer [16] evaluated various EOS by comparing them with the Maxwell construction. To incorporate various EOS, a modified effective mass equation is required. From (3.36), an effective mass equation that maintains consistency regardless of the EOS can be derived.

$$\psi(\mathbf{x}) = \sqrt{\frac{2(p - \rho c_s^2)}{c_0 G}}, \tag{3.39}$$

The following are the EOS that were compared.

1. Shan and Chen (S-C) EOS:

$$p = \frac{\rho}{3} + \frac{c_0}{2g}\rho_0^2[1 - \exp(-\rho/\rho_0)]^2, \tag{3.40}$$

   where $\rho_c = \rho_0 \ln 2$, $g_c = -2/(9\rho_0)$.
2. van der Waals (vdW) EOS:

$$p = \frac{\rho RT}{1 - b\rho} - a\rho^2, \tag{3.41}$$

   where $\rho_c = 1/(3b)$, $p_c = a/(27b^2)$, $T_c = 8a/(27Rb)$.
3. Redlich and Kwong (R-K) EOS:

$$p = \frac{\rho RT}{1 - b\rho} - \frac{a\rho^2}{\sqrt{T}(1 + b\rho)}, \tag{3.42}$$

   $a = 0.42748R^2T_c^{2.5}/p_c$, $b = 0.08664RT_c/p_c$
4. Redlich, Kwong, and Soave (R-K-S) EOS:

$$p = \frac{RT\rho}{1 - b\rho} - \frac{a\varepsilon(T)\rho^2}{1 + b\rho}, \tag{3.43}$$

   where $\varepsilon(T) = \left[1 + (0.480 + 1.574\omega - 0.176\omega^2)\left(1 - \sqrt{T/T_c}\right)\right]^2$.
5. Peng and Robinson (P-R) EOS:

$$p = \frac{\rho RT}{1 - b\rho} - \frac{a\varepsilon(T)\rho^2}{1 + 2b\rho - b^2\rho^2}, \tag{3.44}$$

   where $a = 0.45724R^2T_c^2/p_c$, $b = 0.0778RT_c/p_c$, $\varepsilon(T) = \left[1 + (0.37464 + 1.54226\omega - 0.26992\omega^2)\left(1 - \sqrt{T/T_c}\right)\right]^2$.
6. Carnahan and Starling (C-S) EOS:

$$p = \rho RT\frac{1 + b\rho/4 + (b\rho/4)^2 - (b\rho/4)^3}{(1 - b\rho/4)^3} - a\rho^2, \tag{3.45}$$

   where $a = 0.4963R^2T_c^2/p_c$, $b = 0.187RT_c/p_c$.

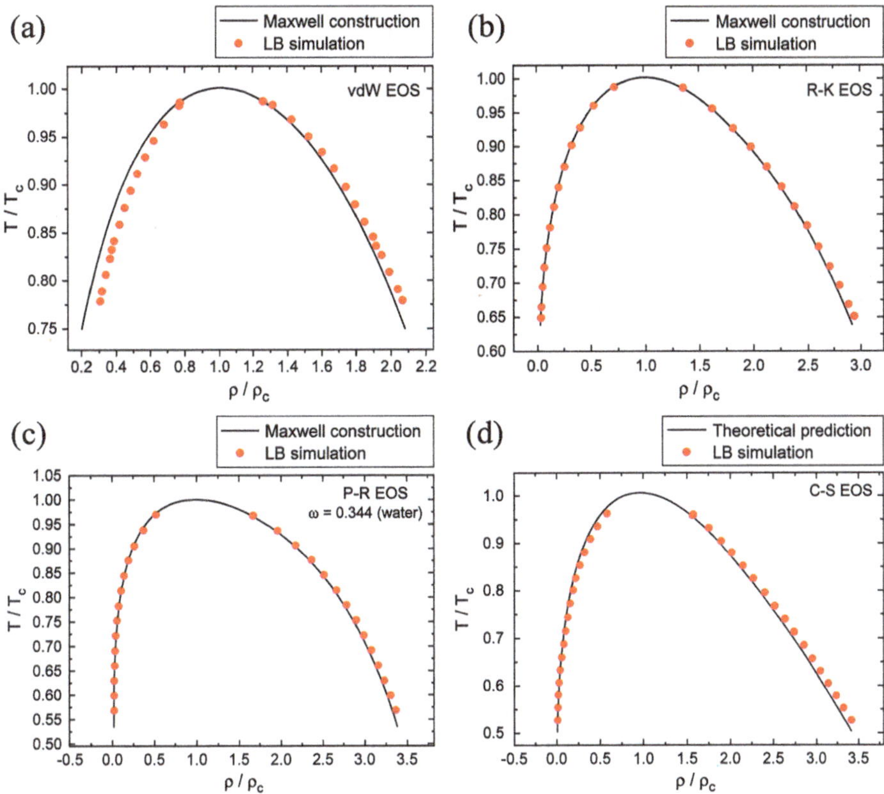


**Fig. 3.4** Comparison of coexistence curves obtained from simulations with theoretical values for different EOS: (**a**) vdW, (**b**) R-K, (**c**) P-R, and (**d**) C-S EOS [16]

Figure 3.4 shows the comparison of the coexistence curves for four different EOS. The vdW EOS exhibited increasing error relative to the Maxwell construction as the temperature decreased, i.e., as the density ratio increased, resulting in a maximum relative error of 33.83%. In contrast, the R-K, P-R, and C-S EOS showed results that were relatively consistent with theoretical values even at low temperatures. Additionally, the R-K and P-R EOS were compared with the real properties of saturated water. The results revealed that the P-R EOS is the most suitable for simulating water. These findings demonstrate the importance of selecting an appropriate EOS to overcome the thermodynamic inconsistency of the pseudopotential model.

In addition to the EOS, improvements in the interparticle force can also reduce thermodynamic inconsistency. Gong and Cheng [17] identified that the interparticle force model in (3.33) is thermodynamically inconsistent and proposed an improved interparticle force model. (3.31) can be rewritten as

$$\mathbf{F}_{int} = -\frac{1}{2} c_0 g \nabla \psi^2(\mathbf{x}). \tag{3.46}$$

Similarly, when (3.46) is discretized, it can be expressed as

$$\mathbf{F}_{int} = -\frac{3c_0 g}{2} \sum_i \omega_i \psi^2(\mathbf{x} + \mathbf{e}_i \Delta t)\mathbf{e}_i. \tag{3.47}$$

where (3.31) and (3.46) represent the same expression in their continuous form, but their discretized forms differ. Equation (3.33) uses both the local and neighboring effective mass, whereas (3.47) utilizes only the neighboring effective mass. Therefore, the new interparticle force combining the two equations can be formulated as follows

$$\mathbf{F}_{int} = -3c_0 g \left[ \beta \psi(\mathbf{x}) \sum_i \omega_i \psi(\mathbf{x} + \mathbf{e}_i \Delta t)\mathbf{e}_i + 0.5(1-\beta) \sum_i \omega_i \psi^2(\mathbf{x} + \mathbf{e}_i \Delta t)\mathbf{e}_i \right], \tag{3.48}$$

where $\beta$ is the weighting factor. The introduction of $\beta$ allows weighting between (3.33) and (3.47). Figure 3.5 shows the comparison of the interparticle force schemes by Shan and Chen (S-C) [2], Zhang and Chen (Z-C) [18], and Gong and Cheng (G-C) for three different EOS. For each EOS, the weighting factors used were $\beta_{SC} = 0.886$, $\beta_{vdW} = 0.55$, and $\beta_{PR} = 1.16$. As shown in the figures, all three schemes exhibited relatively high accuracy in high-temperature environments. However, as the temperature decreased, the S-C and Z-C schemes showed increasing errors, whereas the G-C scheme provided thermodynamically consistent results. These results demonstrate that the appropriate selection of the interparticle force term can reduce thermodynamic inconsistency.

To summarize, the overall algorithm of the pseudopotential model is structured as follows:

1. Find the fluid density and velocity at every node.
2. Calculate the EOS using (3.40)–(3.45).
3. Calculate the effective mass using (3.39).

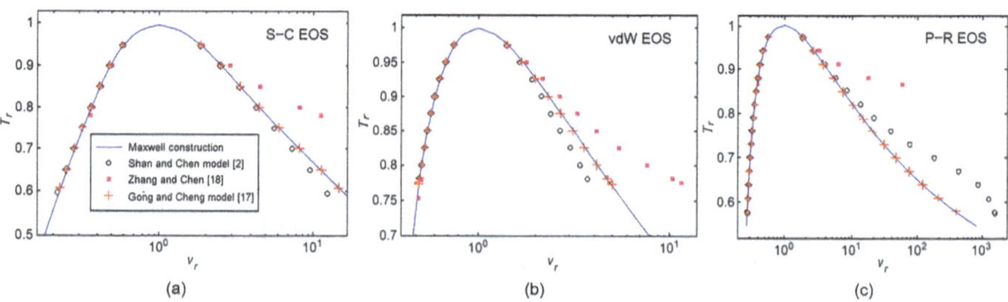


**Fig. 3.5** Coexistence curves of **(a)** S-C EOS, **(b)** vdW EOS, and **(c)** P-R EOS for different schemes of interparticle interaction force term [17]

4. Calculate the interparticle force using (3.48). If additional forces, such as gravity, act on the fluid, sum up all force contributions.
5. Compute the equilibrium distribution.
6. Collide and propagate as usual.
7. Go back to step 1.

As observed in the algorithm, the pseudopotential model does not differ significantly from the general LBM in methodology, except for the introduction of the interparticle force. This is one of the reasons why the pseudopotential model is the most efficient and widely used among multiphase flow models.

**Exercise 3.3 Pseudopotential Model Basics**

(a) Describe the physical meaning of the interparticle force in (3.31).
(b) Explain the role of the effective mass function $\psi(\rho)$ and why different forms are used.
(c) Discuss why pseudopotential LBM is computationally efficient compared to the color-gradient model.

**Solution**

(a) In the Shan–Chen pseudopotential model, the interparticle force is typically written as

$$\mathbf{F}(\mathbf{x}) = -G\psi(\rho(\mathbf{x}))\sum_i \omega_i \psi(\rho(\mathbf{x} + \mathbf{e}_i))\mathbf{e}_i$$

This force:

- Acts as a mean-field attractive (or repulsive) interaction between neighboring lattice sites.
- Drives phase separation by pulling fluid from low-density regions into high-density regions (for attractive $G < 0$).
- Generates a non-ideal pressure tensor, so that the equation of state deviates from the ideal gas law and supports liquid–gas coexistence and surface tension.

Physically, it mimics cohesive molecular forces at a coarse-grained level.

(b) The effective mass function $\psi(\rho)$ controls how the interparticle force depends on density:

- It determines the shape of the non-ideal EOS and the location of coexistence densities.
- Different choices of $\psi(\rho)$ lead to different pressure–density relations.
- Proper design of $\psi(\rho)$ can improve thermodynamic consistency, density ratio, and numerical stability.

Therefore, various forms are used to match a target equation of state or to obtain better phase separation and stability characteristics for a given application.

(c) Pseudopotential LBM is computationally efficient because:

- It uses a single distribution function for the fluid, unlike color-gradient models which require two or more.
- The interparticle force is a local or near-local stencil operation (nearest neighbors), which is cheap to evaluate.
- Surface tension and phase separation are obtained implicitly from the force term, without solving additional PDEs or interface-tracking equations.
- The algorithm remains a simple streaming–collision scheme with a force term, which maps well to high-performance architectures (GPU, multi-core).

As a result, for the same lattice size, pseudopotential simulations typically require less memory and fewer floating-point operations per time step than color-gradient counterparts.

## 3.5 Thermodynamic Model

In discussing the pseudopotential model, it was observed that the EOS includes a temperature term. According to Maxwell construction, the density ratio varies with temperature. Therefore, in the pseudopotential model, the temperature suitable for the desired density ratio must be set uniformly across all nodes. In this case, since a heat transfer model is not included, the temperature is applied as a constant. If the temperature variation is minimal, this approach may not pose significant issues. However, in scenarios where the fluid properties change or where high heat is applied to the extent that phase change occurs, it becomes essential to incorporate a heat transfer model to compute heat convection and diffusion.

Thermal LB employs two approaches: the hybrid method and the double distribution function (DDF) method. The hybrid method solves the temperature field using traditional numerical techniques such as finite difference or finite volume methods. In contrast, the DDF method introduces an additional distribution function, known as the temperature distribution function, to solve the heat equation using the LB scheme. The DDF method ensures methodological consistency by solving the heat equation within the LB framework. It also retains the high computational efficiency of LB, as partial differential equations are computed locally independently. Numerical analysis using the hybrid method is identical to conventional classical methods and will therefore be omitted from this textbook.

The heat transfer equation, neglecting viscous heat dissipation, is

$$\frac{\partial T}{\partial t} + \nabla \cdot (\mathbf{u}^{eq} T) = \nabla \cdot (\alpha \nabla T) + \phi, \tag{3.49}$$

where $\mathbf{u}^{eq}$ is the equilibrium velocity which is defined as $\mathbf{u}^{eq} = \mathbf{u} + \Delta t\mathbf{F}/(2\rho)$. $\alpha$ is the thermal diffusivity and $\phi$ is the phase change source term. When Equation (3.49) is applied to the LB scheme, it becomes

$$g_i(\mathbf{x}+\mathbf{e}_i\Delta t, t+\Delta t) - g_i(\mathbf{x},t) = -\frac{1}{\tau_T}\left[g_i(\mathbf{x},t) - g_i^{eq}(\mathbf{x},t)\right] + \Delta t\omega_i\phi, \tag{3.50}$$

where $g_i$ is the temperature distribution function, and $\tau_T$ is the thermal relaxation time. The equilibrium temperature distribution function, $g_i^{eq}$ is

$$g_i^{eq}(\mathbf{x},t) = T\omega_i\left[1 + \frac{\mathbf{e}_i\cdot\mathbf{u}^{eq}}{c_s^2} + \frac{(\mathbf{e}_i\cdot\mathbf{u}^{eq})^2}{2c_s^4} - \frac{|\mathbf{u}^{eq}|^2}{2c_s^2}\right]. \tag{3.51}$$

The temperature can be obtained similarly to the fluid density, as the summation of $g_i$

$$T = \sum_i g_i = \sum_i g_i^{eq}. \tag{3.52}$$

Thermal diffusivity can be calculated using thermal relaxation time.

$$\alpha = c_s^2(\tau_T - 0.5)\Delta t. \tag{3.53}$$

$\phi$ can be derived from the entropy balance equation.

$$\phi = \frac{T}{\rho^2 c_v}\left(\frac{\partial p}{\partial T}\right)_\rho \frac{d\rho}{dt} + T\nabla\cdot\mathbf{u}^{eq}, \tag{3.54}$$

where $c_v$ is the specific heat capacity. The $d\rho/dt$ in (3.54) leads to computational inefficiency and is therefore replaced using the mass conservation equation [19]:

$$\frac{\partial\rho}{\partial t} + \nabla\cdot(\rho\mathbf{u}^{eq}) = 0, \tag{3.55}$$

Therefore, (3.54) can be rewritten as follows

$$\phi = T\left[1 - \frac{1}{\rho c_v}\left(\frac{\partial p}{\partial T}\right)_\rho\right]\nabla\cdot\mathbf{u}^{eq}, \tag{3.56}$$

As discussed so far, the DDF thermal model does not differ significantly from fluid dynamics calculations, making it easy to implement. The following is the pseudocode for the collision function of the DDF thermal model:

```
//Thermal LB collision function.
//Within this function, the velocity is the equilibrium velocity
updated by the fluid collision function.
//The EOS used is the Peng-Robinson EOS.
for all (x,y) do
```

```
  if (x,y) is fluid node then
    dudx = dvdy = 0.0
   for all i do
      xtemp = nx[x][y][i]
      ytemp = ny[x][y][i]
    dudx += 3.0 * omega[i] * ux[xtemp][ytemp] * ex[i]
    dvdy += 3.0 * omega[i] * uy[xtemp][ytemp] * ey[i]
    end for
   udotu = ux[x][y] * ux[x][y] + uy[x][y] * uy[x][y]
   dpdT = (1.0 + (0.37464 + 1.54226 * acentric - 0.26992 * acentric *
acentric)
   * (1.0 - sqrt(Temperature[x][y] / Tc))) * pow(Temperature[x][y] * Tc,
-0.5)
    phi = Temperature[x][y] * (1.0 - dpdT / (rho[x][y] * cv)) * (dudx +
dvdy)
    for all i do
      udote = ux[x][y] * ex[i] + uy[x][y] * ey[i]
     geq = omega[i] * Temperature[x][y] * (1.0 + 3.0 * udote + 4.5 * udote
      * udote - 1.5 * udotu)
      g[x][y][i] += - (g[x][y][i] - geq) / tau_T[x][y] + omega[i] * phi
    end for
  end if
end for
```

**Exercise 3.4 Thermal Double Distribution Function (DDF) Model**
Explain the role of the entropy source term in (3.56).

**Solution**
The entropy source term in (3.56) accounts for irreversible processes such as viscous dissipation, heat conduction, and possibly phase-change-related entropy production. In the thermal LB framework:

- It appears as a source term in the temperature (or energy) equation, contributing to local heating or cooling not captured by pure advection–diffusion.
- Including this term ensures that the model is consistent with the second law of thermodynamics, i.e., entropy does not decrease in irreversible processes.
- It is particularly important in flows with strong viscous heating or steep temperature gradients.

  Thus, the entropy source term refines the thermal model so that both energy and entropy balances are correctly represented at the macroscopic level.

## 3.6 Wettability

The contact angle, defined as the angle formed at the junction of solid, liquid, and gas phases, reflects the balance of interfacial tensions and characterizes the wettability, hydrophilicity, or hydrophobicity of a surface. Contact angle simulation provides significant advantages in understanding phenomena that are challenging to measure experimentally. Specifically, it enables the analysis of interfacial behaviors at nanometer or micrometer scales and the study of wetting phenomena on heterogeneous surfaces. By predicting contact angles under various conditions—such as changes in temperature, pressure, surface roughness, or chemical composition—simulations complement experimental limitations and offer deeper insights.

Young's equation describes the balance of forces at the three-phase contact line. It is expressed as

$$\cos\theta = \frac{\sigma_{gs} - \sigma_{ls}}{\sigma_{lg}}, \tag{3.57}$$

where $\sigma_{gs}$ is the surface tension between gas and solid, $\sigma_{ls}$ is the surface tension between liquid and solid, and $\sigma_{lg}$ is the surface tension between liquid and gas. Figure 3.6 is a schematic illustrating Young's equation. Young's equation represents the equilibrium condition at the contact line, where the components of the interfacial tensions balance each other. The contact angle $\theta$ depends on the relative magnitudes of the interfacial tensions and reflects the wettability of the solid surface by the liquid.

To control the contact angle in LBM, it is necessary to classify the nodes into four categories as follows

- $\mathbf{x}_{FB}$: A fluid node that is in contact with at least one solid node.
- $\mathbf{x}_{FI}$: A fluid node that is not in contact with any solid node.
- $\mathbf{x}_{SB}$: A solid node that is in contact with at least one fluid node.
- $\mathbf{x}_{SI}$: A solid node that is not in contact with any fluid node.

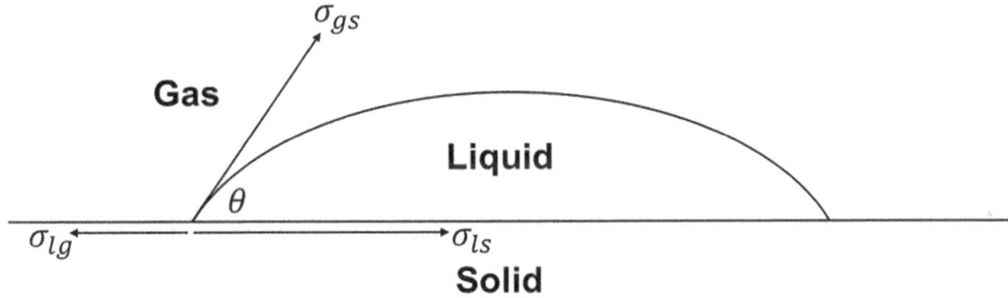


**Fig. 3.6** Schematic illustrating Young's equation

The following chapters will cover the wettability schemes of the color-gradient and pseudopotential models.

### 3.6.1 Wettability: Color-Gradient Model

The simplest wettability scheme in the color-gradient model is the virtual density scheme [20]. This method controls the contact angle by assigning a virtual phase parameter to the $\mathbf{x}_{SB}$ nodes. The virtual phase parameter on the $\mathbf{x}_{SB}$ nodes can be assigned based on the force balance equation in (3.57) as follows

$$\rho^N(\mathbf{x}_{SB}) = \cos\theta. \tag{3.58}$$

Despite its ease of implementation, the virtual density scheme has been reported to cause unphysical flow at the interface where the droplet contacts the solid, leading to inaccurate results [21].

To address this issue, an improved virtual density scheme [22, 23] has been proposed. Figure 3.7 is a schematic of the improved virtual density scheme. $\mathbf{n}_s$ is the unit normal vector of the wall, and $\mathbf{n}^*$ is the unit normal vector to the fluid interface. As shown in the figure, the angle between $\mathbf{n}_s$ and $\mathbf{n}^*$, denoted as $\theta'$, is the current contact angle. The basic approach of the improved virtual density scheme involves rotating $\mathbf{n}^*$ by the desired contact angle, $\theta$. To achieve this, $\mathbf{n}^*$ at $\mathbf{x}_{FB}$ nodes must first be determined. However, since the $\mathbf{x}_{FB}$ nodes lack phase parameter values, (3.13) cannot be directly applied. Therefore, the average value of $\rho^N$ is calculated from the

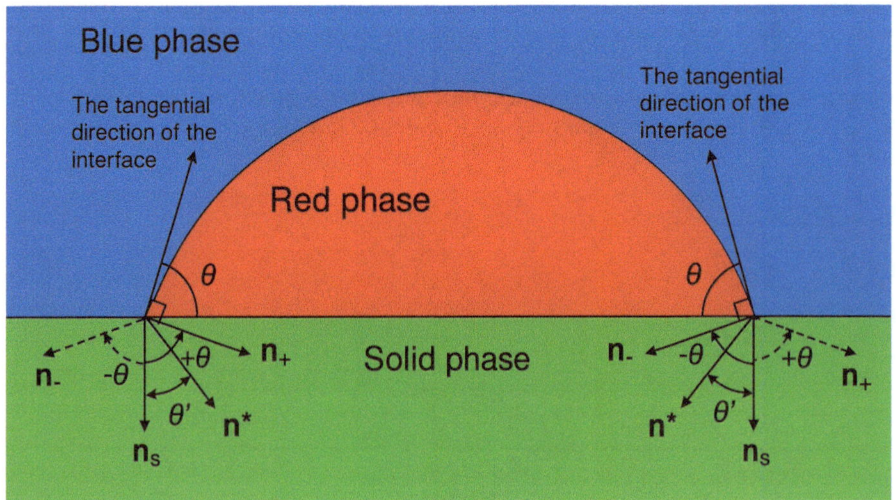


**Fig. 3.7** Schematic of the improved virtual density scheme [22]

surrounding $\mathbf{x}_{FB}$ nodes and assigned to the $\mathbf{x}_{SB}$ nodes. The average $\rho^N$ at $\mathbf{x}_{SB}$ can be calculated as

$$\rho^N(\mathbf{x}_{SB}) = \frac{\sum_i \omega_i \rho^N(\mathbf{x} + \mathbf{e}_i \Delta t)[1 - s_w(\mathbf{x} + \mathbf{e}_i \Delta t)]}{\sum_i \omega_i [1 - s_w(\mathbf{x} + \mathbf{e}_i \Delta t)]}, \tag{3.59}$$

where $s_w$ is the solid node indicator, defined as $s_w(\mathbf{x}_{FB}) = s_w(\mathbf{x}_{FI}) = 0$ and $s_w(\mathbf{x}_{SB}) = s_w(\mathbf{x}_{SI}) = 1$. Since we have determined an arbitrary $\rho^N$ value for the $\mathbf{x}_{SB}$ nodes, $\mathbf{n}^*$ can be calculated.

$$\mathbf{n}^* = -\frac{\nabla \rho^{N*}}{|\nabla \rho^{N*}|}. \tag{3.60}$$

where $\nabla \rho^{N*}$ represents the gradient of the phase parameter before applying the wettability scheme. The angle $\theta'$ between $\mathbf{n}_s$ and $\mathbf{n}^*$ is given by

$$\theta' = \arccos(\mathbf{n}_s \cdot \mathbf{n}^*). \tag{3.61}$$

Now, $\mathbf{n}^*$ is rotated so that the angle between $\mathbf{n}_s$ and $\mathbf{n}^*$ matches the desired contact angle, $\theta$. In the case of 2D, two $\mathbf{n}^*$ vectors are produced, which are denoted as $\mathbf{n}_+$ and $\mathbf{n}_-$.

$$\mathbf{n}_\pm = \left(\cos \pm\theta - \frac{\sin \pm\theta \cos\theta'}{\sin\theta'}\right)\mathbf{n}_s + \frac{\sin \pm\theta}{\sin\theta'}\mathbf{n}^*. \tag{3.62}$$

To select between $\mathbf{n}_+$ and $\mathbf{n}_-$, the one with the shorter Euclidean distance to $\mathbf{n}^*$ is chosen. Finally, $\mathbf{n}^*$ is replaced with $\mathbf{n}$, and the gradient of $\rho^N$ is modified as follows.

$$\nabla \rho^N = |\nabla \rho^{N*}|\mathbf{n}. \tag{3.63}$$

In the improved virtual density scheme, the magnitude of $\nabla \rho^N$ remains unchanged, and only its direction is adjusted. As a result, the contact angle can be controlled without altering the magnitude of the surface tension, as governed by (3.11). The following is the pseudocode for the improved virtual density scheme.

```
//Improved virtual density scheme for D2Q9 color-gradient model.
for all (x,y) do
  if (x,y) is SB node then
    sum1 = sum2 = 0.0
   for all i do
      xtemp = nx[x][y][i]
      ytemp = ny[x][y][i]
```

```
      sum1 += omega[i] * rhoN[xtemp][ytemp] * (1.0 - sw[xtemp][ytemp])
        sum2 += omega[i] * (1.0 - sw[xtemp][ytemp])
      end for
    rhoN[x][y] = sum1 / sum2
  end if
end for

for all (x,y) do
  if (x,y) is FB node then
    grad = sqrt(gradx[x][y] * gradx[x][y] + grady[x][y] * grady[x][y])
    nx_star = - gradx[x][y] / grad
    ny_star = - grady[x][y] / grad
    theta_prime = acos(nsx[x][y] * nx_star + nsy[x][y] * ny_star)
    nx_plus = (cos(theta) - sin(theta) * cos(theta_prime) / sin
(theta_prime))
        * nsx[x][y] + (sin(theta) / sin(theta_prime)) * nx_star
    ny_plus = (cos(theta) - sin(theta) * cos(theta_prime) / sin
(theta_prime))
        * nsy[x][y] + (sin(theta) / sin(theta_prime)) * ny_star
    nx_minus = (cos(theta) + sin(theta) * cos(theta_prime) / sin
(theta_prime))
        * nsx[x][y] - (sin(theta) / sin(theta_prime)) * nx_star
    ny_minus = (cos(theta) + sin(theta) * cos(theta_prime) / sin
(theta_prime))
        * nsy[x][y] - (sin(theta) / sin(theta_prime)) * ny_star
    D_plus = (nx_star - nx_plus) * (nx_star - nx_plus) + (ny_star -
ny_plus) *
        (ny_star - ny_plus)
    D_minus = (nx_star - nx_minus) * (nx_star - nx_minus) + (ny_star -
ny_minus)
        * (ny_star - ny_minus)
    if D_plus < D_minus then
      gradx[x][y] = grad * nx_plus
      grady[x][y] = grad * ny_plus
    else
      gradx[x][y] = grad * nx_minus
      grady[x][y] = grad * ny_minus
    end if
  end if
end for
```

### 3.6.2 Wettability: Pseudopotential Model

In the pseudopotential model, a widely used wettability scheme is the one that applies a fluid-solid interaction force [24].

$$\mathbf{F}_{ads} = -G_w \rho(\mathbf{x}_{FB}) \sum_i \omega_i s(\mathbf{x} + \mathbf{e}_i \Delta t)\mathbf{e}_i, \tag{3.64}$$

where $G_w$ is the adhesive parameter. By adjusting $G_w$, contact angle can be controlled. Benzi et al. [25] proposed an adhesion force that introduces a wall density, $\rho_w$, to model the interaction.

$$\mathbf{F}_{ads} = -G_w \psi(\rho) \sum_i \omega_i \psi(\rho_w) s(\mathbf{x} + \mathbf{e}_i \Delta t)\mathbf{e}_i, \tag{3.65}$$

However, these methods also have the limitation of inducing unphysical flow [26]. The pseudopotential model can also control the contact angle without using an adhesive force term by employing virtual density, similar to the color-gradient model. In this scheme [27], the virtual density at $\mathbf{x}_{SB}$ is defined as

$$\rho(\mathbf{x}_{SB}) = \begin{cases} \varphi \rho_{ave}(\mathbf{x}_{SB}), & \text{for decreasing } \theta, \\ \rho_{ave}(\mathbf{x}_{SB}) - \Delta\rho, & \text{for increasing } \theta, \end{cases} \tag{3.66}$$

where$\varphi$ and $\Delta\rho$ are the constants used to adjust the contact angle. The ranges for $\varphi$ and $\Delta\rho$ are $\varphi \geq 1$ and $\Delta\rho \geq 0$, respectively. To increase hydrophobicity, $\varphi$ is increased, and to enhance hydrophilicity, $\Delta\rho$ is increased. $\rho_{ave}$ can be calculated similarly to (3.59), as follows

$$\rho_{ave}(\mathbf{x}_{SB}) = \frac{\sum_i \omega_i \rho(\mathbf{x} + \mathbf{e}_i \Delta t)[1 - s_w(\mathbf{x} + \mathbf{e}_i \Delta t)]}{\sum_i \omega_i [1 - s_w(\mathbf{x} + \mathbf{e}_i \Delta t)]}. \tag{3.67}$$

For numerical stability, when $\rho(\mathbf{x}_{SB}) > \rho_l$ or $\rho(\mathbf{x}_{SB}) < \rho_g$, $\rho(\mathbf{x}_{SB})$ is set as $\rho(\mathbf{x}_{SB}) = \rho_l$ and $\rho(\mathbf{x}_{SB}) = \rho_g$. The following is the pseudocode for the improved virtual density scheme.

```
//Improved virtual density scheme for D2Q9 pseudopotential model.
for all (x,y) do
  if (x,y) is SB node then
    sum1 = sum2 = 0.0
   for all i do
      xtemp = nx[x][y][i]
      ytemp = ny[x][y][i]
```

```
      sum1 += omega[i] * rho[xtemp][ytemp] * (1.0 - sw[xtemp][ytemp])
        sum2 += omega[i] * (1.0 - sw[xtemp][ytemp])
      end for
    rho_ave = sum1 / sum2
    if surface is hydrophobic then
      rho[x][y] = phi * rho_ave
    else
      rho[x][y] = rho_ave - incrho
    end if
    if rho[x][y] > rho_l then
      rho[x][y] = rho_l
    else if rho[x][y] < rho_g then
      rho[x][y] = rho_g
    end if
  end if
end for
```

## 3.7 Project: Droplet Wettability

Wettability control is a critical factor that determines the interaction between liquid and solid surfaces, playing a key role in various scientific and engineering applications. Even with the same liquid, changes in the substrate can alter the interaction between the liquid and the solid surface, resulting in a different contact angle (Fig. 3.8). By leveraging such changes in liquid-solid interactions, patterned wettability substrates can be fabricated for diverse applications. Figure 3.9 presents a

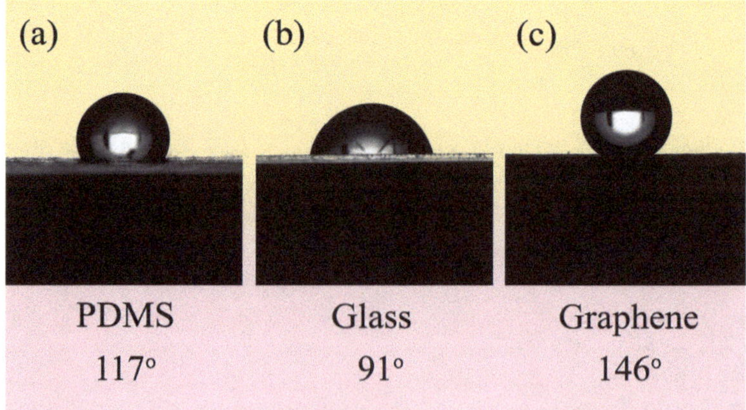


**Fig. 3.8** Water contact angle measurement of (**a**) PDMS, (**b**) glass, and (**c**) graphene, respectively [28]

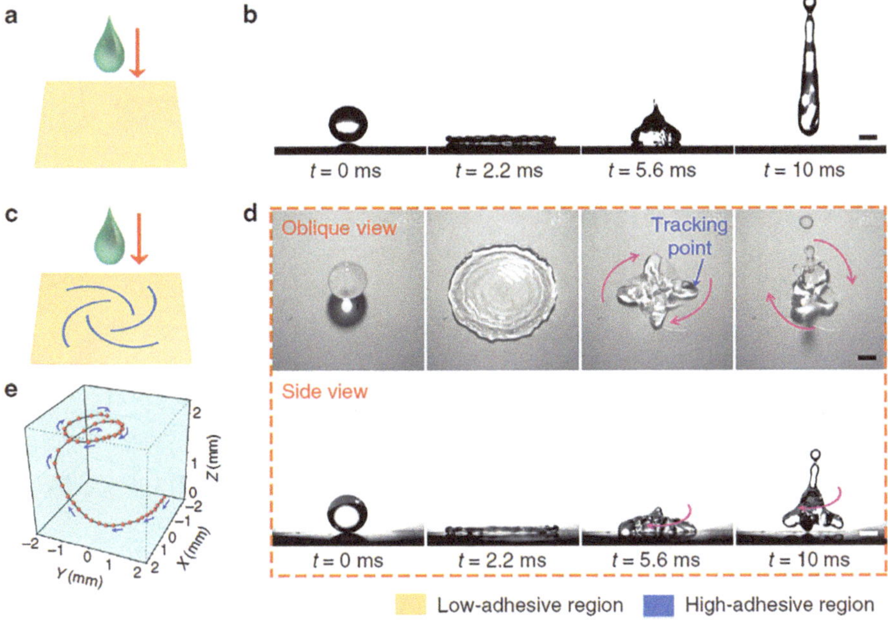


**Fig. 3.9** (**a**, **b**) Scheme and sequenced images of a water droplet impacting on a homogeneous hydrophobic and low-adhesive surface. (**c**) Scheme of a water droplet impacting on a low-adhesive substrate with high-adhesive patterns. (**d**) Selected snapshots to show the synchronized oblique and side views of the droplet gyrating by impacting on the deliberately-designed adhesion-patterned substrate. (**e**) Three-dimensional trajectory of the tracking point [29]

study analyzing droplet jumping dynamics on a patterned wettability substrate. Figure 3.9a and b show the case where a droplet impacts a homogeneous hydrophobic substrate. In this case, the droplet undergoes a spreading process upon collision with the substrate and subsequently rebounds. However, as shown in Fig. 3.9c, when the substrate features a high-adhesive spiral pattern, the droplet is influenced by the spiral pattern during the rebound process, causing it to rotate as it jumps off the surface. In addition to this, patterned wettability substrates can be applied in various fields, including evaporation [8], boiling [30], and droplet transport [31].

In this project, as shown in Fig. 3.10, the pseudopotential model is used to model a sessile droplet placed on a substrate. By analyzing the interaction between the droplet and the substrate, various contact angles are observed. The conditions and objectives of the project are as follows:

Project Conditions:

- The LBM multiphase model uses the D2Q9 pseudopotential model.
- The P-R EOS is applied and $\beta$ is set to 1.16.
- The wettability model uses (3.66).
- The heat transfer model is not applied, and the temperature across the entire domain is fixed at $T = 0.95T_c$.

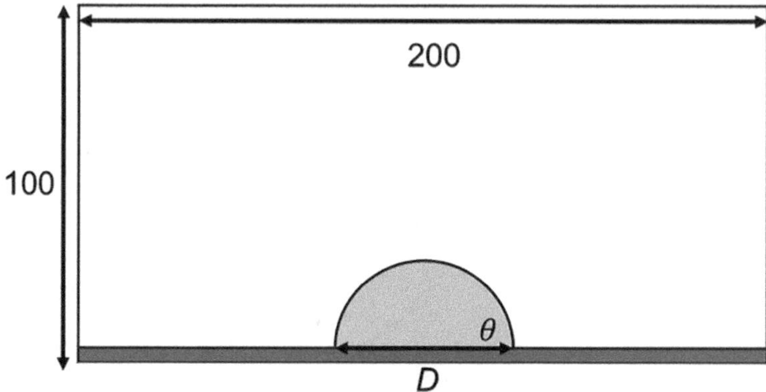


**Fig. 3.10** Schematic of simulation domain for droplet wettability control

- The densities of the liquid and gas are $\rho_l = 4.927$ and $\rho_g = 1.013$, respectively, with the relaxation time set to 1 for both.
- The simulation domain has a width of 200 *lu* and a height of 100 *lu*.
- The left and right surfaces of the domain are periodic, the bottom surface is solid, and the top surface applies a constant pressure boundary condition.
- The droplet diameter is 60 *lu*.

Project Objectives:

- Verify the appropriate liquid and gas densities at $T = 0.95T_c$ using the Maxwell construction for the P-R EOS.
- Compare the simulated liquid and gas densities with theoretical values to ensure consistency and validate the accuracy of the pseudopotential model.
- Measure the contact angle by varying $\varphi$ and $\Delta\rho$, and present the results as a graph to analyze the relationship between $\varphi$, $\Delta\rho$, and the contact angle.
- Adjust the parameters to achieve contact angles of 60°, 90°, and 120°, and perform simulations.

**Summary**

1. **Multiphase flow simulation challenges**
   - Multiphase systems involve interface tension, curvature, density contrast, and topological changes.
   - Traditional CFD requires explicit interface tracking, leading to numerical diffusion and complex geometry handling.
2. **LBM as a mesoscopic multiphase framework**
   - LBM solves the Boltzmann equation using discrete velocities, enabling natural interface representation.
   - Local collision–streaming operations simplify computation and improve parallelism.

- Mesoscopic interactions allow emergence of macroscopic surface tension and phase separation.

3. **Three major multiphase LBM model categories**

   (a) Color-gradient model

   - Uses red/blue distribution functions and a color field.
   - Collision, perturbation (surface tension), and recoloring (sharpening) operators.
   - Independent tuning of viscosity, density ratio, and surface tension.
   - Sharp interface and robust control of wettability.

   (b) Pseudopotential (Shan–Chen) model

   - Introduces interparticle force based on a pseudopotential function $\psi(\rho)$.
   - Phase separation occurs naturally if interaction strength $G$ exceeds critical value.
   - Surface tension, coexistence densities depend on equation of state (EOS).
   - Simple implementation but spurious currents and thermodynamic inconsistency can occur.

   (c) Free-energy model

   - Based on minimization of a thermodynamic free-energy functional.
   - Governs interface thickness, curvature, and surface tension through variational derivative.
   - Highly accurate thermodynamic consistency; ideal for wetting and contact-line modeling.
   - Requires solving Poisson-type equations $\rightarrow$ higher computational cost.

4. **High-density-ratio considerations**

   - Numerical stability decreases as density ratio increases.
   - CG-MRT and free-energy LBM enable higher density ratios than the classical Shan–Chen model.
   - D3Q19 lattice symmetry and isotropic gradient operators critical for stable interface evolution.

5. **Model selection guidelines**

   - Color-gradient: best for sharp interfaces, independent tuning, high-density ratios, droplet deformation.
   - Pseudopotential: simplest for phase separation, porous media, boiling/evaporation.
   - Free-energy: preferred for contact-line physics, wetting, thermodynamic accuracy.
   - Trade-offs among accuracy, stability, computational overhead, and interface fidelity.

## References

1. D. H. Rothman, and J. M. Keller, "Immiscible cellular-automaton fluids," J Stat Phys **52**, 1119-1127 (1988).
2. X. Shan, and H. Chen, "Lattice boltzmann model for simulating flows with multiple phases and components," Physical review E **47**, 1815 (1993).
3. M. R. Swift, E. Orlandini, W. Osborn, and J. Yeomans, "Lattice boltzmann simulations of liquid-gas and binary fluid systems," Physical Review E **54**, 5041 (1996).
4. T. Inamuro, T. Ogata, S. Tajima, and N. Konishi, "A lattice boltzmann method for incompressible two-phase flows with large density differences," J. Comput. Phys. **198**, 628-644 (2004).
5. H. M. Lee, and J. S. Lee, "Effects of heat transfer on particle suspended drop-on-demand inkjet printing using lattice boltzmann method," Appl. Therm. Eng. **213**, 118637 (2022).
6. M. Monteferrante, A. Montessori, S. Succi, D. Pisignano, and M. Lauricella, "Lattice boltzmann multicomponent model for direct-writing printing," Phys. Fluids **33**, (2021).
7. S. Sohrabi, and Y. Liu, "Modeling thermal inkjet and cell printing process using modified pseudopotential and thermal lattice boltzmann methods," Phys. Rev. E **97**, 033105 (2018).
8. H. M. Lee, and J. S. Lee, "Suppression of the coffee-ring effect using a contact-angle radial-band-patterned substrate: A lattice boltzmann study," Phys. Fluids **36**, (2024).
9. H. M. Lee, and J. S. Lee, "Particle deposition dynamics in evaporating droplets using lattice boltzmann and magnetic particle simulation," Phys. Fluids **35**, (2023).
10. T. Reis, and T. N. Phillips, "Lattice boltzmann model for simulating immiscible two-phase flows," Journal of Physics A: Mathematical and Theoretical **40**, 4033-4053 (2007).
11. H. Liu, A. J. Valocchi, and Q. Kang, "Three-dimensional lattice boltzmann model for immiscible two-phase flow simulations," Phys. Rev. E **85**, 046309 (2012).
12. M. Latva-Kokko, and D. H. Rothman, "Diffusion properties of gradient-based lattice boltzmann models of immiscible fluids," Phys. Rev. E **71**, 056702 (2005).
13. Q. Li, K. H. Luo, Y. L. He, Y. J. Gao, and W. Q. Tao, "Coupling lattice boltzmann model for simulation of thermal flows on standard lattices," Physical Review E **85**, 016710 (2012).
14. Y. Ba, H. H. Liu, Q. Li, Q. J. Kang, and J. J. Sun, "Multiple-relaxation-time color-gradient lattice boltzmann model for simulating two-phase flows with high density ratio," Physical Review E **94**, (2016).
15. Z. X. Wen, Q. Li, Y. Yu, and K. H. Luo, "Improved three-dimensional color-gradient lattice boltzmann model for immiscible two-phase flows," Phys. Rev. E **100**, 023301 (2019).
16. P. Yuan, and L. Schaefer, "Equations of state in a lattice boltzmann model," Phys. Fluids **18**, 042101 (2006).
17. S. Gong, and P. Cheng, "Numerical investigation of droplet motion and coalescence by an improved lattice boltzmann model for phase transitions and multiphase flows," Comput. Fluids **53**, 93-104 (2012).
18. R. Zhang, and H. Chen, "Lattice boltzmann method for simulations of liquid-vapor thermal flows," Physical Review E **67**, 066711 (2003).
19. S. Gong, and P. Cheng, "A lattice boltzmann method for simulation of liquid–vapor phase-change heat transfer," Int. J. Heat Mass Transf. **55**, 4923-4927 (2012).
20. M. Latva-Kokko, and D. H. Rothman, "Static contact angle in lattice boltzmann models of immiscible fluids," Physical Review E—Statistical, Nonlinear, and Soft Matter Physics **72**, 046701 (2005).
21. S. Leclaire, K. Abahri, R. Belarbi, and R. Bennacer, "Modeling of static contact angles with curved boundaries using a multiphase lattice boltzmann method with variable density and viscosity ratios," International Journal for Numerical Methods in Fluids **82**, 451-470 (2016).
22. T. Akai, B. Bijeljic, and M. J. Blunt, "Wetting boundary condition for the color-gradient lattice boltzmann method: Validation with analytical and experimental data," Advances in Water Resources **116**, 56-66 (2018).
23. Z. Xu, H. Liu, and A. J. Valocchi, "Lattice boltzmann simulation of immiscible two-phase flow with capillary valve effect in porous media," Water Resources Research **53**, 3770-3790 (2017).

24. N. S. Martys, and H. Chen, “Simulation of multicomponent fluids in complex three-dimensional geometries by the lattice boltzmann method,” Physical review E **53**, 743 (1996).
25. R. Benzi, L. Biferale, M. Sbragaglia, S. Succi, and F. Toschi, “Mesoscopic modeling of a two-phase flow in the presence of boundaries: The contact angle,” Physical Review E—Statistical, Nonlinear, and Soft Matter Physics **74**, 021509 (2006).
26. H. Huang, Z. Li, S. Liu, and X. Y. Lu, “Shan-and-chen-type multiphase lattice boltzmann study of viscous coupling effects for two-phase flow in porous media,” International journal for numerical methods in fluids **61**, 341-354 (2009).
27. Q. Li, Y. Yu, and K. H. Luo, “Implementation of contact angles in pseudopotential lattice boltzmann simulations with curved boundaries,” Phys. Rev. E **100**, 053313 (2019).
28. N. Jiang, Y. Wang, K. C. Chan, C. Y. Chan, H. Sun, and G. Li, “Additive manufactured graphene coating with synergistic photothermal and superhydrophobic effects for bactericidal applications,” Global Challenges **4**, 1900054 (2020).
29. H. Li, W. Fang, Y. Li, Q. Yang, M. Li, Q. Li, X.-Q. Feng, and Y. Song, “Spontaneous droplets gyrating via asymmetric self-splitting on heterogeneous surfaces,” Nature Communications **10**, 950 (2019).
30. J. S. Lee, and J. S. Lee, “Numerical study of hydrophobic-island shapes with patterned wettability for pool boiling," Appl. Therm. Eng. **127**, 1632-1641 (2017).
31. J. S. Lee, J. Y. Moon, and J. S. Lee, "Study of transporting of droplets on heterogeneous surface structure using the lattice boltzmann approach," Appl. Therm. Eng. **72**, 104-113 (2014).

# Chapter 4
# Particle Dynamics

## 4.1 Overview of Particle Dynamics in LBM

Particle dynamics simulations play a critical role in a wide range of industrial applications, offering invaluable insights into the motion, interaction, and behavior of particles in fluid environments. In the pharmaceutical industry, simulations aid in optimizing drug delivery systems [1] by analyzing nanoparticle transport within biological systems. In the energy sector, particle dynamics help improve the performance of lithium-ion batteries [2] by studying ion transport and dendrite suppression. Environmental engineering applications include pollutant removal systems [3] where sedimentation and particle separation are crucial for efficiency. Additionally, particle dynamics simulations are widely used in manufacturing processes such as 3D printing [4] and powder coating [5], where the distribution and aggregation of particles directly influence product quality. These diverse applications demonstrate the significance of particle dynamics in advancing technology and optimizing industrial processes.

The LBM provides a unique mesoscopic framework for fluid simulation, making it an excellent tool for coupling with particle dynamics models. By simulating fluid flow at the particle level, LBM seamlessly integrates the interactions between particles and surrounding fluids. The flexibility of LBM allows for the incorporation of external forces such as drag, lift, Brownian motion, and electrokinetic effects, enabling accurate modeling of complex particle behaviors. Furthermore, LBM efficiently handles dynamic boundaries and irregular geometries, making it particularly suitable for simulating particle-laden flows, sedimentation, and particle interactions in multiphase environments. This integration enhances the capability to predict particle behavior under various conditions, bridging the gap between theoretical studies and real-world applications.

Many researchers have attempted to combine LBM with particle dynamics models. In 1994, Ladd [6, 7] was the first to propose a particle dynamics model

 
H. M. Lee, J. S. Lee, *Lattice Boltzmann Methodology for Single-Phase and Multiphase Nanoparticle Modeling*,
https://doi.org/10.1007/978-981-95-9117-6_4

within the LBM framework. The main idea of Ladd's method is to define each particle as a moving boundary. Fluid near the particle boundary interacts according to the bounce-back rule. However, Ladd's method has limitations, including numerical instability in simulating interactions at the fluid-particle boundary and a strong dependence of particle resolution on grid size.

In 2005, Nakayama and Yamamoto [8] first introduced the smoothed profile method (SPM). Later, Jafari et al. [9] proposed a model combining SPM with LBM and verified its accuracy. In SPM, particles are modeled on a fluid domain composed of an Eulerian grid using a profile function. One of the key features of SPM is that the fluid-particle boundary is represented with a finite thickness. This results in high simulation stability and eliminates the need for complex boundary conditions compared to Ladd's method. Thanks to these advantages, SPM has been widely used for analyzing capillary flow and particle interactions, including applications combined with color-gradient [10] and pseudopotential [11] models. However, the numerical stability of the SPM method may degrade as the particle boundary becomes thicker.

The immersed boundary method (IBM) is one of the CFD methodologies designed to simulate complex or moving boundaries, and it can be utilized in LBM to simulate moving particles. In IBM, particles with surfaces represented by a Lagrangian mesh float within a fluid domain composed of an Eulerian mesh. IBM allows for the simulation of complex particle shapes and deformable particles due to the Lagrangian mesh on the particle surface. Moon et al. [12] used IBM to model red blood cell (RBC) and analyzed the degree of deformation of RBCs in a microchannel by varying the elastic modulus. However, since numerous Lagrangian nodes are distributed on the particle surface, modeling particle interactions may require significant computational effort. As a result, IBM may not be suitable for systems with a large number of suspended particles.

If the size of the particles reaches the nanoscale, particle modeling becomes more complex. For particles larger than the microscale, inertial forces such as drag, lift, or gravity predominantly influence particle dynamics. However, for nanoparticles, interaction forces like van der Waals attraction and electrostatic repulsion, which are negligible for micro/macroparticles, must be considered. Nanoparticles can stabilize or aggregate depending on the balance of these attractive and repulsive forces. This balance of interaction forces can be explained by the DLVO theory, which assumes that the total interaction energy is the sum of all interaction energies. Jiang et al. [13] applied DLVO theory to the SPM-LB model to describe interactions between nanoparticles. In addition to these interaction forces, the Brownian motion of nanoparticles, a random movement caused by thermal fluctuations, must also be taken into account. The Langevin equation can be used to describe the motion of nanoparticles influenced by both deterministic forces and random fluctuations. Langevin dynamics have been employed in LBM to simulate the Brownian motion of nanoparticles [14], and this research has revealed deposition patterns of nanoparticles.

This chapter introduces particle modeling in LBM through explanations of SPM and IBM. Additionally, it covers methodologies for nanoparticle modeling by introducing the DLVO theory and Langevin equation combined with LBM.

## 4.2 Smoothed Profile Method

In the SPM, the boundary of solid particles is composed of a continuous boundary with thickness, rather than a thin boundary, as shown in Fig. 4.1. By using this method to handle boundaries, higher numerical stability can be achieved compared to particles with thin boundaries. The profile function for the $i$th 2D spherical particle $\phi_i$ can be determined as follows

$$\phi_i(\mathbf{x},t)=\begin{cases}0, & h<-\dfrac{1}{2}\xi,\\ 1, & h>\dfrac{1}{2}\xi,\\ \dfrac{1}{2}\left[\sin\left(\dfrac{\pi h}{\xi}\right)+1\right], & \text{else,}\end{cases} \tag{4.1}$$

where $\xi$ is the interface thickness, and $h = R_P - |\mathbf{X}_i - \mathbf{x}|$. $R_P$ is the radius, and $\mathbf{X}_i$ is the center position of particle. If there are $N_P$ particles in total, the overall profile function $\phi$ can be calculated as the sum of the individual profile functions.

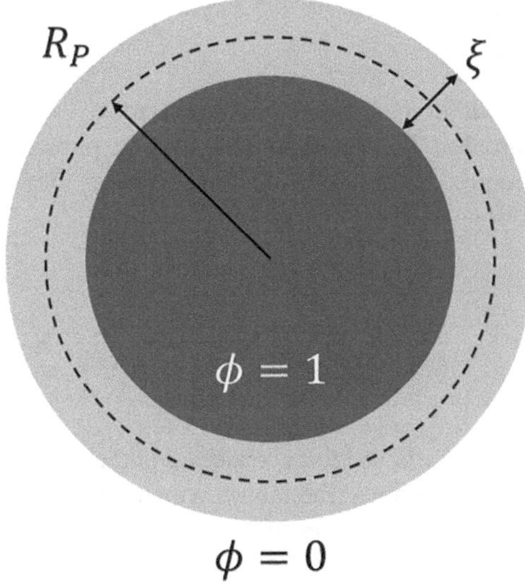


**Fig. 4.1** Profile diagram of a sphere rigid particle in SPM

$$\phi(\mathbf{x},t) = \sum_{i=1}^{N_P} \phi_i(\mathbf{x},t), \tag{4.2}$$

Within the particle interior region, an interaction force $\mathbf{f}_P$ acts between the fluid and the particle. This force is calculated based on the velocity difference between the fluid and the velocity inside the rigid particle.

$$\mathbf{f}_P(\mathbf{x},t) = \phi(\mathbf{x},t)[\mathbf{u}(\mathbf{x},t) - \mathbf{u}_P(\mathbf{x},t)]/\Delta t, \tag{4.3}$$

where $\mathbf{u}$ is the fluid velocity, and $\mathbf{u}_P$ is the velocity inside the rigid particle. $\mathbf{u}_P$ can be calculated as

$$\phi(\mathbf{x},t)\mathbf{u}_P(\mathbf{x},t) = \sum_{i=1}^{N_P} \phi_i(\mathbf{x},t)[\mathbf{U}_i + \mathbf{\Omega}_i \times (\mathbf{x} - \mathbf{X}_i)], \tag{4.4}$$

where $\mathbf{U}_i$ is the linear velocity of particle and $\mathbf{\Omega}_i$ is the angular velocity of particle. The solid–fluid interaction force calculated in (4.3) is applied to the fluid in the opposite direction.

$$\mathbf{f}_H(\mathbf{x},t) = -\mathbf{f}_P(\mathbf{x},t), \tag{4.5}$$

where $\mathbf{f}_H$ is applied as an external force to the fluid nodes. The hydrodynamic force and torque acting on the $i$th particle can be calculated using $\mathbf{f}_P$ as follows.

$$\mathbf{F}_{P,i} = \sum_{\mathbf{x} \in V_P} \rho(\mathbf{x})\mathbf{f}_P(\mathbf{x},t), \tag{4.6}$$

$$\mathbf{T}_{P,i} = \sum_{\mathbf{x} \in V_P} (\mathbf{x} - \mathbf{X}_i) \times \rho(\mathbf{x})\mathbf{f}_P(\mathbf{x},t). \tag{4.7}$$

In addition to hydrodynamic forces, various interparticle forces can act on the particles. To prevent overlapping of particles, an interparticle collision force $\mathbf{F}_C$ is required. The collision force acting between the $i$th and $j$th particles is as follows [15].

$$\mathbf{F}_{C,i} = \begin{cases} 0, & |\mathbf{X}_i - \mathbf{X}_j| > 2R_P + \zeta, \\ \dfrac{C_P}{\varepsilon_P}\left(\dfrac{|\mathbf{X}_i - \mathbf{X}_j| - 2R_P - \zeta}{\zeta}\right)^2 \dfrac{\mathbf{X}_i - \mathbf{X}_j}{|\mathbf{X}_i - \mathbf{X}_j|}, & |\mathbf{X}_i - \mathbf{X}_j| \le 2R_P + \zeta, \end{cases} \tag{4.8}$$

where $C_P$ is the force scale, $\varepsilon_P$ is the stiffness scale factor, and $\zeta$ is the threshold. Having obtained the total force and torque acting on the particle, the linear velocity and angular velocity for the next timestep can be calculated using Newton's second law.

$$\mathbf{U}_i^{n+1} = \mathbf{U}_i^n + \frac{\mathbf{F}_{i,total}}{M_P}\Delta t, \tag{4.9}$$

$$\boldsymbol{\Omega}_i^{n+1} = \boldsymbol{\Omega}_i^n + \frac{\mathbf{T}_{i,total}}{I_P}\Delta t, \tag{4.10}$$

where $M_P$ and $I_P$ denote the mass and moment of inertia of the particle, respectively. For a 2D circular particle, $M_P = \rho_P \pi R_P^2$, $I_P = 0.5 M_P R_P^2$, and for a 3D spherical particle, $M_P = \rho_P (4/3)\pi R_P^3$, $I_P = (2/5) M_P R_P^2$. $\rho_P$ is the density of particle. Finally, the position of the particle's center at the next timestep is calculated as

$$\mathbf{X}_i^{n+1} = \mathbf{X}_i^n + 0.5\left(\mathbf{U}_i^n + \mathbf{U}_i^{n+1}\right)\Delta t. \tag{4.11}$$

The algorithm of the SPM can be summarized as follows.

1. At the initial timestep, set the particle's center position, linear velocity, and angular velocity.
2. Calculate the moment terms of the fluid.
3. From the current center position, determine the particles' $\phi_i$ and $\mathbf{u}_P$ (4.1) and (4.4).
4. Calculate the $\mathbf{f}_P$ and $\mathbf{f}_H$ using (4.3) and (4.5).
5. Using (4.6) and (4.7), calculate the $\mathbf{F}_{P,\ i}$ and $\mathbf{T}_{P,\ i}$ acting on the particle.
6. Calculate the particle interaction force including (4.8).
7. Update the linear velocity, angular velocity, and position of the particles for the next timestep.
8. Perform the collision and streaming steps for the fluid and return to step 2.

## 4.3 Immersed Boundary Method

In IBM, particles are immersed in a fluid composed of Eulerian nodes, as shown in Fig. 4.2. The surfaces of the particles are distributed with Lagrangian nodes. For circular or spherical particles, the center of particle is denoted as $\mathbf{X}_i$, and the positions of the points distributed on the particle surface are represented as $\mathbf{X}_B$. The movement of the particles can be determined through the hydrodynamic forces acting on the $\mathbf{X}_B$. To achieve this, the velocity at the $\mathbf{X}_B$ positions, $\mathbf{U}_B^*$ must first be calculated. Since $\mathbf{X}_B$ are Lagrangian points, the velocity is derived from the velocity values of the surrounding fluid nodes.

$$\mathbf{U}_B^*(\mathbf{X}_B, t+\Delta t) = \sum_{\mathbf{x}} \mathbf{u}^*(\mathbf{x}, t+\Delta t) D(\mathbf{x} - \mathbf{X}_B)\Delta x \Delta y, \tag{4.12}$$

where $D$ is the Dirac delta function [16]. When the domain is 2D, the Dirac delta function is defined as

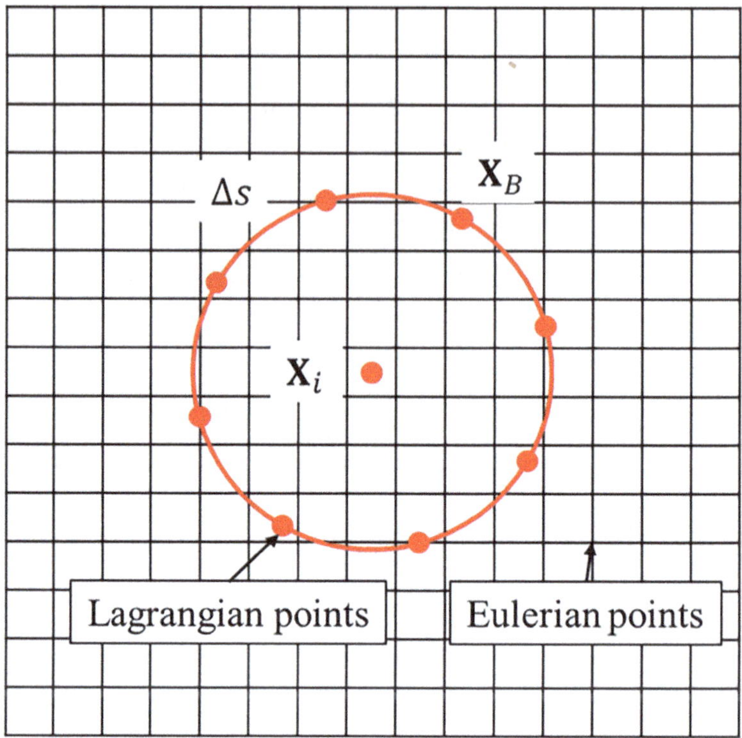


**Fig. 4.2** Schematic of the IBM

$$D(\mathbf{x}-\mathbf{X}_B)=\frac{1}{h^2}\delta_h\left(\frac{x_i-X_B}{h}\right)\delta_h\left(\frac{y_i-Y_B}{h}\right), \tag{4.13}$$

and $\delta_h$ is

$$\delta_h(r)=\begin{cases}\frac{1}{4h}\left[1+\cos\frac{\pi r}{2}\right], & |r|\le 2,\\ 0, & |r|>2.\end{cases} \tag{4.14}$$

where $h = \Delta x = \Delta y$ is lattice spacing. $\mathbf{u}^*$ is the intermediate velocity of fluid. As discussed in Chap. 2, when using Guo et al.'s [17] forcing scheme, the equilibrium velocity is modified by the force term.

$$\mathbf{u}^{eq}=\frac{1}{\rho}\sum_i f_i\mathbf{e}_i+\frac{\mathbf{F}\Delta t}{2\rho}. \tag{4.15}$$

The intermediate velocity, which does not account for the force term, can be expressed as follows.

$$\mathbf{u}^* = \frac{1}{\rho}\sum_i f_i \mathbf{e}_i. \tag{4.16}$$

The hydrodynamic force, $\mathbf{F}_B$, can also be calculated using the velocity at the Lagrangian points obtained from (4.12).

$$\mathbf{F}_B(\mathbf{X}_B, t+\Delta t) = 2\rho \frac{\mathbf{U}_B - \mathbf{U}_B^*(\mathbf{X}_B, t+\Delta t)}{\Delta t}, \tag{4.17}$$

where $\mathbf{U}_B$ is the real velocity at Lagrangian points. Since the hydrodynamic force at the Lagrangian points has been calculated, this force must also be reflected in the fluid domain. To achieve this, the forces at the Lagrangian points are distributed to the Eulerian points using the Dirac delta function.

$$\mathbf{F}_B(\mathbf{x}, t+\Delta t) = \sum_{N_P}\sum_{N_B} \mathbf{F}_B(\mathbf{X}_B, t+\Delta t) D(\mathbf{x}-\mathbf{X}_B)\Delta s, \tag{4.18}$$

where $N_B$ is the number of boundary points distributed on the particle surface, and $N_P$ is the total number of particles. $\Delta s$ is the arc length between two adjacent Lagrangian points. As the hydrodynamic force is added to the Eulerian points, the velocity of fluid should also be updated as $\mathbf{u}^* = \mathbf{u}^* + \mathbf{F}_B/2\rho$. Subsequently, repeat (4.12)–(4.18) until numerical convergence is achieved.

Having calculated the hydrodynamic force at all Lagrangian points, the total force and torque acting on the entire particle can be computed.

$$\mathbf{F}_{P,i} = \sum_{N_B} \mathbf{F}_{B,i}(\mathbf{X}_B)\Delta s, \tag{4.19}$$

$$\mathbf{T}_{P,i} = \sum_{N_B} (\mathbf{X}_{B,i} - \mathbf{X}_i) \times \mathbf{F}_B(\mathbf{X}_B)\Delta s. \tag{4.20}$$

The linear velocity and angular velocity of the particle center at the next timestep are

$$\mathbf{U}_i^{n+1} = \mathbf{U}_i^n + \frac{\mathbf{F}_{i,total}}{M_P}\Delta t, \tag{4.21}$$

$$\boldsymbol{\Omega}_i^{n+1} = \boldsymbol{\Omega}_i^n + \frac{\mathbf{T}_{i,total}}{I_P}\Delta t. \tag{4.22}$$

The position of the particle center is also updated.

$$\mathbf{X}_i^{n+1} = \mathbf{X}_i^n + 0.5\left(\mathbf{U}_i^n + \mathbf{U}_i^{n+1}\right)\Delta t, \tag{4.23}$$

If the particle is rigid, the particle velocity at the Lagrangian points can be calculated using $\mathbf{U}_i$ and $\boldsymbol{\Omega}_i$.

$$\mathbf{U}_B = \mathbf{U}_i + \boldsymbol{\Omega}_i \times (\mathbf{X}_B - \mathbf{X}_i), \tag{4.24}$$

The IBM algorithm can be summarized as follows.

1. Initialize the particle information.
2. Calculate the moment terms of the fluid.
3. Calculate $\mathbf{U}_B^*$ at each Lagrangian point (4.12).
4. Calculate the hydrodynamic force at the Lagrangian points, $\mathbf{F}_B(\mathbf{X}_B)$ and distribute it to the Eulerian points, $\mathbf{F}_B(\mathbf{x})$ (4.17) and (4.18).
5. Update $\mathbf{u}^*$ as $\mathbf{u}^* = \mathbf{u}^* + \mathbf{F}_B/2\rho$.
6. Repeat steps 3–5 util convergence is achieved.
7. Calculate $\mathbf{F}_{P,\ i}$ and $\mathbf{T}_{P,\ i}$. If there is an interaction force, add it to the total force.
8. Update $\mathbf{U}_i$, $\boldsymbol{\Omega}_i$, $\mathbf{X}_i$, and $\mathbf{U}_B$ for the next timestep (4.21)–(4.24).

**Exercise 4.1 SPM vs IBM**

Using the descriptions in Sects. 4.2 and 4.3, compare the SPM and the IBM in the following five aspects:

1. Boundary representation
2. Computational cost
3. Stability characteristics
4. Suitability for large-scale particle suspensions
5. Accuracy for irregular particle shapes

**Solution**

1. **Boundary representation**

   SPM: The particle boundary is represented by a continuous profile function with finite thickness, producing a smooth transition between solid and fluid. The boundary is smeared over several lattice nodes.

   IBM: The fluid is discretized on an Eulerian grid, while the particle surface is represented by Lagrangian nodes. A discrete Dirac delta function interpolates between the two grids.
2. **Computational cost**

   SPM: Boundary handling is performed through a volume-force formulation, making the method relatively efficient, especially for systems with many particles.

   IBM: The presence of numerous Lagrangian nodes on the particle surface significantly increases computational cost. Interpolation and spreading operations further increase the cost, especially with many particles.
3. **Stability characteristics**

   SPM: The finite thickness boundary provides high numerical stability, mitigating sharp discontinuities that can cause instabilities in thin-boundary methods.

   IBM: Stability is sensitive to the accuracy of the interpolation kernel and the coupling between Eulerian and Lagrangian meshes. Timestep restrictions are often stricter, especially for deformable bodies.

4. **Suitability for large-scale particle suspensions**
   SPM: Well suited for large numbers of rigid particles (suspensions, sedimentation), since each particle is represented through a simple profile function.
   IBM: Not well suited for large particle counts due to computational overhead from many Lagrangian surface nodes.
5. **Accuracy for irregular particle shapes**
   SPM: Best suited for spherical or simple geometries; extending the profile function to highly irregular shapes may be difficult.
   IBM: Very effective for arbitrary and deformable shapes (e.g., RBCs), since the Lagrangian mesh can capture complex surface deformation.

## 4.4 Nanoparticle Dynamics

### *4.4.1 DLVO Theory*

The DLVO theory posits that the interactions between nanoparticles are represented by the balance between electrostatic repulsive forces and van der Waals attractive forces. The stability of a colloid indicates that the particles do not aggregate. Particle aggregation occurs when the van der Waals attractive forces overcome the electrostatic repulsive forces, causing the particles to adhere to each other. Thus, the stability of the colloid can be determined by controlling the electrostatic repulsion, and the DLVO theory provides a quantitative framework for evaluating this stability.

Electrostatic repulsion arises from the overlap of diffused electrical double layers. According to the Stern model illustrated in Fig. 4.3a, the electrical double layer consists of an inner Stern layer and an outer diffusion layer. The Stern layer is a thin layer near the solid surface where the potential decreases rapidly with distance from the surface. The diffusion layer is relatively thicker than the Stern layer, where the potential gradually decreases as the distance from the surface increases. As depicted in Fig. 4.3b, when the diffusion layers of charged nanoparticles overlap, electrostatic repulsion acts between the particles, while van der Waals forces simultaneously act as attractive inputs. To reflect the characteristics of nanoparticles, various LBM studies [13, 18–21] have utilized the DLVO theory as an interaction model for nanoparticles. The following discusses the interaction forces required to apply the DLVO theory in the LBM nanoparticle model.

In 1937, Hamaker [22] proposed a van der Waals attraction model. The van der Waals force acting between two spherical particles of equal radius $R$ is

$$\mathbf{F}_A^{p-p} = \frac{A_H}{3R}\left[\frac{2(R_c+1)}{R_c^2+2R_c} - \frac{R_c+1}{\left(R_c^2+2R_c\right)^2} - \frac{2}{R_c+1} - \frac{1}{\left(R_c+1\right)^3}\right]\mathbf{n}_{ij}, \tag{4.25}$$

where $A_H$ is the Hamaker constant, and $R_c = h/2R$. $h$ is the minimum gap between the particle surfaces. $\mathbf{n}_{ij} = (\mathbf{X}_j - \mathbf{X}_i)/|\mathbf{X}_j - \mathbf{X}_i|$ is the unit vector from particle $i$ to $j$.

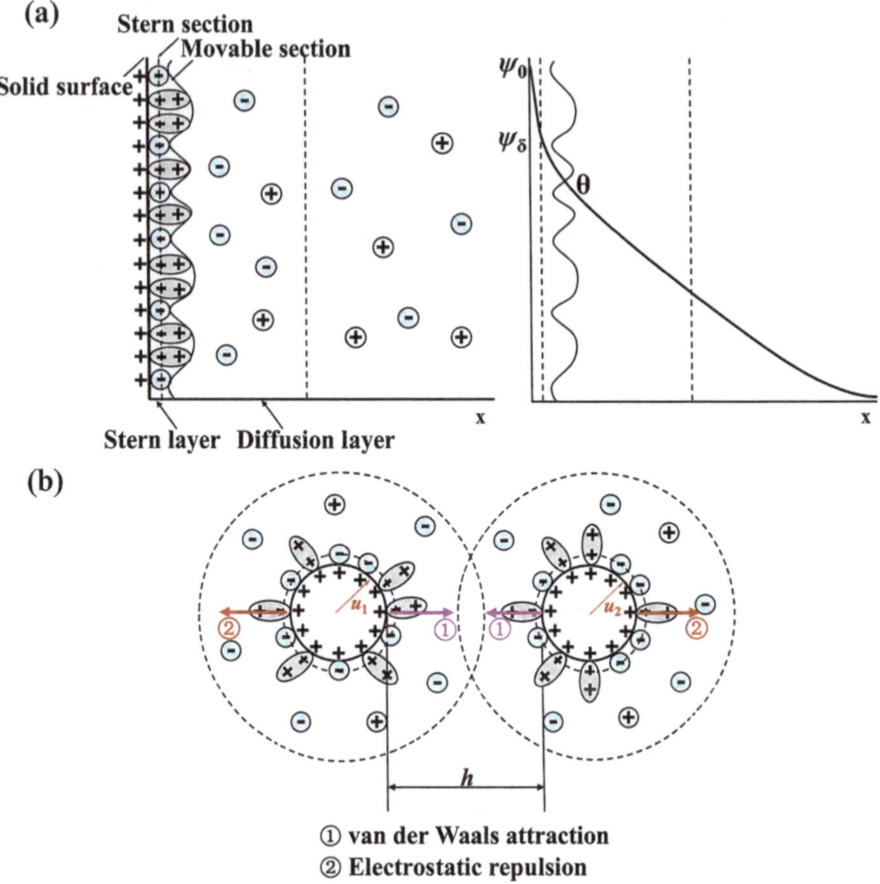


**Fig. 4.3** (**a**) Stern model of the electrical double layer and (**b**) schematic of the overlapped diffusion layers [13]

Van der Waals force can also act between a spherical particle and an infinitely flat plane. The van der Waals force acting between a spherical particle and a flat plane is

$$\mathbf{F}_A^{p-w} = -\frac{A_H}{6R}\left[\frac{2}{R_c} - \frac{1}{R_c^2} - \frac{2}{R_c + 1} - \frac{1}{(R_c + 1)^2}\right]\mathbf{n}_w, \tag{4.26}$$

where $\mathbf{n}_w$ is the normal vector of wall.

Similarly, the electrostatic repulsion force acting between spherical particles is [23]

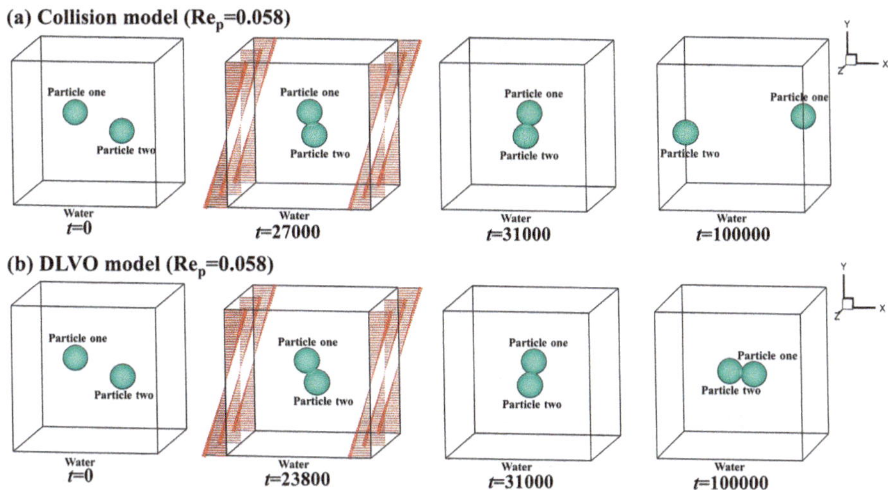


**Fig. 4.4** Dynamic motion of two nanoparticles in shear flow from the (**a**) collision model and (**b**) DLVO model [13]

$$\mathbf{F}_{ELE}^{p-p} = -\frac{64\pi\eta_{\infty}k_BTR}{\kappa}\tanh^2\left(\frac{ze\psi_{\delta}}{4k_BT}\right)\exp(-\kappa h)\mathbf{n}_{ij}, \tag{4.27}$$

where $\eta_{\infty}$, $T$, and $\psi_{\delta}$ are the number density of the ions, absolute temperature, and the zeta potential, respectively. $k_B = 1.38 \times 10^{-23}$ J/K is the Boltzmann constant. $\kappa^{-1} = \sqrt{\epsilon k_BT/2z^2e^2\eta_{\infty}}$ is the Debye length, where $\epsilon$ is the medium permittivity, $z$ is the valency of ions, and $e = 1.6 \times 10^{-19}$ C is the electrical charge. The electrostatic repulsion force acting between a spherical particle and a wall is

$$\mathbf{F}_{ELE}^{p-w} = \frac{128\pi\eta_{\infty}k_BTR}{\kappa}\tanh^2\left(\frac{ze\psi_{\delta}}{4k_BT}\right)\exp(-\kappa h)\mathbf{n}_{w}. \tag{4.28}$$

In the DLVO theory, the total potential energy acting on a particle is expressed as the sum of the attraction potential and the repulsion potential. Therefore, the DLVO force acting on a particle can be expressed as follows.

$$\mathbf{F}_{DLVO} = \mathbf{F}_A^{p-p} + \mathbf{F}_A^{p-w} + \mathbf{F}_{ELE}^{p-p} + \mathbf{F}_{ELE}^{p-w}. \tag{4.29}$$

Applying the DLVO model enables a more realistic representation of interactions between nanoparticles. Typically, collision models like (4.8) are used to prevent overlap between particles. Jiang et al. [13] combined the SPM with the DLVO model to compare the collision model and the DLVO model. In their study, as shown in Fig. 4.4, two particles collided within a shear flow, and the models were compared. In the collision model, the particles adhered to each other upon collision, rotated around the domain center, and then separated. However, in the DLVO model, the

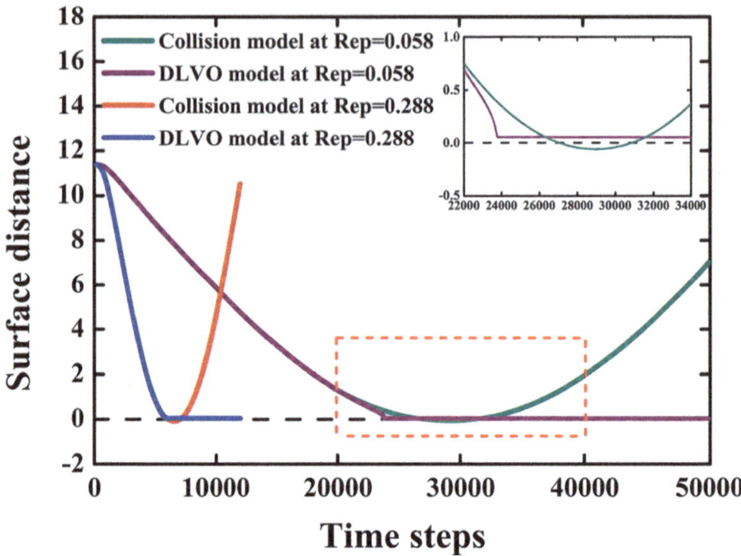


**Fig. 4.5** Surface distance between two particles in water and enlarged view of the region indicated by the red dashed box [13]

particles rotated after the collision but did not separate due to the strong van der Waals attraction.

To further analyze the difference in results, the particle spacing over time was examined, as shown in Fig. 4.5. The analysis revealed that, as previously observed, the collision model resulted in particle separation after collision, whereas the DLVO model maintained the same surface distance after the collision. Additionally, during particle collision (red dashed box), the collision model exhibited negative surface distance, indicating overlapping, while the DLVO model maintained a very small surface distance. These results demonstrate that the DLVO model provides a more realistic representation of particle interactions.

By applying the DLVO model, the aggregation process of nanoparticles within a suspension can also be observed. Figure 4.6 illustrates the particle aggregation process when the DLVO model is applied. The Reynolds number $Re_P = 0.1$ was kept constant, and the Debye length $\kappa$ was nondimensionalized by multiplying it with the particle radius. As shown (4.27) and (4.28), the electrostatic force is inversely proportional to the Debye length. In other words, as $\kappa R_{hyb}$ increases, the influence of van der Waals attraction becomes more dominant.

In all cases, the nanoparticles were commonly aligned near the top and bottom walls due to the attractive forces acting with respect to the walls. However, in the central region of the domain, away from the walls, as the magnitude of the electrostatic force decreased, the particles aggregated in a less aligned and more mixed form. Additionally, as the volume concentration increased, particle mobility

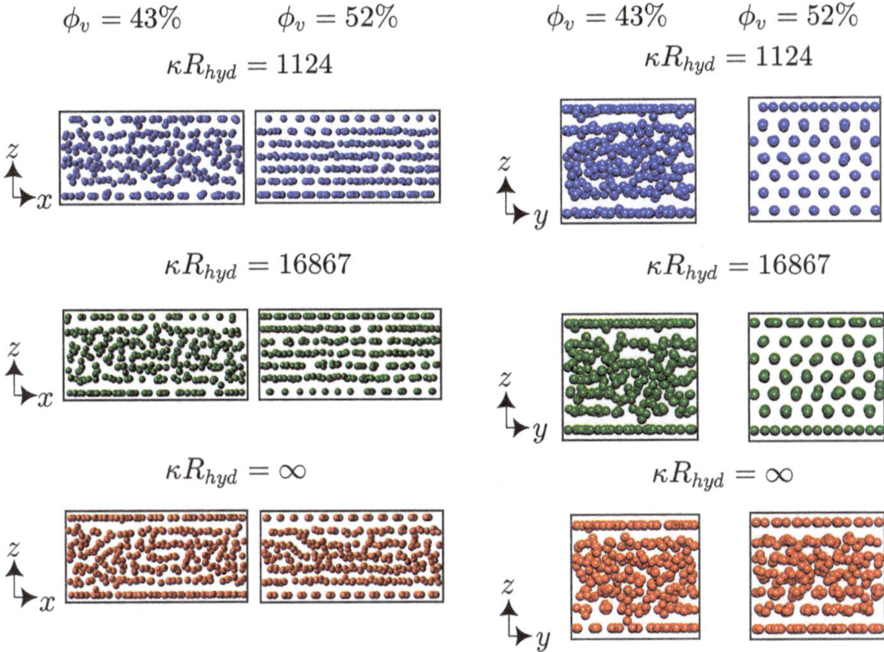


**Fig. 4.6** Nanoparticle aggregation process based on volume concentration and $\kappa R_{hyd}$ [19]

decreased, and the aggregated particles formed sliding layers due to the electrostatic force pushing other particles away.

These results demonstrate that the LBM combined with the DLVO model can be used to observe the stability of nanoparticle suspensions under various environmental conditions.

**Exercise 4.2 DLVO Forces: Interpretation and Scaling**
Based on (4.25)–(4.29):

(a) Describe the physical origin of van der Waals attraction and electrostatic repulsion.
(b) Determine whether each force decays algebraically or exponentially with particle separation.
(c) Explain how the Debye length affects nanoparticle stability, referencing Fig. 4.6.

**Solution**

(a) Van der Waals forces arise from fluctuating or induced dipoles in atoms and molecules. Even electrically neutral particles exhibit instantaneous charge fluctuations, which induce dipoles in neighboring particles and result in a net attractive interaction. At the continuum scale for colloidal particles, these microscopic interactions are integrated into an effective attractive force, often characterized by the Hamaker constant [as in (4.25)].

Electrostatic repulsion originates from the overlap of electrical double layers surrounding charged particle surfaces. According to the Stern model (Fig. 4.3), each particle is surrounded by a compact Stern layer and a diffuse layer of counterions. When two such diffuse layers overlap, an osmotic and electrostatic repulsion develops between the particles, which resists approach and aggregation. This repulsion is described in (4.27) and (4.28).

(b) The van der Waals attraction between spherical particles or between a sphere and a wall typically decays algebraically with separation distance. For example, (4.25) shows a dependence of the form $\propto 1/h^2$ (or similar power laws), where $h$ is the surface-to-surface separation.

The electrostatic repulsion decays approximately exponentially with distance, controlled by the Debye length $\kappa^{-1}$. In (4.27) and (4.28), the repulsive force contains terms like $\exp(-\kappa h)$, indicating that the range of the repulsion is set by the thickness of the diffuse layer.

Thus, attraction is long-ranged on an algebraic scale, while repulsion is short-to medium-range and decays exponentially.

(c) The Debye length $\kappa^{-1}$ characterizes the thickness of the diffuse layer and thus the effective range of electrostatic repulsion:

- A larger Debye length (smaller $\kappa$) means that electrostatic repulsion extends over a longer distance, which increases the energy barrier preventing particles from approaching each other. This enhances colloidal stability and suppresses aggregation.
- A smaller Debye length (larger $\kappa$) implies that the double layer is more strongly screened; electrostatic repulsion decays rapidly, allowing van der Waals attraction to dominate at shorter separations.

In Fig. 4.6, as $\kappa R_{hyb}$ increases (effectively decreasing the relative Debye length for a given particle radius), electrostatic repulsion becomes weaker, and the aggregates become more compact and less ordered in the bulk region. At higher volume concentrations, reduced mobility and layered structures emerge due to the modified balance of attraction and repulsion.

### *4.4.2 Brownian Dynamics*

Brownian motion refers to the random movement of nanoparticles within a fluid. It occurs as fluid molecules, driven by thermal energy, move rapidly and collide with the particles, causing them to undergo random motion. Since these collisions are random, the particles do not move continuously in one direction but follow an irregular path.

Brownian motion explains the random movement of particles on very short time scales. On such scales, the particle motion exhibits probabilistic rather than deterministic characteristics. When simulating systems on microscopic time scales, considering Brownian motion is essential to ensure the accuracy of the model. The major time scales considered in Brownian dynamics include the viscous diffusion time scale

$$\tau_\nu = \frac{R_P^2}{\nu}, \tag{4.30}$$

which accounts for the time for the hydrodynamic momentum to diffuse over a distance of the particle radius, the particle velocity relaxation time scale

$$\tau_r = \frac{M_P}{\gamma}, \tag{4.31}$$

over which the particle velocity decays to the algebraic long-time tail regime, and the Brownian diffusion time scale

$$\tau_B = \frac{R_P^2 \gamma}{k_B T}, \tag{4.32}$$

which measures the time the particle has diffused its own radius [24].

The Langevin equation (LE) is a stochastic differential equation used to describe probabilistic phenomena, such as Brownian motion, by combining deterministic forces and random fluctuations acting on particles. Liu et al. [24] proposed the LB-LE model, which integrates LBM with LE to simulate the Brownian motion of nanoparticles. In the LB-LE model, the hydrodynamic force acting on the nanoparticles is replaced solely by the Stokes drag force, and the particles are assumed to be points. The equation of motion for particles in the LB-LE model is as follows.

$$M_P \frac{d\mathbf{u}_P}{dt} = \mathbf{F}_I + \mathbf{F}_D + \mathbf{F}_B, \tag{4.33}$$

where $\mathbf{F}_I$ is the interparticle force, $\mathbf{F}_D$ is the viscous drag force, and $\mathbf{F}_B$ is the Brown force. $\mathbf{F}_D$ is obtained by

$$\mathbf{F}_D = -\gamma[\mathbf{u}_P(t) - \mathbf{u}(\mathbf{X}_i, t)], \tag{4.34}$$

where $\gamma = 6\pi\mu R_P$ is the Stokes' drag coefficient, $\mathbf{u}_P$ is the nanoparticle velocity, and $\mathbf{u}(\mathbf{X}_i, t)$ is interpolated fluid velocity at the position where the center of nanoparticle. $\mathbf{F}_B$ is calculated by $\sqrt{2D_P}d\mathbf{W}(t)$, where $D_P = k_B T/\gamma$ denotes the diffusion coefficient. $\mathbf{W}(t)$ is Gaussian random variable with variance $\langle|\mathbf{W}(t + \Delta t) - \mathbf{W}(t)|^2\rangle = N\Delta t$, where $N$ is two (three) in two (three) dimensions [18].

Using the above equations, all forces acting on the nanoparticles can be determined, enabling the simulation of particle behavior at the next timestep. However, depending on the magnitude of the inertial force acting on the particle, the system is

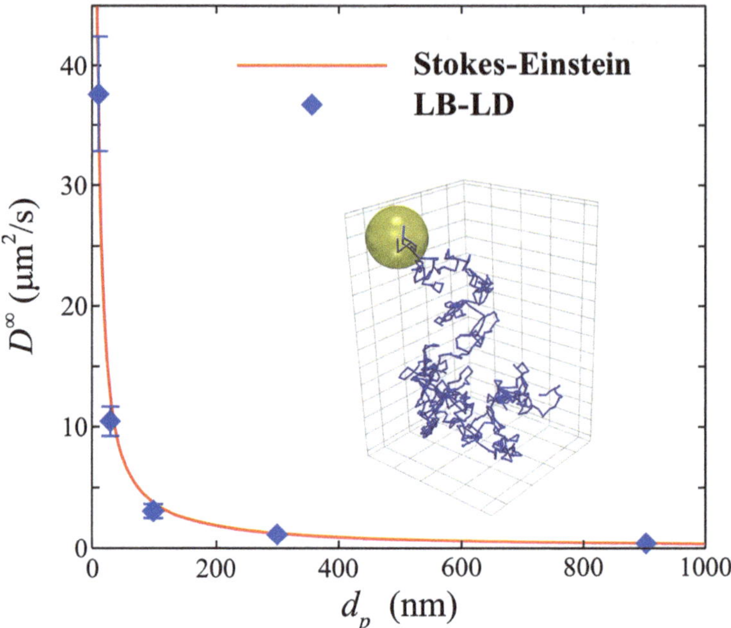


**Fig. 4.7** Normalized self-diffusivity of colloidal particles as a function of particle size at infinite dilution [24]

classified as either an overdamped or underdamped LE system. For this purpose, the Stokes number $St = \tau_r/\Delta t$ can be introduced. If $St < 1$, the inertial force has little effect on the particle, and the particle behavior can be solved using the overdamped LE. Conversely, if $St \geq 1$, the LE is updated with a time interval similar to or shorter than $\tau_r$, requiring the underdamped LE. Therefore, the equation describing nanoparticle behavior based on $St$ is as follows.

$$\mathbf{u}_P(t+\Delta t) = \mathbf{u}_P(t) + \frac{1}{\gamma}[\mathbf{F}_I + \mathbf{F}_B], \quad (St < 1) \tag{4.35}$$

$$\mathbf{u}_P(t+\Delta t) = \mathbf{u}_P(t) + \frac{\Delta t}{M_P}[\mathbf{F}_I + \mathbf{F}_B + \mathbf{F}_D], \quad (St \geq 1) \tag{4.36}$$

Figure 4.7 shows the comparison of particle long-time diffusivity $D^{\infty}$ as a function of particle size simulated using the LB-LE model with the Stokes-Einstein relation. $D^{\infty}$ can be calculated as

$$D^{\infty} = \frac{1}{6}\frac{d}{dt}\left\langle [\mathbf{X}_i(t) - \mathbf{X}_i(0)]^2 \right\rangle\Big|_{t\to\infty}. \tag{4.37}$$

The theoretical Brownian diffusion obtained through the Stokes-Einstein relation is

$$D^B = \frac{k_B T}{\gamma}. \tag{4.38}$$

Figure 4.7 also illustrates the random trajectories of particles due to Brownian motion. The diffusivity $D^{\infty}$ obtained from the simulations showed a high level of agreement with the theoretical values. These results demonstrate that the LB-LE model, which incorporates thermal fluctuations, can effectively simulate the Brownian motion of nanoparticles.

**Exercise 4.3 LB–LE Coupling**
Section 4.4.2 introduces the LB–LE model.

(a) Explain why the LB–LE model replaces the full hydrodynamic force with only the Stokes drag in (4.34).
(b) Discuss the advantages and limitations of assuming the particle is a point particle.
(c) Explain how thermal fluctuations are introduced and why a Gaussian random force is required.

**Solution**
(a) In the LB–LE model, nanoparticles are assumed to be very small and suspended in a low-Reynolds-number flow, where inertial and complex nonlinear hydrodynamic effects are negligible compared with viscous drag. Under these conditions, the dominant hydrodynamic contribution on each particle can be well approximated by the Stokes drag, which is linear in the slip velocity between the particle and the surrounding fluid. Replacing the full hydrodynamic force with the Stokes drag greatly simplifies the coupling between LBM and particle dynamics: it avoids resolving detailed flow disturbances around each particle and allows the use of a simple closed-form expression for the hydrodynamic force. This simplification reduces computational cost while retaining the essential physics of nanoparticle–fluid momentum exchange in the Brownian regime.
(b) Treating nanoparticles as point particles offers several advantages:

- The particle size is not explicitly resolved on the lattice, so there is no need for a resolved particle boundary or surface mesh, which significantly reduces memory and computational cost.
- Hydrodynamic coupling is performed only through interpolated local fluid velocity and a drag law, making the algorithm simple and efficient.
- It becomes feasible to simulate large numbers of nanoparticles in complex flows.

However, this assumption has important limitations:

- Near-field hydrodynamics (lubrication forces, flow disturbance around finite-size particles) are not resolved accurately.
- Finite-size effects such as steric exclusion, geometric confinement, or detailed particle–wall interactions cannot be captured.

- The model is not suitable for larger particles or for dense suspensions where particle–particle hydrodynamic interactions and excluded volume become significant.

  Thus, the point-particle assumption is appropriate mainly for dilute suspensions of nanoscale particles where Brownian and drag forces dominate.

(c) In the LB–LE framework, thermal fluctuations are introduced through a stochastic force term—often called the Brownian force—added to the deterministic forces (drag and interparticle forces) in the Langevin equation. This random force is modeled as a Gaussian white noise process with zero mean and a variance chosen to satisfy the fluctuation–dissipation theorem, ensuring consistency with the target temperature and diffusion coefficient. A Gaussian random force is required because:

- The underlying molecular collisions are the result of many independent microscopic interactions; by the central limit theorem, their cumulative effect on the particle is well approximated by a Gaussian distribution.
- The Langevin equation with Gaussian noise yields Brownian motion whose statistical properties (mean-squared displacement, long-time diffusivity) match the theoretical Stokes–Einstein relation, as confirmed in Fig. 4.7.

Therefore, the Gaussian random force provides a thermodynamically consistent and mathematically tractable way to model thermal fluctuations in nanoparticle dynamics.

## 4.5 Project: Particle Sedimentation

Sedimentation, the process by which particles settle under the influence of gravity in a fluid, is a fundamental phenomenon in various natural and industrial systems. Understanding sedimentation dynamics is crucial for optimizing processes such as wastewater treatment, chemical separation, pharmaceutical manufacturing, and sediment transport in environmental engineering. Simulating particle sedimentation provides valuable insights into the underlying physics, including drag forces, particle–particle interactions, and the influence of fluid flow, enabling the design and optimization of efficient systems. Particle sedimentation due to gravity can also be simulated using LBM. Figure 4.8 shows the velocity contour for 30 particles during sedimentation [25]. During the sedimentation process of multiple particles, various physical phenomena, such as drafting, kissing, and tumbling, occur.

In this project, the SPM model is used to simulate the sedimentation of a single spherical particle under gravity, and the results for particle position and sedimentation velocity will be compared with experimental data. The simulation domain is shown in Fig. 4.9. The project conditions and objectives are as follows.

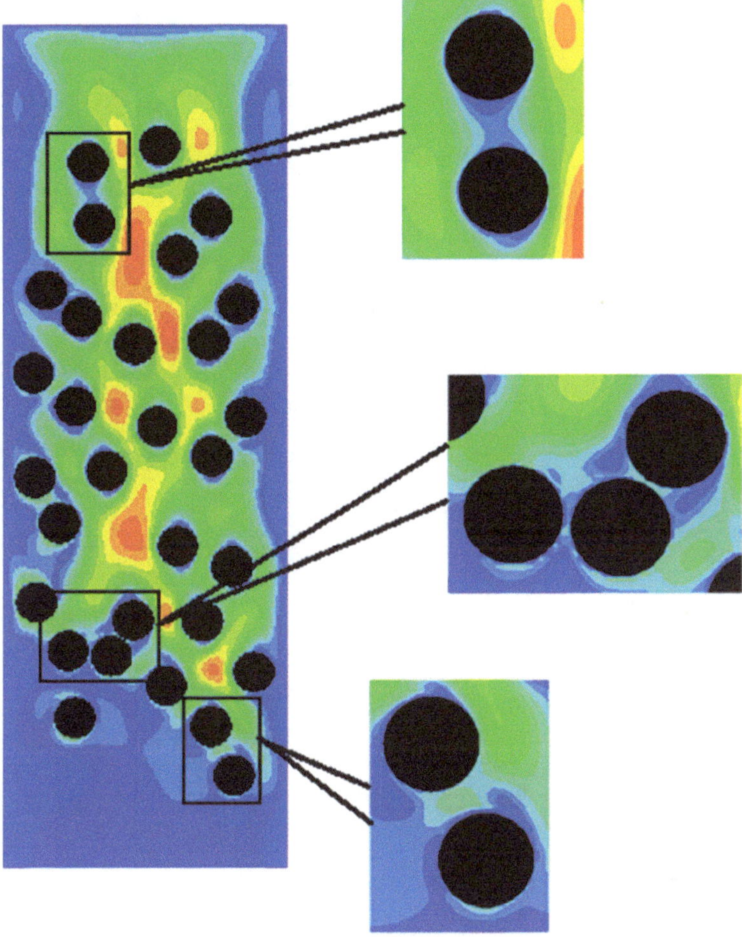

**Fig. 4.8** Velocity contour of 30 settling spherical particles at the middle plane [25]

Project Conditions:

- The particle model uses SPM.
- The size of the simulation domain is 10 × 10 × 16 cm.
- The particle has a diameter of 15 mm, a density of 1120 kg/m$^3$, and is positioned 12 cm above the bottom surface.
  The gravitational acceleration is 9.81 m/s.
- The fluid properties for each case are shown in Table 4.1.

Project Objectives:

- Measure the particle's position and sedimentation velocity over time and compare the results with the experimental data [27] shown in Fig. 4.10.

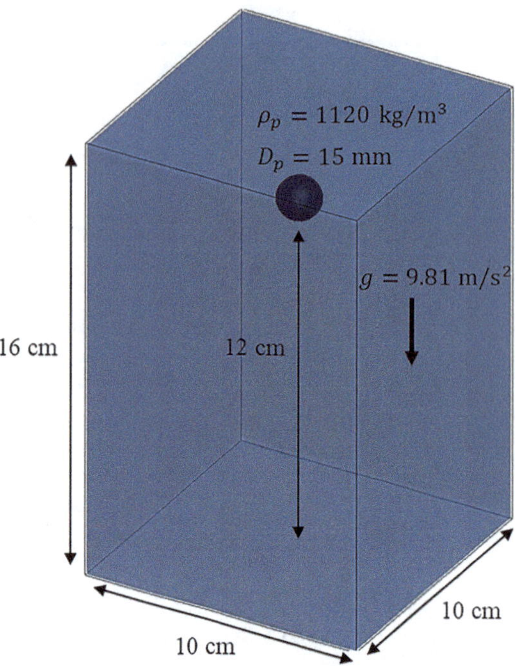


**Fig. 4.9** Simulation domain of a sphere settling in a container [26]

**Table 4.1** Fluid properties of sedimentation experiments

| | Fluid density, $\rho$ (kg/m$^3$) | Viscosity, $\mu$ (kg/m · s) | Reynolds number, $Re$ $(=\rho V_{fall} D_p/\mu)$ |
|---|---|---|---|
| Case 1 | 970 | 0.373 | 1.5 |
| Case 2 | 965 | 0.212 | 4.1 |
| Case 3 | 962 | 0.113 | 11.6 |
| Case 4 | 960 | 0.058 | 31.9 |

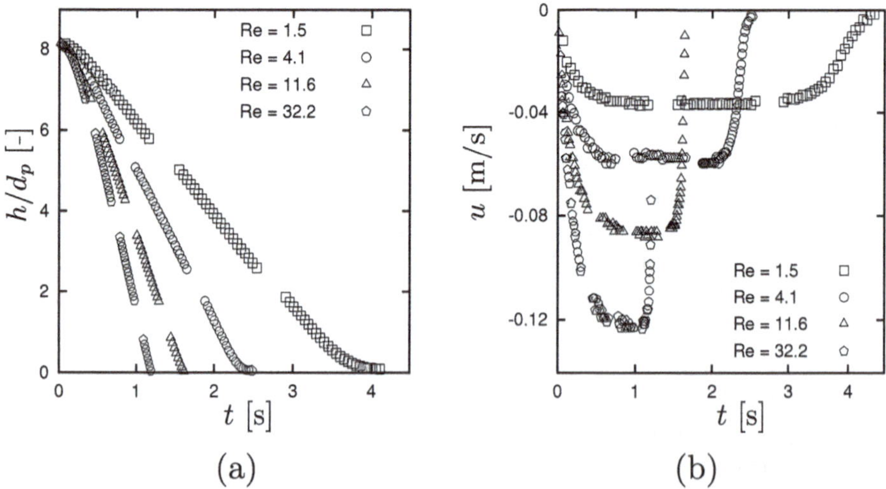


**Fig. 4.10** Experimental data on (**a**) sphere trajectory and (**b**) sedimentation velocity of four cases [27]

- Perform simulations by varying the interface thickness $\xi$ that determines the particle's boundary and compare the results.
- Measure the error rate between the simulation and experimental results, and discuss the possible reasons for these discrepancies.

**Summary**

1. **Smoothed Profile Method (SPM)**
   - Replaces sharp particle boundary with a continuous smooth profile $\phi_P(\mathbf{x})$.
   - Advantages:
     - Eliminates sharp interface discontinuities
     - Avoids staircase artifacts
     - Enables efficient and stable simulation of many particles
   - LBM computes fluid velocity; particle velocity is reconstructed from rigid-body motion.
2. **Hydrodynamic coupling**
   - SPM computes hydrodynamic force and torque via integrated momentum exchange:

$$\mathbf{F}_{P,i} = \sum_{\mathbf{x} \in V_P} \rho(\mathbf{x}) \mathbf{f}_P(\mathbf{x}, t), \mathbf{T}_{P,i} = \sum_{\mathbf{x} \in V_P} (\mathbf{x} - \mathbf{X}_i) \times \rho(\mathbf{x}) \mathbf{f}_P(\mathbf{x}, t)$$

   - Ensures Newton's third law consistency.
3. **Immersed Boundary Method (IBM)**
   - Represents boundary using Lagrangian markers embedded in Eulerian fluid grid.
   - Interaction mediated by delta kernel smearing:
     - Spreads boundary forces to fluid
     - Interpolates fluid velocity to boundary points
   - Suitable for flexible/elastic membranes or moving solid structures.
4. **DLVO interactions**
   - van der Waals attraction:

$$F_{vdW} \sim -\frac{A_H}{6\pi d^3}$$

   - Electrostatic double-layer repulsion:

$$F_{el} \sim e^{-\kappa d}$$

   - Overall DLVO potential determines:

   - Agglomeration vs. stable dispersion
   - Particle adhesion to the substrate
   - Distance at which particles stick irreversibly

5. **Numerical implementation considerations**

   - Small nanoparticle length scale → strong stiffness in Langevin equation → need for small $\Delta t$.
   - DLVO force singularity near surfaces requires regularization or minimum cutoff distance.

6. **Applications**

   - Inkjet printing inks with functional nanoparticles.
   - Colloidal deposition and assembly on substrates.
   - Control of coffee-ring pattern through DLVO engineering or surface modification.
   - General microfluidic and nanofluidic transport of colloids.

## References

1. R. Karnik, F. Gu, P. Basto, C. Cannizzaro, L. Dean, W. Kyei-Manu, R. Langer, and O. C. Farokhzad, "Microfluidic platform for controlled synthesis of polymeric nanoparticles," Nano letters **8**, 2906-2912 (2008).
2. L. Bläubaum, F. Röder, C. Nowak, H. S. Chan, A. Kwade, and U. Krewer, "Impact of particle size distribution on performance of lithium-ion batteries," ChemElectroChem **7**, 4755-4766 (2020).
3. G. Xu, Y. Zhang, and J. Gregory, "Different pollutants removal efficiencies and pollutants distribution with particle size of wastewater treated by cept process," Water Practice and Technology **1**, wpt2006047 (2006).
4. P. Calvert, "Inkjet printing for materials and devices," Chemistry of materials **13**, 3299-3305 (2001).
5. S. Adhikari, S. Selvaraj, and D. H. Kim, "Progress in powder coating technology using atomic layer deposition," Adv. Mater. Interfaces **5**, 1800581 (2018).
6. A. J. Ladd, "Numerical simulations of particulate suspensions via a discretized boltzmann equation. Part 1. Theoretical foundation," Journal of fluid mechanics **271**, 285-309 (1994).
7. A. J. Ladd, "Numerical simulations of particulate suspensions via a discretized boltzmann equation. Part 2. Numerical results," Journal of fluid mechanics **271**, 311-339 (1994).
8. Y. Nakayama, and R. Yamamoto, "Simulation method to resolve hydrodynamic interactions in colloidal dispersions," Phys. Rev. E **71**, 036707 (2005).
9. S. Jafari, R. Yamamoto, and M. Rahnama, "Lattice-boltzmann method combined with smoothed-profile method for particulate suspensions," Physical Review E—Statistical, Nonlinear, and Soft Matter Physics **83**, 026702 (2011).
10. Y. K. Lee, and K. H. Ahn, "Particle dynamics at fluid interfaces studied by the color gradient lattice boltzmann method coupled with the smoothed profile method," Physical Review E **101**, 053302 (2020).
11. W. Hu, T. Lin, C. Yang, C. Tu, X. Li, F. Xu, F. Bao, X. Gao, and Y. Zhang, "Particle dynamics at fluid interface by pseudo-potential lattice boltzmann method coupled with smoothed profile method," Phys. Fluids **36**, (2024).

12. J. Y. Moon, R. I. Tanner, and J. S. Lee, "A numerical study on the elastic modulus of volume and area dilation for a deformable cell in a microchannel," Biomicrofluidics **10**, (2016).
13. L. Jiang, M. Rahnama, B. Zhang, X. Zhu, P.-C. Sui, D.-D. Ye, and N. Djilali, "Predicting the interaction between nanoparticles in shear flow using lattice boltzmann method and derjaguin–landau–verwey–overbeek (dlvo) theory," Phys. Fluids **32**, 043302 (2020).
14. L. Zhang, and X. Wang, "Numerical coffee-ring patterns with new interfacial schemes in 3d hybrid lb-le model," Powder Technol. **392**, 130-140 (2021).
15. Z.-G. Feng, and E. E. Michaelides, "The immersed boundary-lattice boltzmann method for solving fluid–particles interaction problems," J. Comput. Phys. **195**, 602-628 (2004).
16. C. S. Peskin, "The immersed boundary method," Acta numerica **11**, 479-517 (2002).
17. Z. Guo, C. Zheng, and B. Shi, "Discrete lattice effects on the forcing term in the lattice boltzmann method," Physical review E **65**, 046308 (2002).
18. R. Verberg, J. M. Yeomans, and A. C. Balazs, "Modeling the flow of fluid/particle mixtures in microchannels: Encapsulating nanoparticles within monodisperse droplets," The Journal of chemical physics **123**, (2005).
19. S. Srinivasan, H. E. Van den Akker, and O. Shardt, "Inclusion of dlvo forces in simulations of non-brownian solid suspensions: Rheology and structure," International Journal of Multiphase Flow **149**, 103929 (2022).
20. H. M. Lee, and J. S. Lee, "Particle deposition dynamics in evaporating droplets using lattice boltzmann and magnetic particle simulation," Phys. Fluids **35**, (2023).
21. H. M. Lee, and J. S. Lee, "Suppression of the coffee-ring effect using a contact-angle radial-band-patterned substrate: A lattice boltzmann study," Phys. Fluids **36**, (2024).
22. H. C. Hamaker, "The london—van der waals attraction between spherical particles," Physica **4**, 1058-1072 (1937).
23. J. H. Masliyah, and S. Bhattacharjee, *Electrokinetic and colloid transport phenomena* (John Wiley & Sons, 2006).
24. Z. Liu, Y. Zhu, J. R. Clausen, J. B. Lechman, R. R. Rao, and C. K. Aidun, "Multiscale method based on coupled lattice-boltzmann and langevin-dynamics for direct simulation of nanoscale particle/polymer suspensions in complex flows," International Journal for Numerical Methods in Fluids **91**, 228-246 (2019).
25. Z. Hashemi, O. Abouali, and R. Kamali, "Three dimensional thermal lattice boltzmann simulation of heating/cooling spheres falling in a newtonian liquid," International Journal of Thermal Sciences **82**, 23-33 (2014).
26. H. M. Lee, and J. S. Lee, "Effects of heat transfer on particle suspended drop-on-demand inkjet printing using lattice boltzmann method," Appl. Therm. Eng. **213**, (2022).
27. A. ten Cate, C. H. Nieuwstad, J. J. Derksen, and H. E. A. Van den Akker, "Particle imaging velocimetry experiments and lattice-boltzmann simulations on a single sphere settling under gravity," Phys. Fluids **14**, 4012-4025 (2002).

# Chapter 5
# Introduction to Inkjet Printing

## 5.1 Introduction

Inkjet printing is a precision deposition technology that digitally ejects liquid ink onto a substrate in a non-contact manner to form desired patterns. Unlike conventional manufacturing processes, inkjet printing offers enhanced flexibility and efficiency. Since ink can be directly deposited at specific locations in response to digital signals without requiring masks or molds, it is considered a highly suitable method for low-cost, high-resolution, and small-batch production. These advantages have opened up a wide range of applications in areas such as electronic devices, displays, biosensors, tissue engineering scaffolds, energy materials, and polymer-based devices, and both academic and industrial research on this technology is actively underway.

Fundamentally, inkjet printing relies on a series of fluid dynamic phenomena involving the formation and flight of droplets and their deposition onto substrates. In particular, to ensure that fine droplets are formed stably and reach the target location with precision, various physical parameters such as ink viscosity, surface tension, and density must be carefully controlled in conjunction with geometric features of the nozzle, waveform characteristics of the actuation signal, actuation frequency, and environmental conditions. The interaction among these complex factors results in highly nonlinear behavior, making it difficult to predict printing success through simple static analysis.

Inkjet printing technologies are broadly classified into continuous inkjet (CIJ) and drop-on-demand (DOD) systems depending on the ejection mechanism. Among these, DOD systems are widely used due to their ability to achieve high resolution and printing precision. DOD inkjet systems can be further categorized into piezoelectric, thermal, and electrohydrodynamic actuation types, each of which is optimized for specific fluid properties and offers distinct advantages and limitations.

 
H. M. Lee, J. S. Lee, *Lattice Boltzmann Methodology for Single-Phase and Multiphase Nanoparticle Modeling*,
https://doi.org/10.1007/978-981-95-9117-6_5

Therefore, understanding the physical principles and constraints of each mechanism is essential for designing application-specific printing conditions.

In practical applications, inks often consist of complex mixtures containing nanoparticles, polymers, and biomolecules. Compared to simple pure-liquid inks, such particle-laden inks exhibit more complex fluid behaviors. Increases in viscosity, emergence of non-Newtonian characteristics, inhomogeneous particle distributions, and the potential formation of satellite droplets can all hinder high-resolution printing. To effectively handle such complex inks, conventional theoretical frameworks must often be adapted, or new numerical analysis approaches must be introduced.

Further complexities arise in the latter stages of inkjet processes, where droplet evaporation and particle deposition occur. One of the most notable issues is the coffee-ring effect, in which excessive particle accumulation near the droplet's contact line hinders uniform pattern formation. Various physical and chemical strategies have been explored to suppress this phenomenon, including electrowetting, Marangoni flow control, modulation of interparticle interactions, and application of external magnetic fields.

In recent years, the integration of artificial intelligence (AI) technologies with inkjet printing has become increasingly prominent. Approaches using machine learning to predict droplet morphology, automatically optimize waveform design, and monitor droplet behavior in real time for multi-nozzle synchronization suggest a promising future for enhancing both the precision and throughput of inkjet systems. In particular, AI models offer powerful alternatives to physics-based approaches when dealing with complex systems, positioning AI-assisted printing as a key enabler for next-generation manufacturing.

This chapter provides a comprehensive overview of inkjet printing technologies, covering both fundamental principles and recent research trends. Special emphasis is placed on the physical mechanisms of DOD systems, fluid dynamic behavior during jetting, the influence of complex inks, droplet evaporation and particle deposition mechanisms, and AI-based process control strategies. Numerical simulations and experimental validations are introduced throughout the chapter to support a well-rounded understanding of both theoretical and practical aspects of inkjet printing.

## 5.2 Classification of Inkjet Printing

As shown in Fig. 5.1, inkjet printing technology is classified into continuous inkjet (CIJ) and drop-on-demand (DOD) inkjet. CIJ continuously ejects ink from the nozzle, forming a steady stream of ink droplets, while DOD precisely deposits ink droplets only when needed using either piezoelectric actuators or heated ink bubbles.

Figure 5.2 illustrates the jetting mechanisms of CIJ and DOD inkjet systems. In CIJ, a high-pressure pump forces ink through the nozzle, generating a continuous ink jet. This jet is broken into uniform droplets due to Rayleigh-Plateau instability. The formed ink droplets pass through charging electrodes, gaining an electric charge.

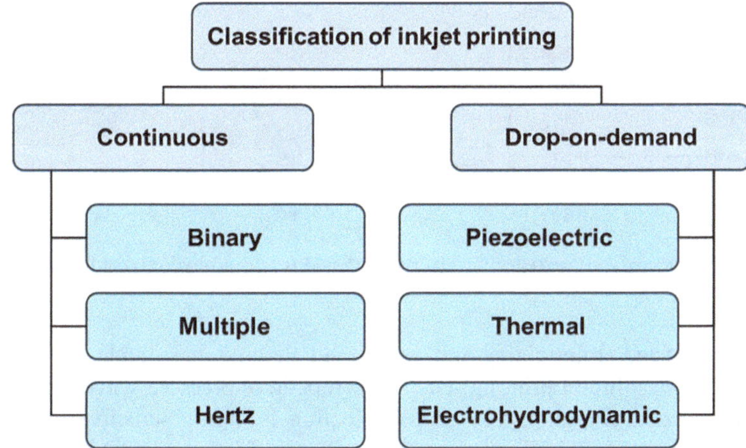


**Fig. 5.1** Classification of inkjet printing technologies [1]

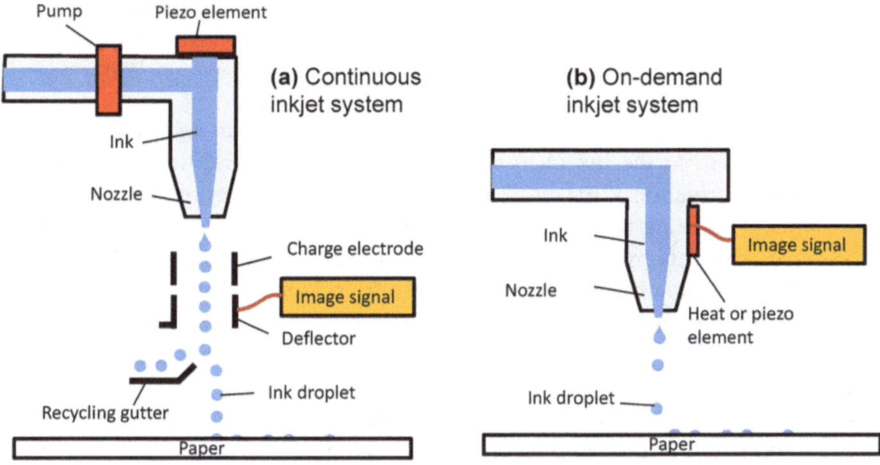


**Fig. 5.2** Schematic of (**a**) continuous inkjet and (**b**) drop-on-demand inkjet [2]

The charged droplets are deflected within an electric field and deposited at the desired locations on the substrate, while uncharged droplets are collected by the recycling gutter. In DOD inkjet, an actuator is activated based on specific signals, generating pressure within the ink chamber. As the pressure increases, a small ink droplet is ejected from the nozzle and deposited at a designated location on the substrate.

CIJ can eject thousands of droplets per second, offering high printing speeds and the ability to use inks with various viscosity characteristics, making it widely applicable across industries. However, it requires precise control of droplet deflection and has lower deposition accuracy compared to DOD. DOD generates droplets

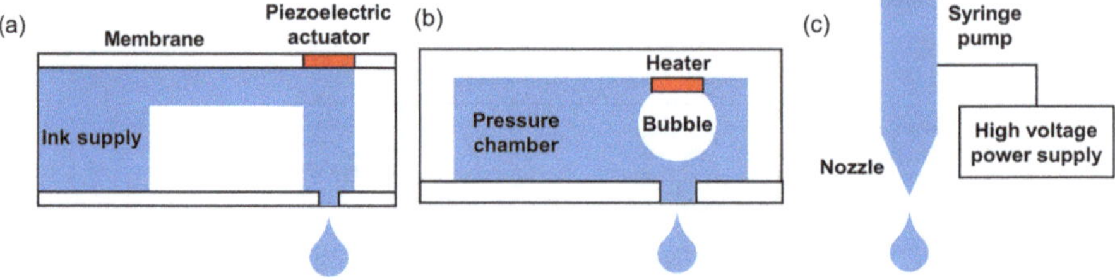


**Fig. 5.3** Schematic of (**a**) piezoelectric, (**b**) thermal, and (**c**) electrohydrodynamic inkjet printing [3]

only when needed, minimizing ink waste and making it suitable for industries requiring high-resolution printing. However, repetitive pressure waveform changes may cause nozzle clogging, and droplet ejection is highly sensitive to the ink's physical properties. In this chapter, we will primarily focus on DOD inkjet printing, which is utilized in applications demanding high precision, such as cell printing and electronic printing.

DOD inkjet printing can be further classified based on the droplet jetting mechanism into piezoelectric inkjet printhead (PIP), thermal inkjet printing (TIJ), and electrohydrodynamic (EHD) printing. Figure 5.3 illustrates these three types of DOD printing schematically.

PIP uses a piezoelectric element that deforms in response to an electrical signal, applying pressure to the ink chamber and ejecting droplets. Since the degree of mechanical deformation of the piezoelectric element varies depending on the voltage waveforms, optimizing these waveforms is essential for precise droplet ejection.

TIJ ejects droplets by heating the ink to form vapor bubbles. A resistor located inside the nozzle rapidly heats up when an electrical signal is applied, vaporizing a portion of the ink. The resulting vapor bubble creates pressure, pushing the ink droplet out of the nozzle. TIJ primarily uses low-viscosity inks, such as water-based inks.

EHD utilizes a strong electric field to generate ink droplets. By applying the electric field to the nozzle or an open surface, extremely small droplets are ejected from the ink interface. This technique can produce droplets on the nanometer scale, making it suitable for high-resolution and nanoscale applications.

Inkjet printing technology requires the selection of an appropriate methodology based on specific requirements, such as the desired resolution and the physical properties of the ink. Properties like viscosity and surface tension of the ink significantly influence droplet formation, jet stability, and printing quality on the substrate. In general, inks often contain various particles intended for deposition on the substrate. When particle-laden inks are used, complex interactions between the particles and the fluid may arise, necessitating more precise control. For this reason, the design and optimization of inkjet printing systems must be accompanied by fluid dynamic analysis, ensuring stable and precise printing quality.

**Exercise 5.1 Classification and Operating Principles of Inkjet Printing**

1. Summarize the fundamental differences between CIJ and DOD inkjet printing in terms of ejection mechanism, resolution, and operation.
2. Explain why DOD is preferred for high-resolution applications such as cell printing and electronic printing.
3. Describe the physical principles behind Rayleigh–Plateau instability in CIJ.

**Solution**

(a) CIJ:

- Ejection mechanism: A continuous liquid jet is ejected from a nozzle under constant pressure and then periodically broken into droplets by an external perturbation (e.g., piezoelectric vibration). Only some droplets are used for printing; the others are deflected and recycled.
- Resolution: Limited by continuous jet diameter, perturbation frequency, and deflection accuracy. Typically lower spatial resolution than DOD.
- Operation: Requires high-speed droplet deflection and ink recirculation system (charging electrodes, deflection plates, gutter), with relatively high ink consumption.

DOD:

- Ejection mechanism: A droplet is generated only when required, by a transient pressure pulse (piezoelectric, thermal, or EHD actuation) inside the chamber.
- Resolution: Higher, because droplet volume and timing can be more precisely controlled and there is no continuous jet.
- Operation: No recirculation; each actuation corresponds to one (or a small number of) droplets. System is structurally simpler near the nozzle exit but more demanding in waveform and actuation control.

(b) DOD is preferred for high-resolution cell printing and electronic printing because:

- Each droplet is individually addressed in time and space, enabling precise patterning.
- Droplet volume and velocity can be tuned by waveform design, allowing control over feature size and film thickness.
- No need to recycle unused droplets, which is important for expensive or sensitive materials such as biomaterials, functional polymers, or nanoparticle inks.
- Reduced risk of contamination or mixing in recirculation lines, improving pattern fidelity and device reliability.

(c) In CIJ, a cylindrical liquid jet is inherently unstable to axisymmetric perturbations due to Rayleigh–Plateau instability:

- Surface tension tends to reduce surface area; small sinusoidal perturbations with wavelength larger than the circumference grow over time.

- As these perturbations grow, the jet necks down at some locations and bulges at others, eventually pinching off into droplets.
- External forcing (e.g., piezo-driven pressure oscillations) sets the frequency and amplitude of the initial perturbation, so that the jet breaks into droplets of controlled size and spacing.

## 5.3 Fluid Dynamics in DOD Inkjet Printing

Figure 5.4 illustrates the DOD inkjet printing process schematically over time. The physical phenomena that can occur during each stage of the printing process are summarized below.

① **Acoustics in printhead:** To eject the ink, the piezoelectric element receives an electrical signal and undergoes mechanical deformation. This deformation generates a pressure wave within the chamber, and when sufficient pressure change occurs, a droplet is ejected from the nozzle. In this process, a complex energy conversion sequence takes place, involving electrical energy, mechanical deformation, acoustic energy, and fluid dynamics. This complexity highlights the importance of optimizing the voltage waveform for precise droplet ejection.

② **Nozzle clogging:** To ensure high droplet reproducibility, repetitive pressure waves are generated within the chamber at a high frequency. However, these pressure waves can lead to nozzle clogging. Nozzle clogging disrupts the flow around the printhead, reducing ejection accuracy and, in severe cases, preventing ejection altogether. Clogging can occur due to the accumulation of particles adhering to the printhead walls and cavitation-induced bubbles trapped within the chamber by the particles.

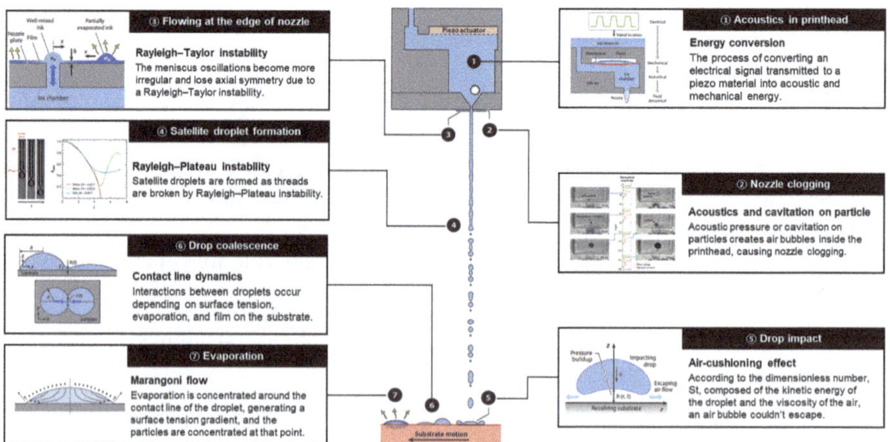


**Fig. 5.4** Physical phenomena occurring in each printing stage of DOD inkjet [4–6]

③ **Flowing at the edge of nozzle:** A thin ink film can form on the nozzle plate due to the oscillating meniscus. Particles within this thin film may be transported to the meniscus, potentially causing bubble formation. Additionally, particle concentration differences within the ink film can trigger Rayleigh–Taylor instability, further intensifying the irregularity of meniscus oscillation.

④ **Satellite droplet formation:** The ejected ink separates from the nozzle and travels toward the substrate. The droplet size is determined by the timing of the separation. During flight, Rayleigh–Plateau instability can cause the breakup of the liquid thread, leading to the formation of small satellite droplets. Due to their small size, satellite droplets are highly susceptible to ambient air, making them difficult to control and potentially reducing printing precision. The droplet separation is influenced by factors such as applied pressure, pressure wave frequency, ink viscosity and surface tension, and nozzle geometry, necessitating a thorough optimization process.

⑤ **Drop impact:** The flying droplet eventually collides with the substrate. Before impact, an air layer forms between the droplet and the substrate, causing a dimple formation on the droplet's impacting surface. This deformation can trap the air layer between the droplet and the substrate. This trapped air negatively affects printing quality and should be minimized. The amount of trapped air is influenced by the droplet's impact velocity, and this relationship can be described using the Stokes number, $St$.

⑥ **Drop coalescence:** Through the coalescence of adhered droplets, conductive lines or films can be formed on the substrate. Controlling drop coalescence requires precise management of the distance between droplets and the patterned wettability of the substrate. Patterned wettability is particularly important as it enables easy control of the droplet's contact line, making it a critical factor to consider when designing droplet interactions.

⑦ **Evaporation:** Finally, particle-laden droplet evaporates and adheres the particles onto the substrate. During evaporation, the highest evaporation rate occurs near the contact line, causing particles to migrate and accumulate around it, forming a ring-like deposition. This phenomenon is undesirable in industries where uniform particle depositions are required.

As reviewed so far, various physical phenomena must be considered in the inkjet printing process. The following sections will provide a more detailed discussion of the key processes involved in inkjet printing and review related previous studies.

### 5.3.1 Jetting Process

In the jetting process, several key requirements must be met to ensure stable printing quality. First, sufficient pressure must be delivered within the chamber for ink ejection. Second, the ejected droplet should avoid splitting into small satellite droplets. Third, the droplet must achieve an appropriate flight velocity to prevent splashing upon impact with the substrate. Lastly, the ink viscosity should not be too high to allow effective transmission of pressure waves. To determine a printable environment that satisfies these requirements, the following dimensionless numbers can be utilized.

$$Re = \frac{\rho U R}{\mu}, \quad We = \frac{\rho U^2 R}{\sigma}, \quad Oh = \frac{\mu}{\sqrt{\rho \sigma R}} = \frac{We^{1/2}}{Re}, \tag{5.1}$$

where $\rho$, $\mu$, and $\sigma$ are the density, viscosity, and surface tension of ink, respectively. $U$ is the flight velocity of droplet and $R$ is the radius of droplet. The Reynolds number ($Re$) represents the ratio of inertial forces to viscous forces, the Weber number ($We$) represents the ratio of inertial forces to capillary forces, and the Ohnesorge number ($Oh$) is the ratio of the viscous timescale to the capillary timescale. The $Oh$ is composed solely of the ink's physical properties and does not include the droplet velocity term, making it useful for evaluating the printability of the ink. The inverse of $Oh$, known as the $Z$ number [8], is also commonly used, and it is generally known that stable droplet formation occurs when $1 < Z < 10$ [9].

Using these dimensionless numbers, Derby [10] proposed a printing map for stable inkjet printing, as shown in Fig. 5.5. The operating map categorizes the printing environment into five distinct regions based on $Re$ and $Oh$. If the $Re$ is too low ($Re \leq 2/Oh$), insufficient inertial force may prevent droplet ejection. Conversely, if $Re$ is too high ($Re^{5/4} \geq 50/Oh$), the droplet may splash upon impact with the substrate, causing printing defects. The $Oh$ also affects printing quality. If $Oh$ is too low ($Oh \leq 0.1$), the influence of high surface tension can lead to the formation of small satellite droplets. On the other hand, if $Oh$ is too high ($Oh \geq 1$), excessive viscosity can make droplet ejection difficult or reduce the printing speed. Therefore, stable printing can be achieved only when the droplet's kinetic energy and the ink's physical properties are appropriately balanced.

Figures 5.6 and 5.7 [11] illustrate the droplet formation process under varying $Z$ numbers and actuation amplitudes, respectively. Observing the changes in the $Z$ number, it was found that when $Z \leq 1.17$, droplet ejection did not occur due to the dominant effect of high viscosity. At $Z = 4.47$, only a single droplet was ejected, while at $Z = 8.40$, satellite droplet formation began. As $Z$ increased further, the influence of surface tension led to the separation of multiple small satellite droplets. Next, by varying the actuation amplitudes, the droplet flight velocity was increased. As the droplet's inertial energy increased, the tail lengthened, but surface tension caused it to merge into a single droplet. However, at sufficiently high voltages, the tail failed to retail and resulted in satellite droplet formation.

The operating map based on dimensionless numbers provides a solid foundation for achieving stable printing. However, as shown in the experimental results from Figs. 5.6 and 5.7, unstable droplets can still form due to variations in factors such as the voltage waveform or amplitude, even under conditions where satellite droplets should theoretically be avoided. Therefore, achieving stable and high-quality printing requires an optimization process that accounts for complex interactions, but

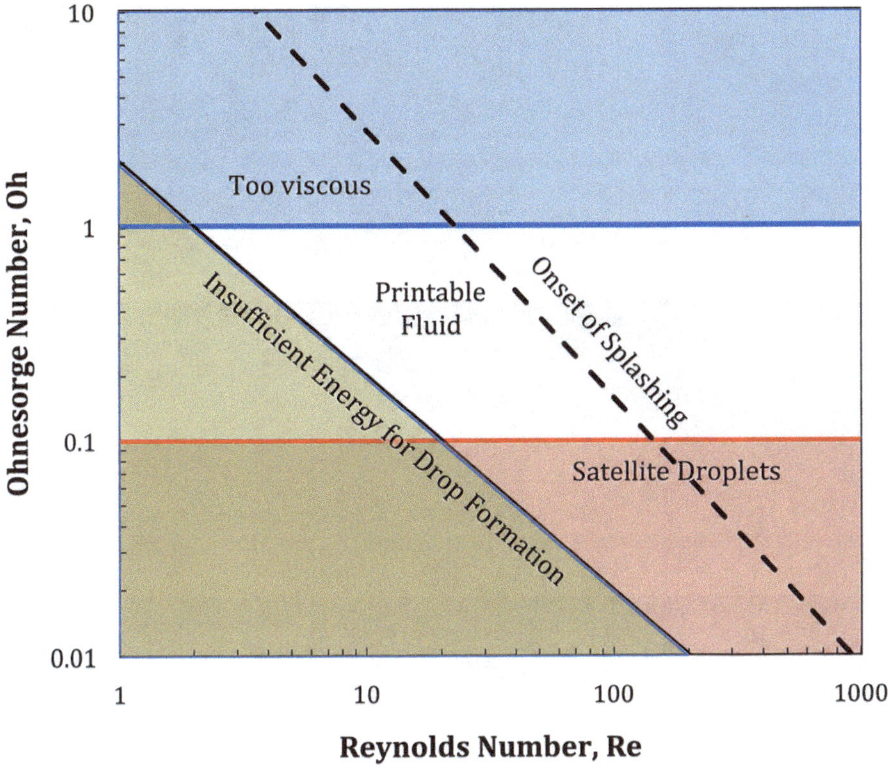


**Fig. 5.5** Operating map for stable DOD inkjet printing [7]

isolating and identifying the independent interactions of each factor through experiments is challenging. Numerical simulations offer a powerful tool to overcome these limitations.

Castrejón-Pita et al. [12] investigated droplet formation and its dynamics in DOD inkjet printing through experiments and Lagrangian finite-element simulations. Using laser Doppler anemometry (LDA), they quantitatively measured the velocity field inside the nozzle, and the obtained waveform was applied as a boundary condition in the simulation. A comparison between the simulation results and experimental data demonstrated that the model accurately predicted droplet shape and trajectory (Fig. 5.8a). Antonopoulou et al. [13] employed the finite-element method to examine droplet formation over a broader range of *Re* and *Oh* numbers. Their findings revealed that as the waveform and amplitude increased, the liquid filament lengthened, raising the likelihood of satellite droplet formation. While previous DOD inkjet studies primarily focused on the ink's physical properties, particularly the *Z* number, this study highlighted the significance of *Re* in satellite droplet control and proposed a phase diagram for stable droplet formation (Fig. 5.8b). The velocity of a flying droplet is a crucial factor in predicting satellite droplet formation. However, most previous studies have primarily measured

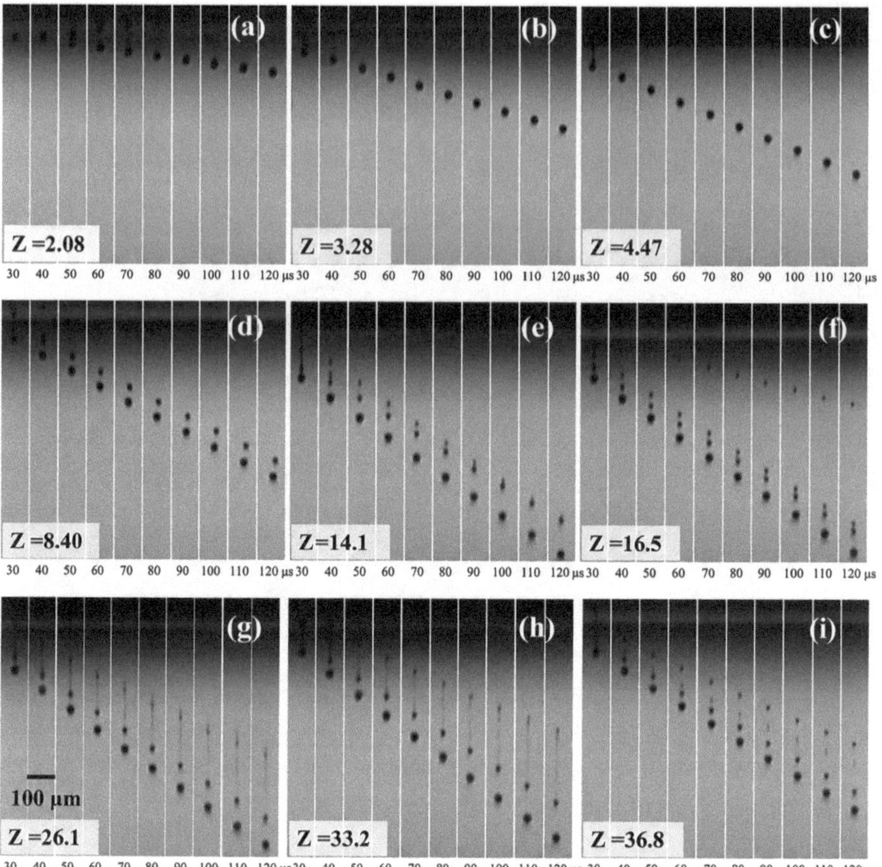


**Fig. 5.6** Images of drop formation for inks with Z ranging from 2.08 to 36.8 at constant dwell voltage of 23 V. (**a**) $Z = 2.08$, (**b**) $Z = 3.28$, (**c**) $Z = 4.47$, (**d**) $Z = 8.40$, (**e**) $Z = 14.1$, (**f**) $Z = 16.5$, (**g**) $Z = 26.1$, (**h**) $Z = 33.2$, (**i**) $Z = 36.8$ [11]

macroscopic properties such as the droplet's average velocity, volume, and the number of satellite droplets. Van der Bos et al. [14] quantitatively measured the internal velocity of droplets using high-speed imaging and short-time laser pulses, comparing the results with a slender jet approximation model (Fig. 5.8c). Several studies have also simulated droplet jetting using LBM. Zhang [15] employed MRT pseudopotential LB model to investigate how nozzle plate wettability and partial nozzle clogging (PNC) affect droplet ejection. The study found that PNC causes the droplet tail to become asymmetrically elongated, reducing jetting accuracy (Fig. 5.8d).

The aforementioned studies on droplet jetting simulations have demonstrated high accuracy when compared with experimental results, showing that not only droplet shape but also internal velocity and pressure distribution within the nozzle can be analyzed. Additionally, numerical models enable the prediction of droplet

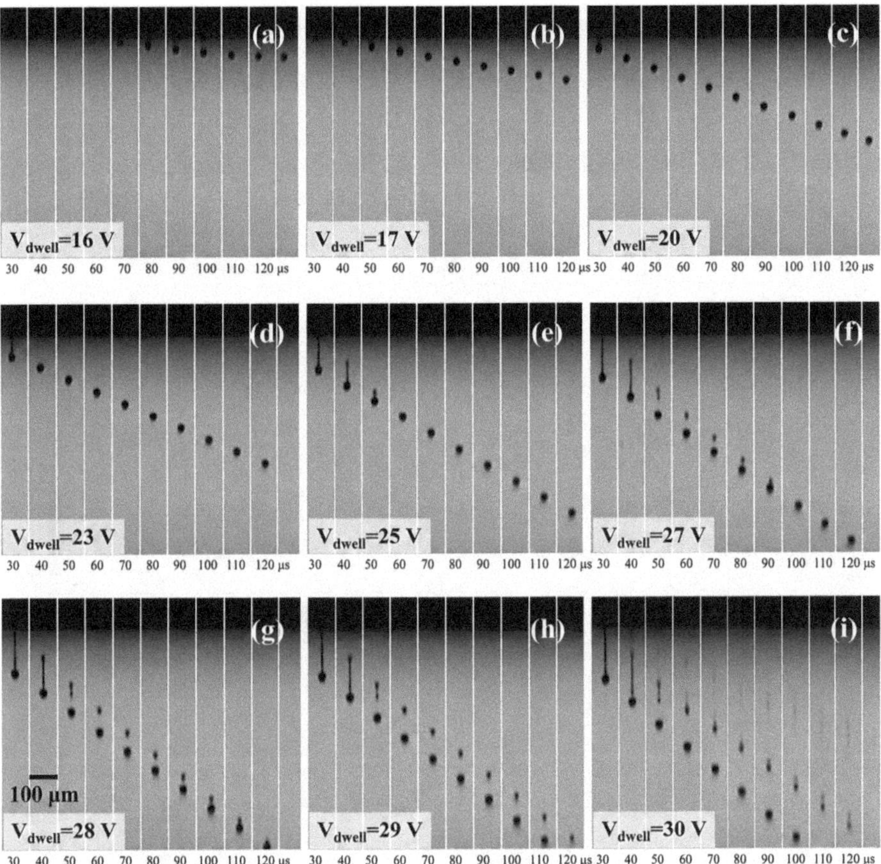


**Fig. 5.7** Images of drop formation with actuation amplitudes of (**a**) 16 V, (**b**) 17 V, (**c**) 20 V, (**d**) 23 V, (**e**) 25 V, (**f**) 27 V, (**g**) 28 V, (**h**) 29 V, (**i**) 30 V ($Z = 4.46$ for all cases) [11]

formation under varying conditions, such as ink properties, waveform characteristics, and printhead geometry, making them highly effective for optimizing the jetting process.

**Exercise 5.2 Dimensionless Numbers for Stable Jetting**

(a) Define *Re*, *We*, *Oh*, and *Z* numbers used in (5.1), and describe the physical meaning of each.

(b) Explain why stable jetting generally occurs when $4 < Z < 10$.

**Solution**

(a) Reynolds number, *Re*: Ratio of inertial to viscous forces. Large *Re*: inertia-dominated, small *Re*: viscous-dominated, strongly damped flow.

Weber number, *We*: Ratio of inertial to capillary forces. Large *We*: inertia overcomes surface tension easily, promoting jetting and breakup.

Ohnesorge number, *Oh*: Ratio of viscous forces to combined inertial–capillary scale. High *Oh*: strong viscous damping, low *Oh*: weak damping and potentially violent breakup.

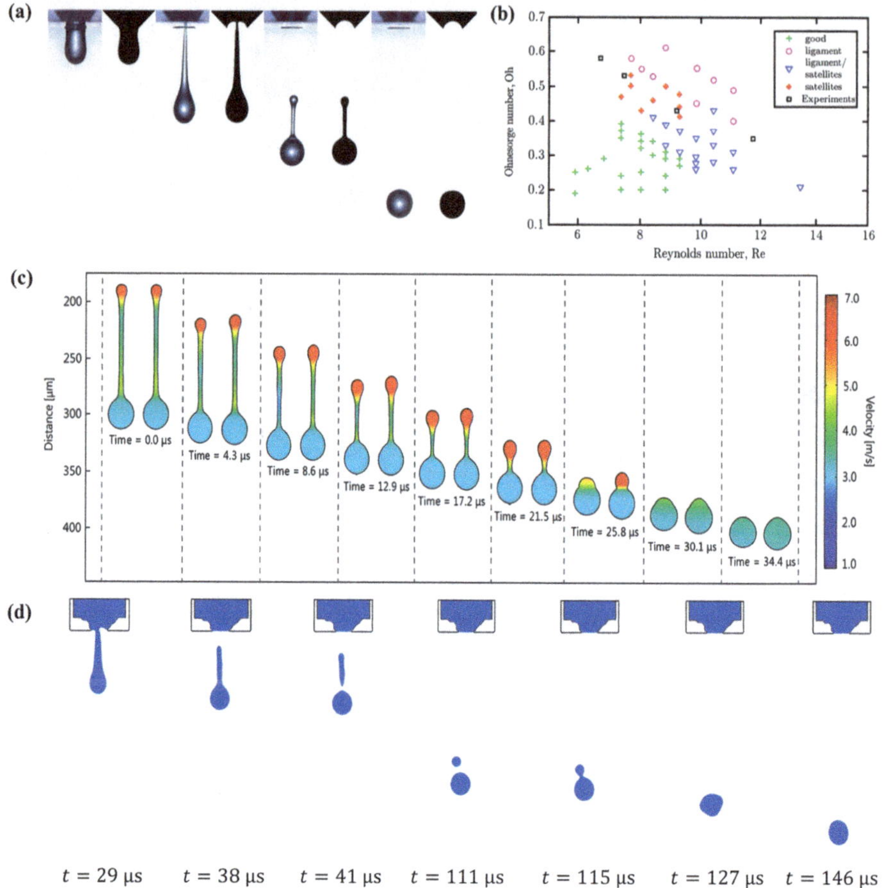


**Fig. 5.8** Simulation studies on droplet jetting. (**a**) Snapshots of experiments (left) and Lagrangian finite-element simulation results (right) [12]. (**b**) Phase diagram of droplet formation plotted using simulations [13]. (**c**) Comparison of droplet formation and velocity field between experiments (left) and simulations (right) [14]. (**d**) LBM simulation of droplet jetting in a printhead with partial nozzle clogging ($\theta_A = 30°$) [15]

*Z* number, *Z*: "Printability" number; large *Z* means low viscosity relative to inertia and surface tension, small *Z* means high viscosity.

(b) For $Z < 4$ (high *Oh*, very viscous):

- Pressure pulses are heavily damped.
- Meniscus motion is small and the liquid cannot form a sufficiently long or fast ligament.
- Result: no jetting or dripping-like behavior.

For $4 < Z < 10$:

- There is a balance between inertial, viscous, and capillary forces.
- Ligaments form with appropriate length and pinch off once, producing mainly a single main droplet with minimal satellites.
- This is the classical "stable operating window."

For $Z > 10$ (very low $Oh$, low viscosity):

- Viscous damping is too weak.
- Ligaments are long and highly unstable to Rayleigh–Plateau breakup.
- Multiple pinch-off events occur, and satellite droplets become frequent.

Hence, $4 < Z < 10$ corresponds to the regime where actuation energy is neither overdamped nor too violent, leading to stable, single-droplet ejection.

### 5.3.2 Jetting of Particle-Laden Ink

The previously introduced jetting studies analyzed the jetting and formation processes of droplets using pure ink. However, in actual inkjet printing, various particles such as conductive particles, dyes, and surfactants are dispersed within the ink. Compared to the relatively simple system of pure ink, particle-laden ink exhibits more complex physical phenomena due to the presence of particles. The inclusion of particles in ink can cause various issues, such as nozzle clogging and the coffee-ring effect, but it can also influence the jetting process itself.

Particle-laden ink exhibits increased viscosity compared to pure ink. The viscosity model [16] for a Newtonian fluid containing noncolloidal spherical particles is given as follows.

$$\mu_p(\phi) = \mu_0 \frac{\exp(-2.34\phi)}{(1 - \phi/\phi_m)^3}, \tag{5.2}$$

where $\mu_p$ and $\mu_0$ represent the viscosity modified by the particles and the viscosity of the Newtonian fluid, respectively. $\phi$ denotes the volume fraction of particle, while $\phi_m = 0.62$ corresponds to the maximum packing fraction. Figure 5.9 illustrates the effective viscosity ($\mu_p/\mu_0$) as a function of particle concentration, showing that the effective viscosity increases exponentially with increasing particle concentration. This viscosity increase significantly alters the pinch-off dynamics of the droplet.

Particle-laden ink undergoes detachment through three distinct stages [17]. In the initial stage, the process is significantly influenced by particle concentration. As particle concentration increases, the viscosity rises, slowing down the thinning process and resulting in a longer filament. In the next stage, the sufficiently thinned filament begins to separate. Since only a small number of particles are present in the filament at this point, the viscosity change is minimal, and the pinch-off proceeds similarly to that of pure ink. In the final stage, the few remaining particles in the filament induce instability, accelerating the detachment process.

Figure 5.10 presents snapshots and minimum neck diameter, $W$ measurements comparing the pinch-off dynamics of particle-laden ink and pure ink. Here, $t_p$

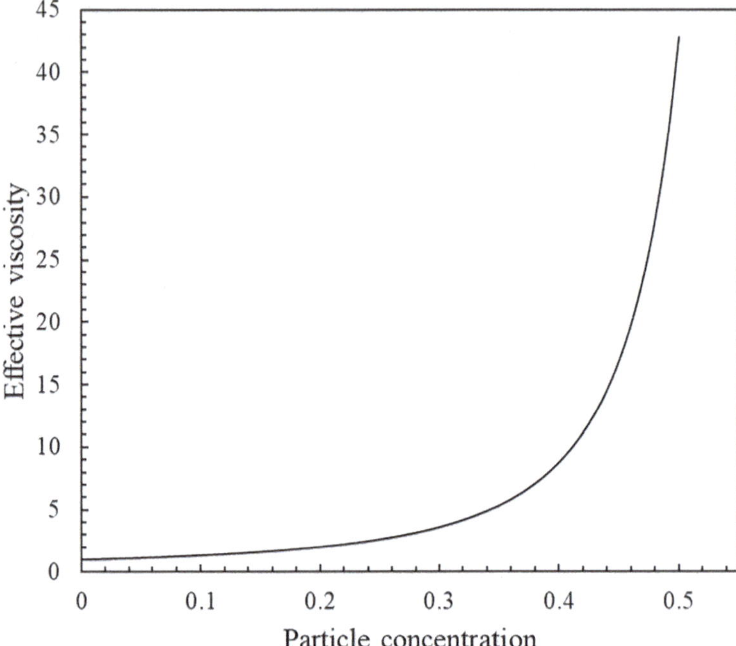


**Fig. 5.9** Effective viscosity as a function of the particle volume fraction

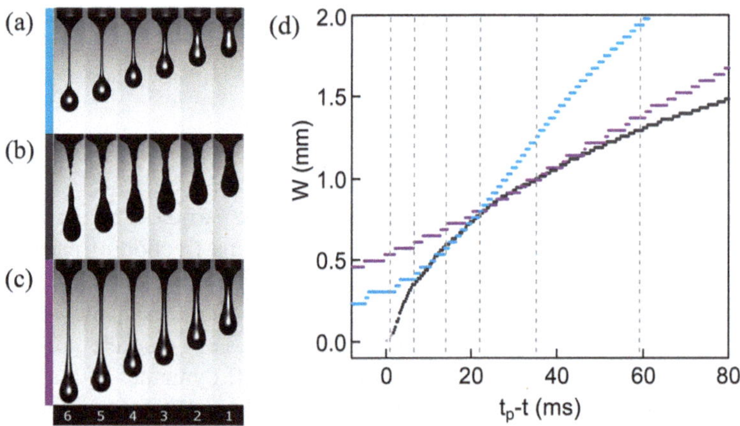


**Fig. 5.10** Comparison of pinch-off dynamics between pure and particle-laden ink [17]. Detachment process of (**a**) pure oil Shin Etsu (cyan), (**b**) particle suspended oil (gray) and (**c**) pure oil AP1000 (purple). The particle diameter is 140 μm and the concentration is 40%. (**d**) Evolution of the minimal neck diameter *W*. The timing of each snapshot in the detachment process is indicated by the gray dashed lines in the graph

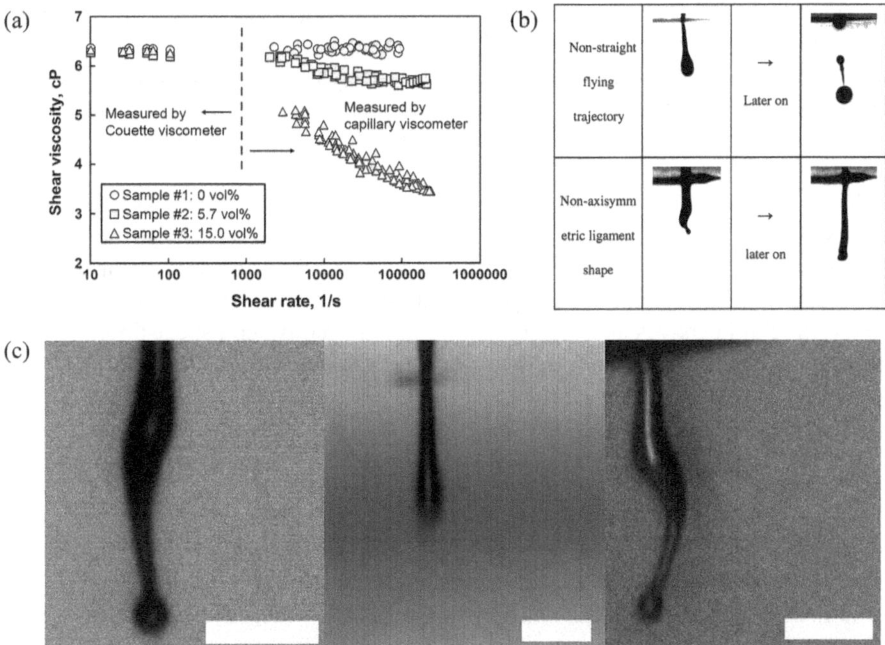


**Fig. 5.11** Inkjet printing of particle-laden non-Newtonian ink. (**a**) Shear viscosity of viscoelastic ink as a function of shear rate [18]. As particle concentration increases, the ink exhibits shear-thinning behavior. (**b**) Images showing inaccurate printing caused by high particle concentration in the ink [18]. (**c**) Snapshots of the jetting process of viscoelastic ink (PVA, 25 wt% in water), captured from three different angles at the same time [19]

represents the pinch-off time, and the $x$ axis in Fig. 5.10d is expressed as $t_p - t$. To examine the effect of viscosity in the initial stage, (a) and (b) use the same oil, Shin Etsu, while (c) uses oil AP1000, which has a viscosity similar to that of particle-laden ink. The results show that, compared to (a), the decrease in minimum neck diameter in the first stage is significantly slower for the droplet containing particles. This can be attributed to the increased viscosity due to the presence of particles, as evidenced by the comparison with the pure ink (c), which has a similar effective viscosity. Subsequently, the minimum neck diameter of the particle-laden droplet decreases similarly to that of pure ink, and in the final stage, it rapidly diminishes. These results demonstrate that the altered viscosity of ink with a high particle concentration has a substantial impact on the pinch-off dynamics of droplets.

In general, pure ink used in inkjet printing is a Newtonian fluid, meaning its viscosity remains constant regardless of shear rate. However, when particles are included in the ink, it can exhibit non-Newtonian fluid behavior. Two representative non-Newtonian properties are shear-thinning, where viscosity decreases as shear rate increases, and shear-thickening, where viscosity increases with shear rate. Figure 5.11a presents viscosity measurements of inks with particle volume concentrations of 0%, 5.7%, and 15.0% under varying shear rates [18]. While the viscosity of pure ink remains relatively unchanged with increasing shear rate, inks with higher particle concentrations exhibit a decrease in viscosity as shear rate increases, demonstrating shear-thinning behavior. This transition to non-Newtonian characteristics in particle-laden ink can negatively impact printing quality.

To predict the viscosity of non-Newtonian fluids, a commonly used model is expressed as $\mu = \mu_0 \dot{\gamma}^{n-1}$, where $\dot{\gamma}$ represents the shear rate, $\mu_0$ is consistency and $n$ is the viscosity exponent. When $n$ is less than 1, the fluid exhibits shear-thinning behavior; when $n$ is greater than 1, it exhibits shear-thickening behavior; and when $n = 1$, the fluid behaves as a Newtonian fluid. Now, consider a shear-thinning fluid flowing through a cylindrical nozzle under steady-state Poiseuille flow. In this case, the maximum shear rate inside the nozzle is given by:

$$\dot{\gamma}_{max} = \frac{\nu_0}{R_0} \frac{3n+1}{n}, \tag{5.3}$$

where $R_0$ is the radius of nozzle, and $\nu_0$ is the characteristic velocity. In inkjet printing, droplet velocity is typically around 10 m/s, meaning that the maximum shear rate inside the nozzle can reach up to the $10^6$ $s^{-1}$ scale. As shear rate increases, viscosity decreases, reducing the flow resistance inside the nozzle and facilitating droplet ejection. However, once the filament forms, viscosity increases again, delaying filament breakup. In elongated filaments, the viscoelasticity of the ink induces asymmetric instabilities, ultimately leading to non-ideal jetting, as shown in Fig. 5.11b and c [18, 19]. Such asymmetric jetting can prevent droplets from being deposited at the desired locations or cause inconsistencies in satellite droplet formation, negatively impacting printing precision and quality.

So far, it has been shown that increasing particle concentration can raise the viscosity of ink or induce a transition to non-Newtonian behavior, thereby affecting droplet formation. To significantly alter the ink's physical properties, a high particle concentration is required, and at concentrations below 15%, changes in the suspension's properties are negligible. However, even inks with low particle concentrations undergo a droplet formation process distinct from that of pure ink.

Figure 5.12a [20] compares the droplet formation process of pure ink and suspensions with low particle concentrations (1% and 6%). As particles are introduced, droplet breakup accelerates, and satellite droplets form within the filament. This indicates that even with minimal changes in ink properties due to low particle concentrations, thinning dynamics can be significantly affected. When particle concentration is high and particle size is at the nanoscale, the suspension can be approximated as a continuous fluid. However, as shown in Fig. 5.12a, such an assumption is not valid for low concentrations, and the influence of individual particles must be considered.

If particles are present in the thread during the thinning process, they migrate toward both ends, as shown in Fig. 5.12b [21]. This migration leads to the formation of a particle-free region, where very few particles remain. The presence of particles

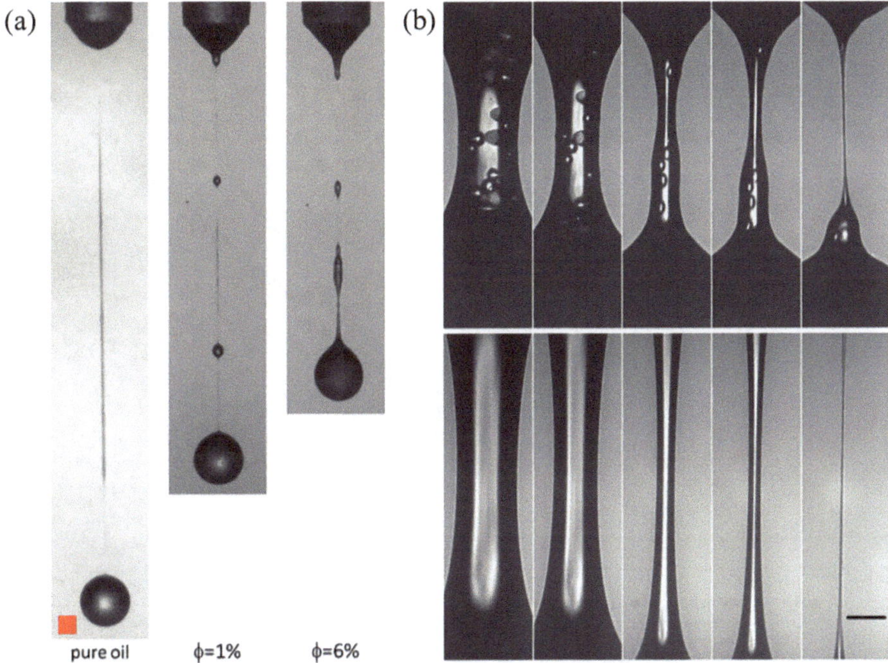


**Fig. 5.12** Filament thinning at low particle concentration. (**a**) Images of droplet separation for pure ink and ink with 1% and 6% particle volume fraction. The diameter of particle is 80 μm [20]. (**b**) Comparison of liquid bridge breakup between fluid containing 3 wt% particles with an 80 μm diameter (up) and pure fluid (down) [21]

in the thread induces a local curvature too large to accommodate the self-similar solution of the Stokes regime, resulting in an instability that accelerates the thinning dynamics compared to pure fluids. Figure 5.13 [21] quantitatively compares the thinning dynamics of a pure fluid and a suspension. The blue line represents the suspension, showing a faster thinning rate than the red line, which corresponds to the pure fluid. As thinning progresses, it becomes dominated by the particle-free region, where no particles are present. In this stage, the thinning process proceeds similarly to that of a pure fluid, as indicated by the black dotted line.

In summary, the inclusion of particles in the jetting process alters ink viscosity and induces a transition to non-Newtonian behavior, leading to differences in droplet formation. Even at low particle concentrations, particles present in the thread can introduce instabilities, accelerating the formation of satellite droplets. As a result, particle-laden ink exhibits more complex physical phenomena compared to pure ink, necessitating adjustments to previously established stable ejection conditions.

Shi et al. [22] extended the operating map shown in Fig. 5.5 to include particle-laden ink, as illustrated in Fig. 5.14. By categorizing ejection scenarios into five types and analyzing droplet formation, they confirmed that pure fluids achieve stable ejection within the range of $0.1 < Oh < 1$. However, for inks with a particle volume

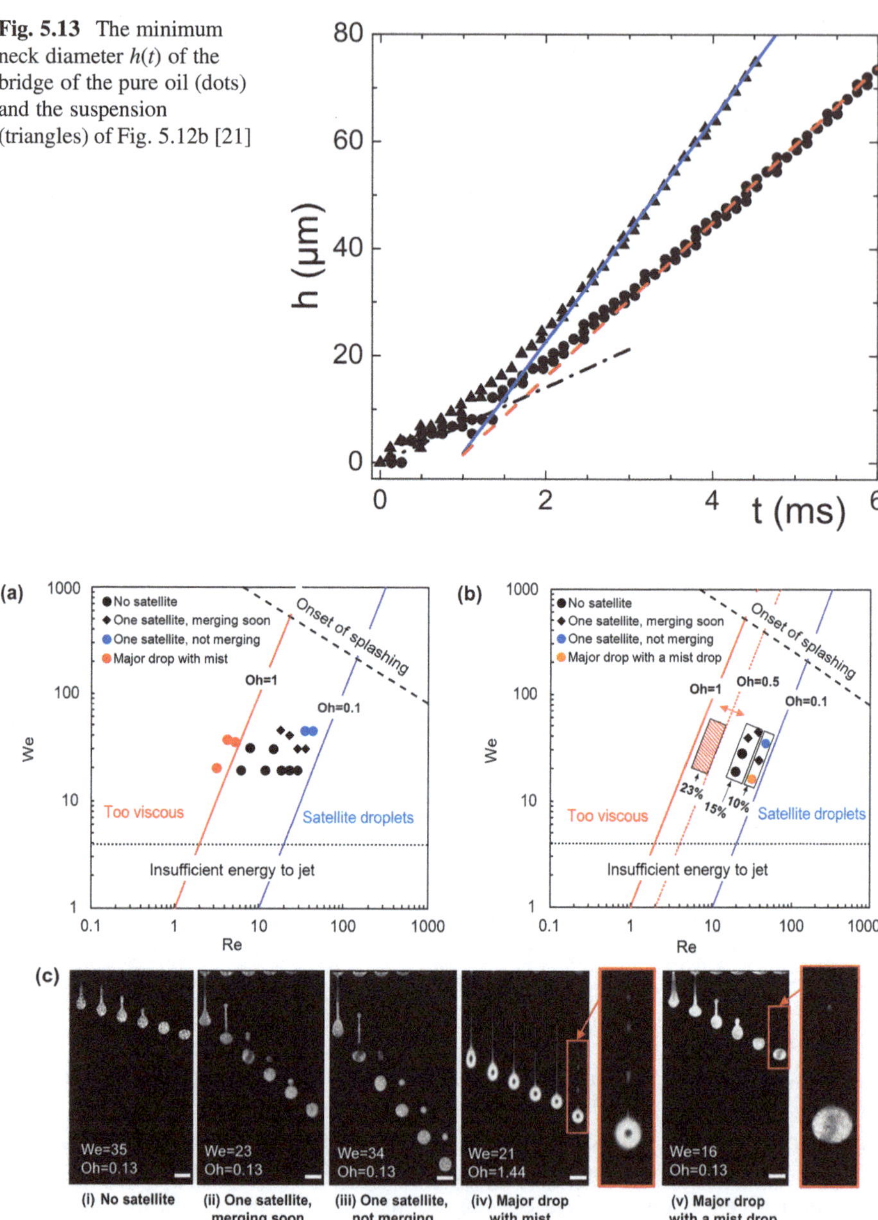


**Fig. 5.13** The minimum neck diameter $h(t)$ of the bridge of the pure oil (dots) and the suspension (triangles) of Fig. 5.12b [21]

**Fig. 5.14** (**a**) Jetting test results for the pure fluids. (**b**) Modified jetting map for particle-laden fluids. The red shaded region represents unstable jetting conditions. (**c**) The five jetting scenarios

fraction below 23%, the stable ejection range narrowed to $0.5 < Oh < 1$. These results indicate that when particles are present in the ink, the ejection conditions must be designed more conservatively to ensure stability.

**Exercise 5.3 Non-Newtonian Behavior of Particle-Laden Inks**

(a) Describe how particle inclusion induces shear-thinning or shear-thickening behavior.

(b) Explain how shear-thinning affects filament breakup and satellite droplet formation.

**Solution**

(a) At moderate volume fractions with deformable microstructure (e.g., soft aggregates, polymer-coated particles), increasing shear can break clusters and align particles with the flow. This reduces flow resistance and leads to shear-thinning: viscosity decreases with shear rate.

In contrast, at high volume fractions or for certain shapes (e.g., rod-like particles), increasing shear can enhance collisions and form transient networks, increasing resistance and producing shear-thickening: viscosity increases with shear rate.

Inks with nanoparticles can therefore exhibit non-Newtonian rheology, especially under the very high shear rates encountered in inkjet nozzles.

(b) For a shear-thinning ink, the local viscosity decreases in regions of high shear, such as inside the nozzle and in the thinning filament. This reduction in viscosity lowers the local Ohnesorge number and effectively increases the $Z$ number, so the flow becomes more susceptible to rapid stretching and Rayleigh–Plateau instability. As a result, the neck region thins faster and pinch-off tends to occur earlier, which can increase the likelihood of satellite droplet formation, especially when inertial effects are significant. In practice, shear-thinning behavior can thus reduce the actuation pressure needed to expel the droplet, but at the same time it may exacerbate filament instability and satellite generation if the waveform and operating conditions are not carefully controlled.

### *5.3.3 Evaporation*

A droplet ejected from a nozzle impacts the substrate and subsequently evaporates, during which the particles contained within the droplet adhere to the substrate. In an evaporating droplet, the highest evaporation rate occurs at the contact line, where the liquid, gas, and solid phases meet. This induces a capillary flow directed toward the contact line within the droplet. As a result, the particles are transported to the contact line and ultimately form a ring-like deposition pattern (Fig. 5.15). This phenomenon is commonly referred to as the "coffee-ring" effect, and it is a critical issue to avoid in industries where uniform particle distribution is required.

Numerous previous studies have been conducted to suppress the coffee-ring effect, and the proposed methodologies can be broadly classified into three categories: (1) control of contact line, (2) control of internal flow within the droplet, and (3) control of particle properties and interparticle interactions. This section

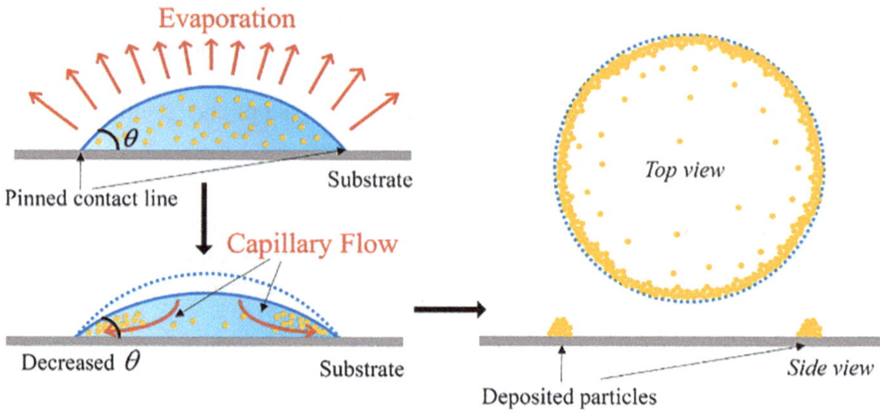


**Fig. 5.15** Schematic of droplet evaporation and particle deposition process [23]

introduces various prior studies that have been carried out to suppress the coffee-ring effect.

The mobility of the contact line significantly influences particle deposition. For a sessile droplet evaporating on a substrate, the contact line remains pinned until the contact angle reaches the receding angle. This results in a continuous capillary flow toward the contact line, which is one of the main causes of the ring-like particle deposition pattern. If the contact line is allowed to move freely, particles become trapped in the Marangoni flow inside the droplet, leading to a "coffee-eye" pattern, where particles accumulate at the center of the droplet. Therefore, the ability to control the movement of the contact line implies the ability to regulate the resulting particle deposition pattern.

Wettability is one of the most effective methods for controlling contact line mobility. Figure 5.16a shows the particle deposition pattern when a superhydrophobic substrate is used to minimize the contact area between the droplet and the substrate. As the influence of the contact line is reduced, the particle adhesion during evaporation is suppressed, resulting in a ball-like particle pattern at the droplet center. Figure 5.16b presents a study in which the contact line was controlled by introducing a hydrophilic patch on a hydrophobic substrate. While the use of a fully hydrophobic substrate alone limits the ability to control deposition patterns, incorporating hydrophilic patches allows for diverse adhesion patterns at relatively low cost. In addition to chemical wettability between the droplet and the surface, contact line pinning can also be suppressed through electrowetting (EW). When an AC voltage exceeding the pinning force is applied to the droplet, repeated depinning of the contact line can occur, thereby preventing the formation of a coffee-ring. Figure 5.16c illustrates the effect of EW on particle deposition, where the application of EW results in a coffee-eye pattern rather than a coffee-ring, in contrast to the case without EW.

As previously mentioned, two main flow types exist within an evaporating droplet: the radial capillary flow directed toward the contact line and the vortex

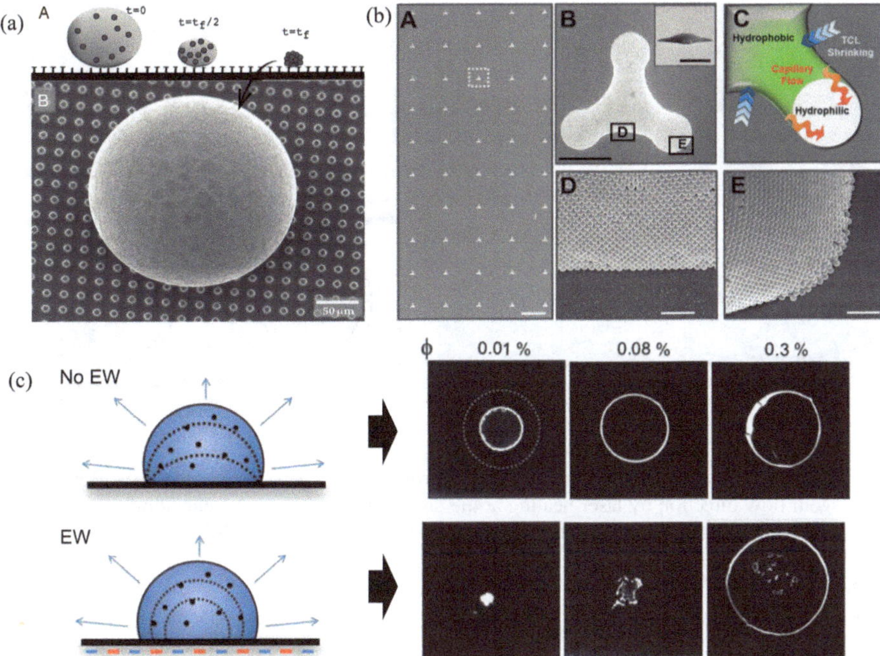


**Fig. 5.16** Suppression of the coffee-ring effect through contact line control. (**a**) Particle distribution on a superhydrophobic substrate [24]. (**b**) Contact line control using a hydrophilic patch [25]. (**c**) Variation in particle deposition pattern induced by electrowetting [26]

Marangoni flow. Capillary flow transports particles to the contact line, promoting deposition, while Marangoni flow traps particles within the internal vortices, thereby delaying their adhesion. Thus, enhancing the Marangoni flow—driven by surface tension gradients—can lead to a more uniform particle distribution.

Marangoni flow can be controlled using either thermal gradients or surfactants. In typical cases where heat is applied from the substrate, the surface temperature of the droplet decreases toward the center, resulting in an outward Marangoni flow. However, as shown in Fig. 5.17a, when a laser is applied to the center of a sessile droplet, the surface temperature gradient is reversed, leading to an inward Marangoni flow. This flow counteracts the capillary flow, enabling a more uniform particle distribution. Surfactants can also induce changes in Marangoni flow. When surfactants are transported to the contact line, the local surface tension is reduced, generating an inward Marangoni flow. This inward flow pushes particles away from the contact line toward the center, suppressing edge deposition and ultimately resulting in a more uniform particle distribution (Fig. 5.17b).

Lastly, the coffee-ring effect can also be suppressed by modifying particle shape and interparticle interactions. Figure 5.18a shows the comparison of the final deposition patterns of ellipsoidal and spherical particles. When ellipsoidal particles reach the air–water interface, they tend to form looser structures compared to

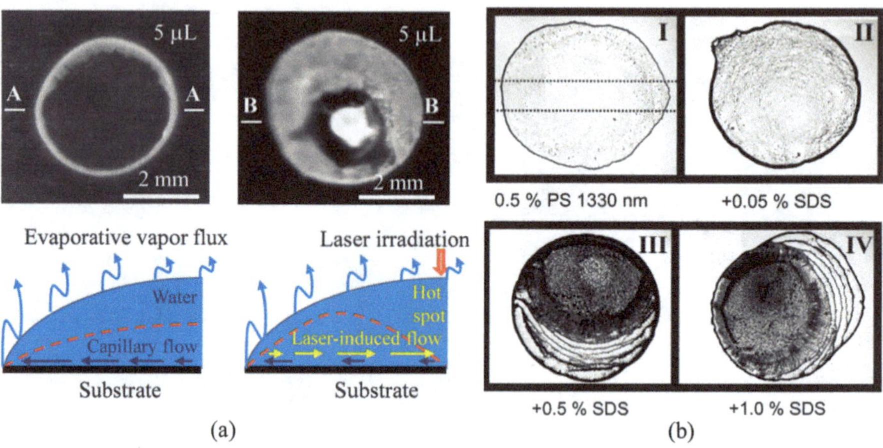


**Fig. 5.17** Suppression of the coffee-ring effect through Marangoni flow control. (**a**) Reversal of Marangoni flow direction by laser heating at the droplet center [27]. (**b**) Generation of Marangoni eddies by surfactants to hinder particle deposition near the contact line [28]

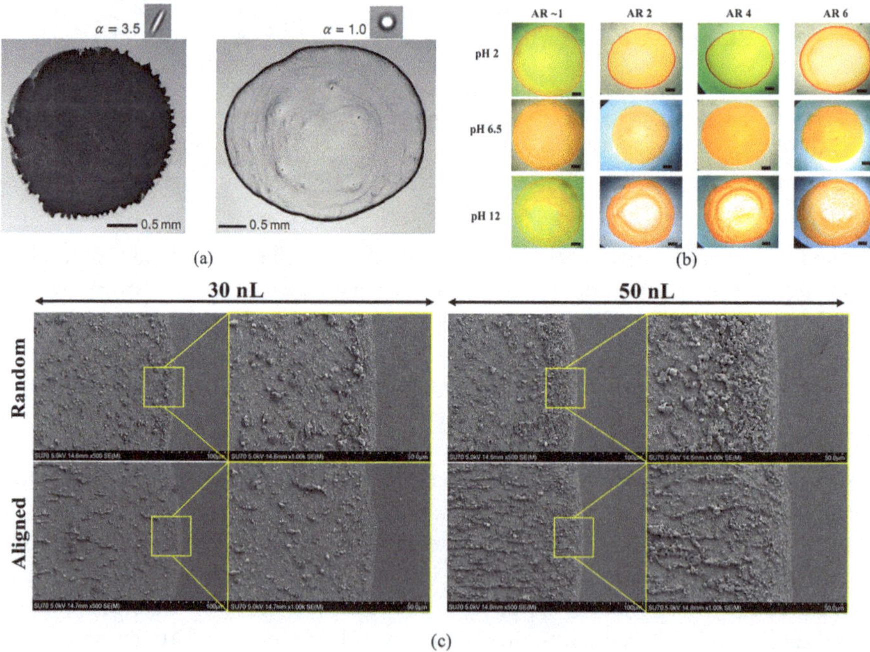


**Fig. 5.18** Particle distribution influenced by particle shape and interactions. (**a**) Comparison of deposition patterns between ellipsoidal and spherical particles [29]. (**b**) Particle distribution under varying aspect ratios and pH levels [30]. (**c**) Effect of an external magnetic field on the deposition pattern of magnetic particles [31]

spherical particles. The authors of this study explained that such loose networks provide greater resistance to external flow, preventing the particles from migrating toward the contact line and thereby enabling a more uniform deposition. However, subsequent studies suggested that particle–particle and particle–substrate interactions may have a greater influence on uniformity than particle shape alone. Figure 5.18b presents a comparison of deposition patterns under varying particle aspect ratios and pH levels of the aqueous solution. Changes in pH affect the interparticle interaction forces. The results showed that while the aspect ratio had a limited effect on deposition patterns, the interaction forces significantly influenced the distribution. In addition, the use of magnetic particles allows for control over the ring pattern (Fig. 5.18c). When exposed to an external magnetic field, magnetic particles align along the direction of magnetic anisotropy. If the particles form chain-like structures parallel to the magnetic field rather than distributing randomly, the number of particles transported by lateral flow can be reduced, thus suppressing the coffee-ring effect.

**Exercise 5.4 Coffee-Ring Formation and Suppression Strategies**

(a) Using Fig. 5.15, describe how capillary flow leads to ring-like deposition.
(b) Compare three methods of coffee-ring suppression: contact line control, Marangoni enhancement, and interparticle interaction tuning.
(c) Explain why electrowetting can induce a coffee-eye pattern instead of a coffee-ring.

**Solution**

(a) Evaporation is strongest near the contact line because vapor can diffuse out more easily there. This non-uniform evaporation causes more liquid to be removed near the edge than at the center. To conserve mass, liquid flows from the interior toward the contact line (outward capillary flow). Suspended particles are carried by this flow and accumulate near the pinned contact line. When the droplet fully evaporates, particles remain concentrated at the perimeter, forming a coffee-ring pattern.
(b) Contact line control:

- By moving or periodically depinning the contact line, the persistent outward capillary flow is disrupted.
- The deposition region is "smeared" over a wider area, reducing the sharpness of the ring.

Marangoni enhancement:

- Inducing strong Marangoni circulation (via thermal or solutal gradients) generates recirculating flows that transport particles back from the edge toward the center.
- This counteracts the outward capillary flux and leads to a more uniform particle distribution.

Interparticle interaction tuning:

- Adjusting DLVO interactions and depletion forces can make particles aggregate in the interior or adhere more strongly to the substrate before reaching the contact line.
- This can produce dot-like or uniformly spread deposits instead of a ring.

(c) Electrowetting dynamically changes the apparent contact angle by applying a voltage to the substrate. During evaporation, applying/releasing voltage can cause the contact line to advance or recede, rather than remaining pinned. This motion redistributes the liquid and particles; the outermost region can be partially depleted while particles accumulate more in an inner annulus. As a result, instead of a maximum at the outer edge, the radial particle density shows a peak at an intermediate radius, with lower deposition at the extreme edge and center—resembling a "coffee-eye" pattern. Thus, electrowetting converts the single outer ring into a ring with a depleted rim and a dense inner band.

### *5.3.4 Inkjet Printing with AI*

With the advancement of artificial intelligence (AI) technologies, inkjet printing has also seen increasing integration of AI-based research. Inkjet printing involves the use of fluid inks, and droplets are ejected simultaneously from multiple print heads, making real-time prediction of droplet morphology essential. In this context, AI technologies are being applied to optimize the electrical signals applied to piezoelectric materials, detect the formation of satellite droplets in real time, and predict the morphology of ejected droplets through image-based analysis.

Figure 5.19a illustrates an artificial intelligence-based closed-loop algorithm designed to automatically generate optimal waveforms capable of producing satellite-free droplets for inks with diverse fluid properties. Newtonian fluids with Z values ranging from 1 to 50 were targeted, and single flash imaging was used to extract drop formation types—non-jetting, single droplet, and multiple droplets—as well as their velocities. Among five machine learning models compared, the multilayer perceptron (MLP) demonstrated the most accurate performance in predicting droplet morphology and velocity. This study showed that the use of machine learning significantly reduces the time required for waveform optimization. Figure 5.19b presents a follow-up to the previous study, focusing on waveform optimization for viscoelastic inks. Compared to Newtonian inks, viscoelastic inks exhibit more sensitivity to waveform variations and require consideration of more parameters, making precise optimization essential. The study proposed a learning framework that combines a Gaussian mixture model (GMM) with MLP to predict the jetting behavior of viscoelastic inks with high accuracy, and utilized the model within a deep reinforcement learning (DRL) environment. Lastly, Fig. 5.19c introduces a sample-efficient Bayesian optimization framework for automated waveform design. Multiphase simulations were conducted using OpenFOAM, with five

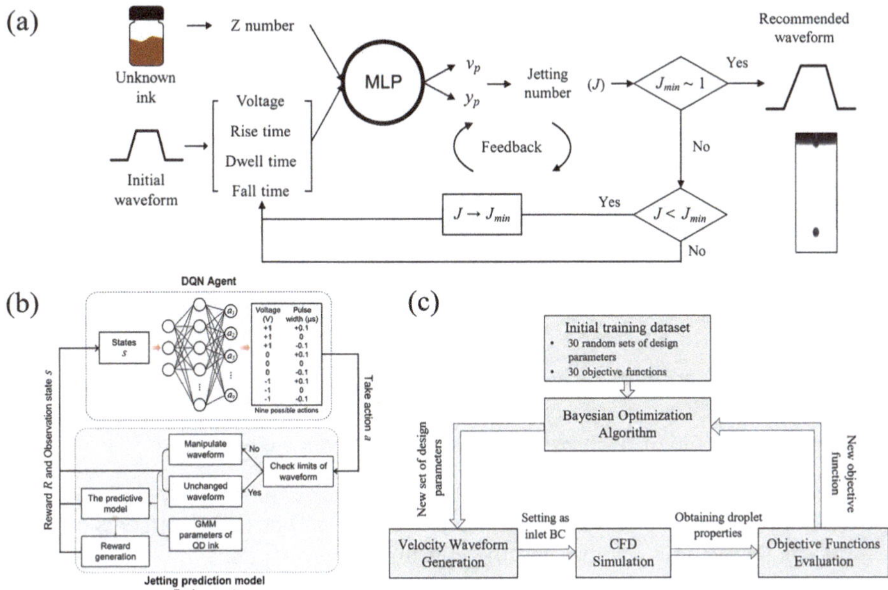


**Fig. 5.19** Schematic of AI-based waveform optimization design. (**a**) MLP algorithm [32]. (**b**) Optimal waveform design for viscoelastic ink [33]. (**c**) Bayesian optimization framework [34]

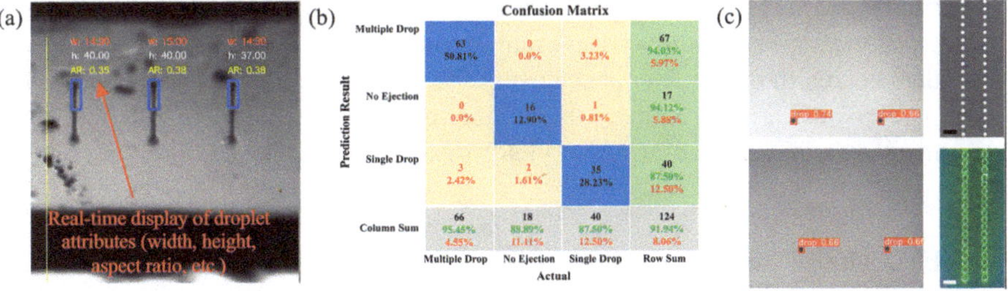


**Fig. 5.20** Inkjet printing in situ monitoring. (**a**) Droplet mode detection results [35]. (**b**) Jetting classification confusion matrix [36]. (**c**) Multijet synchronization from droplet detection [37]

waveform design parameters set. The optimal waveform was identified within 150 simulation trials. As a result, the optimized waveform successfully eliminated satellite droplets and reduced the droplet size to 24.9% of the nozzle diameter.

In inkjet printing, real-time measurement of droplet morphology is essential, and the use of artificial intelligence can significantly reduce the time required for such analysis. Figure 5.20a presents a detection framework that enables real-time monitoring of droplet formation during the inkjet printing process and automatic classification of droplet modes based on image-derived features. The system calculates four input features per frame—droplet size, aspect ratio, velocity, and presence of satellite droplets—and uses them to construct a high-accuracy predictive model. This provides a practical foundation for real-time quality monitoring in industrial settings. Figure 5.20b demonstrates the use of machine learning to predict droplet velocity, radius, and jetting mode under 11 different printing conditions. The general predictive model, trained on approximately 3000 datasets, outperformed traditional dimensionless-number-based approaches, showing the potential to guide ink parameter selection and waveform design without experimental trials. Finally, Figure 5.20c introduces a YOLOv5-based deep learning detection system designed for real-time droplet shape recognition and automated multi-nozzle synchronization in high-throughput inkjet environments. This deep learning framework enables real-time detection, quantitative analysis, and synchronization control, offering a significant improvement over conventional methods that rely on manual observation by skilled operators.

**Exercise 5.5 AI-Assisted Inkjet Printing**
(a) Summarize how machine learning accelerates waveform optimization.
(b) Describe the role of deep learning in multi-nozzle synchronization.

**Solution**
(a) Traditional waveform optimization relies on trial-and-error experiments or high-fidelity simulations, which are time-consuming:

- Machine learning models (e.g., neural networks, surrogate models) are trained on a dataset of waveform parameters → jetting outcomes (droplet volume, velocity, satellites, etc.).
- Once trained, the model can rapidly predict jetting quality for new waveform candidates without running full experiments or simulations.
- Optimization algorithms (Bayesian optimization, gradient-based methods) then search the parameter space using the surrogate model as a fast evaluator.

  Thus, ML greatly reduces the number of costly physical or numerical trials needed to find a waveform that yields desired droplet properties.

(b) In multi-nozzle print heads:

- Each nozzle can have slightly different dynamics (manufacturing tolerances, local temperature, contamination), leading to droplet-to-droplet variability across the array.
- Deep learning models can be trained on large datasets of nozzle state + waveform → droplet outcome, capturing complex, nonlinear relationships.

Thus, deep learning enables synchronized jetting across many nozzles by compensating for non-uniformities, thereby improving print uniformity and throughput.

## 5.4 Conclusions

Inkjet printing has emerged as a versatile and precise deposition technique that enables the patterning of functional materials without the need for physical masks or complex photolithographic processes. In this chapter, we examined the fundamental mechanisms, classification, and application domains of inkjet printing, with a particular focus on drop-on-demand (DOD) systems that are widely employed in high-resolution and high-precision manufacturing.

We first discussed the core principles of DOD inkjet printing, which involve complex interactions between mechanical, acoustic, and fluid dynamic phenomena. The optimization of waveform design, ink properties, and nozzle geometry is essential for achieving stable droplet ejection and minimizing the formation of satellite droplets. The introduction of dimensionless parameters such as Reynolds number, Weber number, Ohnesorge number, and Z number provides a theoretical framework for defining printable regimes and guiding experimental design.

We then addressed the unique challenges associated with particle-laden inks, which introduce additional physical complexities due to increased viscosity, non-Newtonian behavior, and particle-induced instabilities. It was shown that even at low particle concentrations, the presence of solid components within the ink can significantly alter droplet pinch-off dynamics and lead to unintended deposition patterns.

Another critical phenomenon discussed was the evaporation of droplets and the resulting coffee-ring effect, which hampers uniform particle deposition. Multiple suppression strategies—such as contact line manipulation, Marangoni flow control, and tuning of interparticle forces—have been proposed and demonstrated, offering valuable solutions for improving printing uniformity in practical applications.

Finally, we explored recent advancements that integrate artificial intelligence with inkjet printing. Machine learning and deep learning models have shown promise in predicting droplet morphology, optimizing actuation waveforms, and performing real-time monitoring and control of multi-nozzle systems. These developments not only enhance the efficiency and reliability of the printing process but also pave the way for intelligent manufacturing systems capable of adapting to complex material behaviors.

In summary, inkjet printing is a highly interdisciplinary technology that combines principles from fluid mechanics, material science, control engineering, and data-driven modeling. Ongoing innovations in both physical modeling and artificial intelligence are expected to further expand its capabilities, enabling the precise, scalable, and adaptive fabrication of next-generation devices and materials.

**Summary**

1. **Fundamentals of inkjet printing**

   - Inkjet printing ejects liquid ink onto a substrate using a non-contact, digitally controlled mechanism, enabling flexible, mask-free, high-resolution patterning.
   - Applications span electronics, displays, biosensors, tissue scaffolds, and energy materials.
   - Droplet formation, flight stabilization, and deposition involve complex fluid dynamic interactions among viscosity, density, surface tension, nozzle geometry, and actuation waveform.

2. **Classification of inkjet technologies**
   - Inkjet systems are broadly divided into Continuous Inkjet (CIJ) and Drop-on-Demand (DOD) printing.
   - DOD is widely used for high-precision manufacturing and includes piezoelectric, thermal, and electrohydrodynamic (EHD) actuation methods.
   - Each actuation type has its own operating principles, fluid property requirements, and design trade-offs
3. **Physical mechanisms governing droplet jetting**
   - Jetting behavior results from competition between inertia, viscosity, and surface tension, leading to nonlinear and highly sensitive ejection dynamics.
   - Stable jetting requires balanced tuning of ink properties, nozzle geometry, pressure waves, and waveform design.
4. **Particle-laden inks and their complex behaviors**
   - Real inks often contain nanoparticles, polymers, dyes, or biomolecules, making their rheology more complex than pure Newtonian liquids.
   - Particle addition increases viscosity and can induce non-Newtonian behavior, affecting droplet elongation, pinch-off, and satellite droplet formation.
5. **Pinch-off stages in particle-laden jetting**
   - Experiments show a three-stage detachment process:
     1. Viscosity-dominated stage—particle concentration increases effective viscosity and slows neck thinning.
     2. Filament thinning stage—few particles remain, leading to behavior similar to pure ink.
     3. Instability-driven final pinch-off—last remaining particles accelerate final breakup.
   - These stages reveal how particle concentration governs both the elongation and breakup characteristics of DOD jets.
6. **Droplet evaporation and deposition phenomena**
   - Later stages of the printing process involve droplet spreading, evaporation, and particle transport on the substrate.
   - The coffee-ring effect—accumulation of particles at the contact line—remains a major challenge for uniform film formation.
7. **AI-assisted inkjet printing**
   - Recent advances integrate machine learning for tasks such as:
     - Predicting droplet morphology
     - Optimizing waveform design
     - Real-time monitoring and synchronization of multinozzle arrays
   - AI provides new capabilities beyond traditional physics-based approaches, improving both manufacturing throughput and precision.

## References

1. H. Wijshoff, "The dynamics of the piezo inkjet printhead operation," Phys. Rep. **491**, 77-177 (2010).
2. G.-K. Lau, and M. Shrestha, "Ink-jet printing of micro-electro-mechanical systems (mems)," Micromachines **8**, 194 (2017).
3. M. A. Shah, D.-G. Lee, B.-Y. Lee, and S. Hur, "Classifications and applications of inkjet printing technology: A review," IEEE Access **9**, 140079-140102 (2021).
4. D. Lohse, "Fundamental fluid dynamics challenges in inkjet printing," Annual Review of Fluid Mechanics **54**, 349-382 (2022).
5. A. Fraters, M. Van Den Berg, Y. De Loore, H. Reinten, H. Wijshoff, D. Lohse, M. Versluis, and T. Segers, "Inkjet nozzle failure by heterogeneous nucleation: Bubble entrainment, cavitation, and diffusive growth," Physical review applied **12**, 064019 (2019).
6. P. M. Kamat, B. W. Wagoner, A. A. Castrejón-Pita, J. R. Castrejón-Pita, C. R. Anthony, and O. A. Basaran, "Surfactant-driven escape from endpinching during contraction of nearly inviscid filaments," Journal of Fluid Mechanics **899**, (2020).
7. G. H. McKinley, and M. Renardy, "Wolfgang von ohnesorge," Phys. Fluids **23**, (2011).
8. J. Fromm, "Numerical calculation of the fluid dynamics of drop-on-demand jets," IBM Journal of Research and Development **28**, 322-333 (1984).
9. N. Reis, and B. Derby, "Ink jet deposition of ceramic suspensions: Modeling and experiments of droplet formation," MRS Online Proceedings Library (OPL) **625**, 117 (2000).
10. B. Derby, "Inkjet printing of functional and structural materials: Fluid property requirements, feature stability, and resolution," Annual Review of Materials Research **40**, 395-414 (2010).
11. Y. Liu, and B. Derby, "Experimental study of the parameters for stable drop-on-demand inkjet performance," Phys. Fluids **31**, (2019).
12. J. R. Castrejon-Pita, N. F. Morrison, O. G. Harlen, G. D. Martin, and I. M. Hutchings, "Experiments and lagrangian simulations on the formation of droplets in drop-on-demand mode," Phys. Rev. E **83**, 036306 (2011).
13. E. Antonopoulou, O. Harlen, M. Walkley, and N. Kapur, "Jetting behavior in drop-on-demand printing: Laboratory experiments and numerical simulations," Physical Review Fluids **5**, 043603 (2020).
14. A. van der Bos, M.-J. van der Meulen, T. Driessen, M. van den Berg, H. Reinten, H. Wijshoff, M. Versluis, and D. Lohse, "Velocity profile inside piezoacoustic inkjet droplets in flight: Comparison between experiment and numerical simulation," Physical review applied **1**, 014004 (2014).
15. L. Zhang, "Characteristics of drop-on-demand droplet jetting with effect of altered geometry of printhead nozzle," Sensors and Actuators A: Physical **298**, (2019).
16. I. E. Zarraga, D. A. Hill, and D. T. Leighton Jr, "The characterization of the total stress of concentrated suspensions of noncolloidal spheres in newtonian fluids," Journal of Rheology **44**, 185-220 (2000).
17. C. Bonnoit, T. Bertrand, E. Clément, and A. Lindner, "Accelerated drop detachment in granular suspensions," Phys. Fluids **24**, (2012).
18. X. Wang, W. W. Carr, D. G. Bucknall, and J. F. Morris, "Drop-on-demand drop formation of colloidal suspensions," International journal of multiphase flow **38**, 17-26 (2012).
19. M. Bienia, M. Lejeune, M. Chambon, V. Baco-Carles, C. Dossou-Yovo, R. Noguera, and F. Rossignol, "Inkjet printing of ceramic colloidal suspensions: Filament growth and breakup," Chem. Eng. Sci. **149**, 1-13 (2016).
20. M. S. van Deen, T. Bertrand, N. Vu, D. Quéré, E. Clément, and A. Lindner, "Particles accelerate the detachment of viscous liquids," Rheologica Acta **52**, 403-412 (2013).
21. A. Lindner, J. E. Fiscina, and C. Wagner, "Single particles accelerate final stages of capillary break-up," EPL (Europhysics Letters) **110**, (2015).
22. J. Shi, N. Cagney, J. Tatum, A. Condie, and J. R. Castrejón-Pita, "Jetting and droplet formation of particle-loaded fluids," Phys. Fluids **36**, (2024).

23. K. Singh, P. Kumar, H. Raman, H. Sharma, and R. Mangal, "Tailoring the coffee ring effect by chemically active janus colloids," ACS Applied Engineering Materials (2025).
24. A. G. Marin, H. Gelderblom, A. Susarrey-Arce, A. van Houselt, L. Lefferts, J. G. Gardeniers, D. Lohse, and J. H. Snoeijer, "Building microscopic soccer balls with evaporating colloidal fakir drops," Proc Natl Acad Sci U S A **109**, 16455-16458 (2012).
25. L. Wu, Z. Dong, M. Kuang, Y. Li, F. Li, L. Jiang, and Y. Song, "Printing patterned fine 3d structures by manipulating the three phase contact line," Advanced Functional Materials **25**, 2237-2242 (2015).
26. D. Mampallil, H. Eral, D. Van Den Ende, and F. Mugele, "Control of evaporating complex fluids through electrowetting," Soft Matter **8**, 10614-10617 (2012).
27. V. Ta, R. Carter, E. Esenturk, C. Connaughton, T. J. Wasley, J. Li, R. W. Kay, J. Stringer, P. Smith, and J. D. Shephard, "Dynamically controlled deposition of colloidal nanoparticle suspension in evaporating drops using laser radiation," Soft Matter **12**, 4530-4536 (2016).
28. T. Still, P. J. Yunker, and A. G. Yodh, "Surfactant-induced marangoni eddies alter the coffee-rings of evaporating colloidal drops," Langmuir **28**, 4984-4988 (2012).
29. P. J. Yunker, T. Still, M. A. Lohr, and A. G. Yodh, "Suppression of the coffee-ring effect by shape-dependent capillary interactions," Nature **476**, 308-311 (2011).
30. V. R. Dugyala, and M. G. Basavaraj, "Control over coffee-ring formation in evaporating liquid drops containing ellipsoids," Langmuir **30**, 8680-8686 (2014).
31. K. N. Al-Milaji, R. L. Hadimani, S. Gupta, V. K. Pecharsky, and H. Zhao, "Inkjet printing of magnetic particles toward anisotropic magnetic properties," Sci Rep **9**, 16261 (2019).
32. S. Kim, M. Cho, and S. Jung, "The design of an inkjet drive waveform using machine learning," Sci. Rep. **12**, 4841 (2022).
33. S. Kim, M. Cho, and S. Jung, "Reinforcement learning-based dynamic optimization of driving waveforms for inkjet printing of viscoelastic fluids," Langmuir **41**, 10831-10840 (2025).
34. H. Wang, and Y. Hasegawa, "Multi-objective optimization of actuation waveform for high-precision drop-on-demand inkjet printing," Phys. Fluids **35**, (2023).
35. M. Ogunsanya, J. Isichei, S. K. Parupelli, S. Desai, and Y. Cai, "In-situ droplet monitoring of inkjet 3d printing process using image analysis and machine learning models," Procedia Manufacturing **53**, 427-434 (2021).
36. F. P. Brishty, R. Urner, and G. Grau, "Machine learning based data driven inkjet printed electronics: Jetting prediction for novel inks," Flexible and Printed Electronics **7**, 015009 (2022).
37. E. Choi, S. Choi, K. An, and K.-T. Kang, "Deep learning-based inkjet droplet detection for jetting characterizations and multijet synchronization," ACS Applied Materials & Interfaces **16**, 18040-18051 (2024).

# Chapter 6
# Interfacial Behavior of Surfactant-Covered Double Emulsion

## 6.1 Introduction

Double emulsions, often described as droplets encapsulated within larger droplets, are multiphase fluid systems that can take forms such as water-in-oil-in-water (W/O/W) or oil-in-water-in-oil (O/W/O) [1]. Due to their unique compartmentalized structure, they have found widespread use in fields ranging from drug delivery and food encapsulation to cosmetic and biomedical engineering [2–4].

The stability of double emulsions is closely tied to the properties of the interfaces between the inner and outer droplets. To prevent coalescence and preserve droplet integrity, surfactants are typically employed [5]. These surface-active molecules adsorb at fluid interfaces and reduce surface tension, thereby enhancing emulsion stability. However, the distribution of surfactants is rarely uniform, especially under dynamic flow conditions. This nonuniformity can lead to gradients in surface tension, resulting in Marangoni stresses that influence both droplet deformation and internal flow.

While experimental approaches have made significant progress in producing stable double emulsions, they often struggle to resolve fine-scale interface dynamics and surfactant transport. To address this, numerical simulations offer valuable insights. In particular, the LBM provides an efficient framework for modeling multiphase flows with complex interfacial phenomena.

This chapter explores the interplay between surfactant distribution and droplet deformation in a double emulsion subjected to uniaxial extensional flow (Fig. 6.1). Using three-dimensional LBM simulations, we examine how key parameters—such as capillary number, Péclet number, and the radius ratio between inner and outer droplets—affect the interfacial behavior and the stability of the system.

**Exercise 6.1 Role of Surfactants in Double Emulsions**
Explain why surfactants are essential for stabilizing double emulsions such as W/O/W and O/W/O systems.



H. M. Lee, J. S. Lee, *Lattice Boltzmann Methodology for Single-Phase and Multiphase Nanoparticle Modeling*,
https://doi.org/10.1007/978-981-95-9117-6_6

**Solution**
Surfactants are essential in W/O/W and O/W/O double emulsions because they:

- Reduce the interfacial tension between the phases, thereby lowering the thermodynamic driving force for coalescence,
- Provide an interfacial "barrier" that hinders film drainage and rupture when droplets come into close contact,
- Help maintain the concentric double-emulsion structure over long timescales by stabilizing both the inner and outer interfaces.

Without surfactants, the thin films separating inner and outer phases are prone to rupture, leading to rapid coalescence and collapse of the double-emulsion structure.

## 6.2 Numerical Methods

### *6.2.1 Double-Emulsion Modeling*

To model a double emulsion, it is necessary to simulate three immiscible phases, as illustrated in Fig. 6.1. For this purpose, the color-gradient LBM model can be

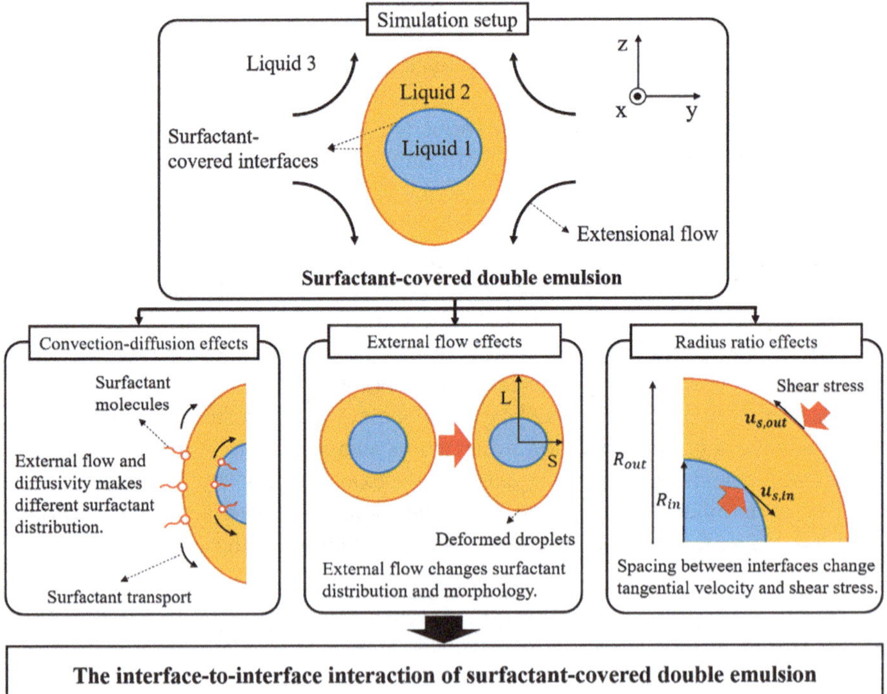


**Fig. 6.1** Schematic summarizing the objectives of this study

employed. Since emulsion systems typically do not exhibit large density ratios between phases, the simplest form of the color-gradient model can be applied to ensure computational efficiency.

$$f_i^k(\mathbf{x}+\mathbf{e}_i\Delta t, t+\Delta t) - f_i^k(\mathbf{x},t) = \Delta t\left(\Omega_i^k\right)^{(3)}\left[\left(\Omega_i^k\right)^{(1)} + \left(\Omega_i^k\right)^{(2)}\right], \tag{6.1}$$

Here, $k$ denotes the three phases, and $\left(\Omega_i^k\right)^{(1)}$, $\left(\Omega_i^k\right)^{(2)}$, and $\left(\Omega_i^k\right)^{(3)}$ correspond to the single-phase collision operator, the perturbation operator, and the recoloring operator, respectively. $\left(\Omega_i^k\right)^{(1)}$ and $\left(\Omega_i^k\right)^{(2)}$ are identical to those presented in Chap. 3, Sect. 3.2. However, because interfacial interactions induced by surfactants are critical in the present simulation, $\left(\Omega_i^k\right)^{(3)}$ adopts Tölke's method [6] to achieve a thinner interface.

$$\max_{j^{k'''}} \mathbf{C}\cdot j^{k'''} = \max_{f_i^{k'''}} \mathbf{C}\cdot\sum_i \mathbf{e}_i f_i^{k'''}, \tag{6.2}$$

where $f_i^{k'''}$ is the density distribution after third operator. $\mathbf{C}$ is a color gradient, which is defined by

$$\mathbf{C} = \frac{1}{\Delta tc}\sum_i \mathbf{e}_i\left(\rho^r(\mathbf{x}+\Delta t\mathbf{e}_i, t) - \rho^b(\mathbf{x}+\Delta t\mathbf{e}_i, t)\right) \tag{6.3}$$

The following presents Tölke's method implemented in pseudocode.

```
//Tölke's method
for all (x,y,z) do
  if (x,y,z) is fluid node then
   sumCx = sumCy = sumCz = 0.0
    for all i do
     xtmp = nx[x][y][z][i]
     ytmp = ny[x][y][z][i]
     ztmp = nz[x][y][z][i]
     sumCx += (rho[xtmp][ytmp][ztmp][1] - rho[xtmp][ytmp][ztmp][2]) *
ex[i]
     sumCy += (rho[xtmp][ytmp][ztmp][1] - rho[xtmp][ytmp][ztmp][2]) *
ey[i]
     sumCz += (rho[xtmp][ytmp][ztmp][1] - rho[xtmp][ytmp][ztmp][2]) *
ez[i]
    end for
    Cx[x][y][z] = sumCx
    Cy[x][y][z] = sumCy
    Cz[x][y][z] = sumCz
```

```
  end if
end for

for all (x,y,z) do
  if (x,y,z) is fluid node then
    for i = 1; i < Q; i += 2 do
     Cdotc = Cx[x][y][z] * ex[i] + Cy[x][y][z] * ey[i] + Cz[x][y][z] * ez
[i]
      tmp = Cdotc * (1.0 / CI) * min(f[x][y][z][i][1], f[x][y][z][i + 1]
[1],
      f[x][y][z][i][2], f[x][y][z][i + 1][2])
      f[x][y][z][i][1] += tmp
      f[x][y][z][i + 1][1] += tmp
      f[x][y][z][i][2] += tmp
      f[x][y][z][i + 1][2] += tmp
    end for
  end if
end for
```

### 6.2.2 Surfactant Transport

Surfactants are commonly added to ensure droplet stability in emulsions and double emulsions. They lower the surface tension, and modeling surfactants is essential for understanding the rheology of an emulsion system. The convection–diffusion equation for an insoluble surfactant is as follows [7].

$$\partial_t \Gamma + \nabla_s \cdot (\mathbf{u}_s \Gamma) + k \Gamma u_n = \alpha_s \nabla^2 \Gamma \tag{6.4}$$

where $\Gamma$ denotes the surfactant concentration, $\mathbf{u}_s$ is the tangential velocity, $k$ is the curvature, $u_n$ is the normal velocity, and $\alpha_s$ is the surfactant diffusivity.

As surfactant is added to a droplet, the surface tension decreases. To express the relationship between surfactant concentration and surface tension, the following Langmuir nonlinear equation [8] is commonly employed.

$$\sigma_s = \sigma_0 + RT\Gamma_\infty \ln(1 - \Gamma/\Gamma_\infty) \tag{6.5}$$

where $\sigma_s$ is the contaminated surface tension, $\sigma_0$ is the surface tension in the absence of surfactants, $R$ is the universal gas constant, and $\Gamma_\infty$ is the saturation concentration of surfactants. In addition to the Langmuir nonlinear equation, the relationship between surfactant concentration and surface tension can also be derived through molecular dynamics simulations [9].

$$\sigma_s/\sigma_0 = 0.9901 - 0.1621(\Gamma/\Gamma_\infty) - 1.0003(\Gamma/\Gamma_\infty)^2 \tag{6.6}$$

The surfactant convection–diffusion equation, (6.4), can be formulated in a three-dimensional domain as follows [10].

$$\begin{aligned}&\partial_t\Gamma + C_1\partial_x\Gamma + C_2\partial_y\Gamma + C_3\partial_z\Gamma + C_4\Gamma + C_5\partial_{xx}\Gamma + C_6\partial_{yy}\Gamma + C_7\partial_{zz}\Gamma\\&\quad + C_8\partial_{xy}\Gamma + C_9\partial_{zx}\Gamma + C_{10}\partial_{yz}\Gamma = 0\end{aligned} \tag{6.7}$$

$C_1$ through $C_{10}$ are defined as follows. First, $C_1$, $C_2$, and $C_3$ are

$$C_1 = u_{sx}, C_2 = u_{sy}, C_3 = u_{sz}. \tag{6.8}$$

The tangential velocity is given by

$$\begin{aligned}u_{sx} &= u_x - n_x^2u_x - n_xn_yu_y - n_zn_xu_z,\\u_{sy} &= u_y - n_xn_yu_x - n_y^2u_y - n_yn_zu_z,\\u_{sz} &= u_z - n_zn_xu_x - n_yn_zu_y - n_z^2u_z.\end{aligned} \tag{6.9}$$

$C_4$ is defined as

$$\begin{aligned}C_4 = {}&\left(n_y^2 + n_z^2\right)\partial_xu_{sx} + \left(n_x^2 + n_z^2\right)\partial_yu_{sy} + \left(n_x^2 + n_y^2\right)\partial_zu_{sz}\\&- n_xn_y\left(\partial_yu_{sx} + \partial_xu_{sy}\right) - n_zn_x\left(\partial_zu_{sx} + \partial_xu_{sz}\right) - n_yn_z\left(\partial_zu_{sy} + \partial_yu_{sz}\right)\\&+ k\left(u_xn_x + u_yn_y + u_zn_z\right).\end{aligned} \tag{6.10}$$

The curvature $k$ can be calculated as

$$\begin{aligned}k = \nabla_s\cdot\mathbf{n} = {}&\left(n_y^2 + n_z^2\right)\partial_xn_x + \left(n_z^2 + n_x^2\right)\partial_yn_y + \left(n_x^2 + n_y^2\right)\partial_zn_z\\&- n_xn_y\left(\partial_yn_x + \partial_xn_y\right) - n_zn_x\left(\partial_zn_x + \partial_xn_z\right) - n_yn_z\left(\partial_zn_y + \partial_yn_z\right).\end{aligned} \tag{6.11}$$

$C_5$ through $C_{10}$ are parameters associated with surfactant diffusion and are defined as follows

$$\begin{aligned}&C_5 = \left(n_x^2 - 1\right)\alpha_s, C_6 = \left(n_y^2 - 1\right)\alpha_s, C_7 = \left(n_z^2 - 1\right)\alpha_s,\\&C_8 = 2n_xn_y\alpha_s, C_9 = 2n_zn_x\alpha_s, C_{10} = 2n_yn_z\alpha_s.\end{aligned} \tag{6.12}$$

Equation (6.7) can be solved using the hopscotch explicit and unconditionally stable finite-difference scheme. This scheme obtains the solution of the differential equation through two successive sweeps, consisting of an explicit part followed by an implicit part. When $i+j+k+n$ is even, the explicit form of (6.7) for calculating $\Gamma_{i,j,k}^{n+1}$ is

$$\begin{aligned}
&\frac{\Gamma_{i,j,k}^{n+1}-\Gamma_{i,j,k}^{n}}{\Delta t}+C_1\frac{\Gamma_{i+1,j,k}^{n}-\Gamma_{i-1,j,k}^{n}}{2\Delta x}+C_2\frac{\Gamma_{i,j+1,k}^{n}-\Gamma_{i,j-1,k}^{n}}{2\Delta y}\\
&\quad+C_3\frac{\Gamma_{i,j,k+1}^{n}-\Gamma_{i,j,k-1}^{n}}{2\Delta z}+C_4\Gamma_{i,j,k}^{n}+C_5\frac{\Gamma_{i+1,j,k}^{n}-2\Gamma_{i,j,k}^{n}+\Gamma_{i-1,j,k}^{n}}{\Delta x^2}\\
&\quad+C_6\frac{\Gamma_{i,j+1,k}^{n}-2\Gamma_{i,j,k}^{n}+\Gamma_{i,j-1,k}^{n}}{\Delta y^2}+C_7\frac{\Gamma_{i,j,k+1}^{n}-2\Gamma_{i,j,k}^{n}+\Gamma_{i,j,k-1}^{n}}{\Delta z^2}+\frac{C_8}{4\Delta x\Delta y}\\
&\quad\times\left(\Gamma_{i+1,j+1,k}^{n}-\Gamma_{i+1,j-1,k}^{n}-\Gamma_{i-1,j+1,k}^{n}+\Gamma_{i-1,j-1,k}^{n}\right)+\frac{C_9}{4\Delta z\Delta x}\\
&\quad\times\left(\Gamma_{i+1,j,k+1}^{n}-\Gamma_{i+1,j,k-1}^{n}-\Gamma_{i-1,j,k+1}^{n}+\Gamma_{i-1,j,k-1}^{n}\right)+\frac{C_{10}}{4\Delta y\Delta z}\\
&\quad\times\left(\Gamma_{i,j+1,k+1}^{n}-\Gamma_{i,j+1,k-1}^{n}-\Gamma_{i,j-1,k+1}^{n}+\Gamma_{i,j-1,k-1}^{n}\right)=0
\end{aligned}\tag{6.13}$$

When $i+j+k+n$ is odd, the implicit form executed during the second sweep is as follows.

$$\begin{aligned}
&\frac{\Gamma_{i,j,k}^{n+1}-\Gamma_{i,j,k}^{n}}{\Delta t}+C_1\frac{\Gamma_{i+1,j,k}^{n+1}-\Gamma_{i-1,j,k}^{n+1}}{2\Delta x}+C_2\frac{\Gamma_{i,j+1,k}^{n+1}-\Gamma_{i,j-1,k}^{n+1}}{2\Delta y}\\
&\quad+C_3\frac{\Gamma_{i,j,k+1}^{n+1}-\Gamma_{i,j,k-1}^{n+1}}{2\Delta z}+C_4\Gamma_{i,j,k}^{n+1}+C_5\frac{\Gamma_{i+1,j,k}^{n+1}-2\Gamma_{i,j,k}^{n+1}+\Gamma_{i-1,j,k}^{n+1}}{\Delta x^2}\\
&\quad+C_6\frac{\Gamma_{i,j+1,k}^{n+1}-2\Gamma_{i,j,k}^{n+1}+\Gamma_{i,j-1,k}^{n+1}}{\Delta y^2}+C_7\frac{\Gamma_{i,j,k+1}^{n+1}-2\Gamma_{i,j,k}^{n+1}+\Gamma_{i,j,k-1}^{n+1}}{\Delta z^2}\\
&\quad+\frac{C_8}{4\Delta x\Delta y}\left(\Gamma_{i+1,j+1,k}^{n+1}-\Gamma_{i+1,j-1,k}^{n+1}-\Gamma_{i-1,j+1,k}^{n+1}+\Gamma_{i-1,j-1,k}^{n+1}\right)+\frac{C_9}{4\Delta z\Delta x}\\
&\quad\times\left(\Gamma_{i+1,j,k+1}^{n+1}-\Gamma_{i+1,j,k-1}^{n+1}-\Gamma_{i-1,j,k+1}^{n+1}+\Gamma_{i-1,j,k-1}^{n+1}\right)+\frac{C_{10}}{4\Delta y\Delta z}\\
&\quad\times\left(\Gamma_{i,j+1,k+1}^{n+1}-\Gamma_{i,j+1,k-1}^{n+1}-\Gamma_{i,j-1,k+1}^{n+1}+\Gamma_{i,j-1,k-1}^{n+1}\right)=0
\end{aligned}\tag{6.14}$$

Finally, the surfactant concentration at the $n+1$ time step is obtained from (6.13) and (6.14). This value is then substituted into (6.6) to calculate the local variation in surface tension. The modified surface tension, $\sigma_s$, is subsequently incorporated into $\left(\Omega_i^k\right)^{(2)}$ in the same manner.

## 6.3 Results

To investigate the interaction between the two interfaces of a surfactant-coated double emulsion, a uniaxial extensional flow is imposed. Three key dimensionless groups are employed for comparison:

- Radius ratio, $\kappa = R_{in}/R_{out}$
- Capillary number, $Ca = \gamma R_{out}\mu/\sigma$
- Péclet number, $Pe = \gamma R_{out}^2/\alpha_s$
- Reynolds number, $Re = \gamma R_{out}^2/\nu$

In these definitions, $R_{in}$ and $R_{out}$ are the radii of the inner and outer droplets, and $\gamma$ denotes the shear rate. The Reynolds number is limited to a maximum value of 0.04, confining the simulation to the creeping-flow regime. All three fluids share a density ratio $\Lambda = \rho^l/\rho^k = 1$ and a viscosity ratio $\lambda = \mu^l/\mu^k = 1$.

The computational domain ($200 \times 200 \times 200\ \text{lu}^3$) contains a continuous phase and concentric outer and inner droplets, with $R_{out} = 40$ lu. Dirichlet conditions are applied at the inlet and outlet, handled with the Hecht–Harting scheme. The imposed velocity field is

$$\mathbf{u}^{\infty} = \gamma \begin{pmatrix} 1 & 0 & 0 \\ 0 & -0.5 & 0 \\ 0 & 0 & -0.5 \end{pmatrix} \cdot \mathbf{x}, \tag{6.15}$$

where $\mathbf{x}$ is the position vector.

At very small $Re$ and $Ca$, both droplets quickly reach steady shapes. Figure 6.2 tracks their deformation index,

$$\mathrm{DI} = (\mathrm{L} - \mathrm{S})/(\mathrm{L} + \mathrm{S}), \tag{6.16}$$

with L and S the major and minor axes. The DI for both droplets stabilizes over time. The inner droplet first stretches in the same direction as the outer one, but eventually takes on a negative DI, indicating reverse deformation. This switch is caused by vortices that form inside the droplet as the internal flow develops.

Figure 6.3 illustrates the corresponding streamline patterns: panel (a) captures the initial inertia-dominated stretching, whereas panel (b) shows the subsequent vortex-induced reversal of the inner droplet's deformation.

**Exercise 6.2 Dimensionless Groups and Flow Regime**
In this chapter, the following dimensionless numbers are introduced: radius ratio $\kappa$, capillary number $Ca$, Péclet number $Pe$, and Reynolds number $Re$.

(a) Briefly interpret the physical meaning of each dimensionless group in the context of a surfactant-covered double emulsion under uniaxial extensional flow.
(b) The simulations are performed with $Re \leq 0.04$. Explain what this implies about the governing flow regime and which terms in the Navier–Stokes equations are negligible.

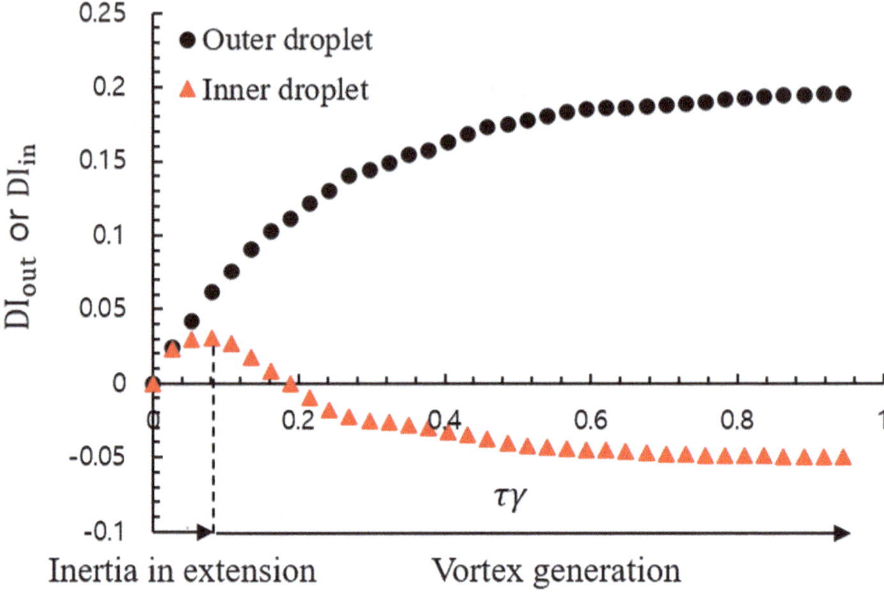


**Fig. 6.2** Time evolution of the deformation of clean outer and inner droplets. Droplets maintain a steady shape below $Ca_{cr}$ ($\kappa = 0.5$, $Ca = 0.09$)

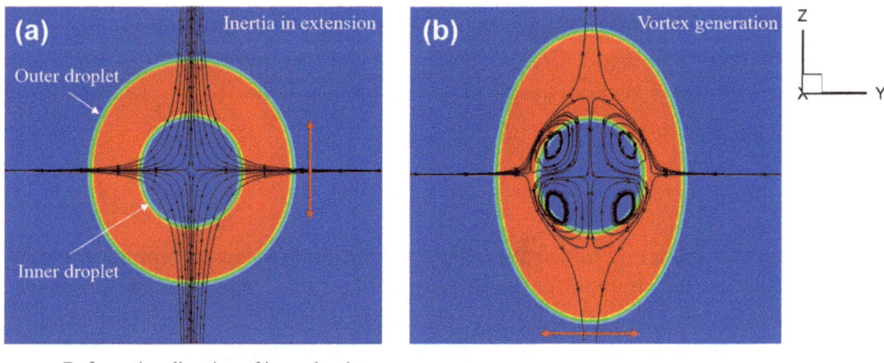


**Fig. 6.3** The streamlines in the region of (**a**) the inertia of extension and (**b**) the vortex generation region

**Solution**

(a) $\kappa$: measures the relative size of the inner droplet to the outer droplet and determines the thickness of the annular gap and the strength of interfacial coupling.

*Ca*: ratio of viscous forces to capillary forces; higher *Ca* corresponds to stronger deformation for the same surface tension.

*Pe*: ratio of convective surfactant transport to diffusive transport; high *Pe* implies convection-dominated surfactant redistribution and greater nonuniformity.

Re: ratio of inertial forces to viscous forces; small *Re* indicates creeping-flow (Stokes) conditions.

(b) Since $Re \leq 0.04$, the simulations lie in the creeping-flow regime where inertia is negligible. In the Navier–Stokes equations, the convective term $\rho(\mathbf{u} \cdot \nabla)\mathbf{u}$ is small compared with the pressure gradient and viscous terms and can be neglected. The flow is thus governed by a quasi-steady balance between viscous and interfacial forces, which simplifies both the physics and the numerical treatment.

### 6.3.1 Model Validation

Validation of the multiphase "clean" double-emulsion model is carried out by comparing its predictions with the numerical data of Stone and Leal [11] (Fig. 6.4). The radius ratio is fixed at $\kappa = 0.5$, and both the viscosity and surface tension ratios are set to unity for all three fluids. In Fig. 6.4, blue circles and triangles represent the outer-droplet deformation index, $DI_{out}$, and the absolute inner-droplet deformation index, $|DI_{in}|$, obtained from the present model. The upper and lower solid curves show $DI_{out}$ and $|DI_{in}|$ reported by Stone and Leal, while red dashed curves correspond to the analytical solutions. Excellent agreement is observed between the present results and the reference data.

The inset on the right of Fig. 6.4 depicts the steady shapes of the double emulsion at different capillary numbers. Once the capillary number exceeds the critical value, $Ca_{cr} \approx 0.11$, both droplets continue to elongate.

### 6.3.2 Effects of Surfactant Transport

The influence of surfactant convection–diffusion on the interaction between the two interfaces of a double emulsion was examined by fixing the capillary number at Ca = 0.06 and varying the Péclet number (Pe). The initial surfactant coverage was $c_{in} = \Gamma_{in}/\Gamma_{\infty} = 0.4$, where $\Gamma_{in}$ is the initial surfactant concentration. Figure 6.5 presents the deformation indices (DI) of both droplets as Pe changes. Increasing Pe lowers the diffusivity $\alpha_s$, so convection becomes the dominant transport mechanism. Two measures were considered: the absolute DI (Fig. 6.5a) and the DI relative to the Pe = 50 case (Fig. 6.5b). The outer droplet's shape shows minimal dependence on Pe, whereas the inner droplet's deformation falls by roughly 40% at high Pe, implying that an uneven surfactant distribution enhances its shape stability.

To clarify how surfactant distribution affects this stability, the XY-plane velocity field between the two interfaces was analyzed (Fig. 6.6). Figure 6.6b illustrates the

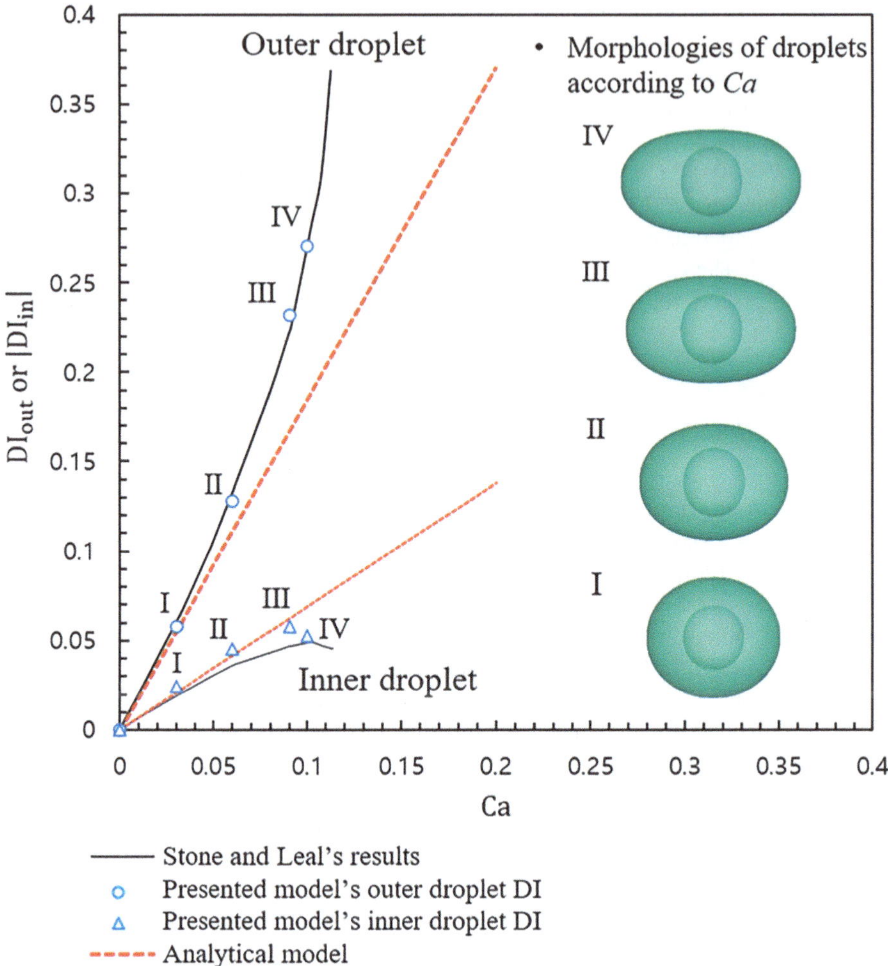


**Fig. 6.4** Deformation indices of a clean double emulsion subjected to uniaxial extensional flow. The panel on the right displays the morphology of the double emulsion at various capillary numbers

field when only the outer interface carries surfactant; for comparison, a clean-interface case with the same average surface tension is also simulated. Surfactant on the outer surface induces vortices in the annular gap, which counteract inward deformation of the inner droplet along its minor axis.

Figure 6.7 illustrates how vortices form between the interfaces. On the outer droplet's XY-plane, stagnation points appear along the inflow axis, limiting convective surfactant transport at those locations (Fig. 6.7a). Elsewhere, surfactant migrates toward the droplet tips, yielding the non-uniform distribution seen in Fig. 6.6c. This uneven coverage sets up a surface tension gradient pointing toward regions of lower

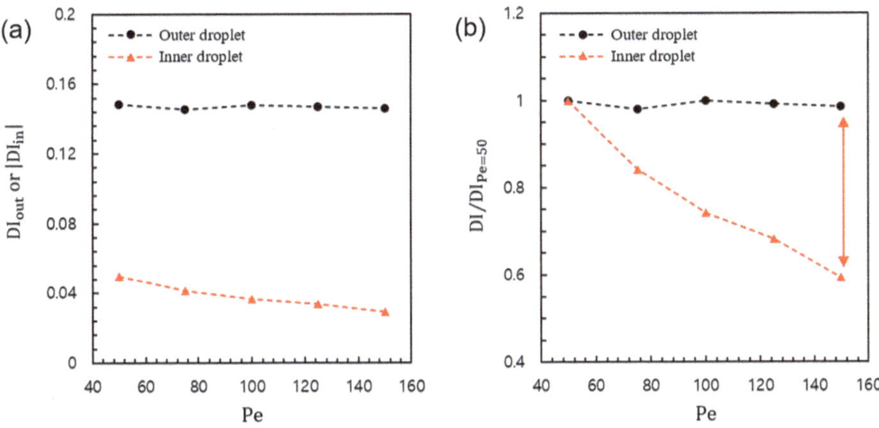


**Fig. 6.5** Variation of deformation index with Péclet number at $Ca = 0.06$. (**a**) Absolute deformation indices. (**b**) Deformation indices normalized by the $Pe = 50$ case. The inner droplet shows strong sensitivity to Pe, whereas the outer droplet is only weakly affected

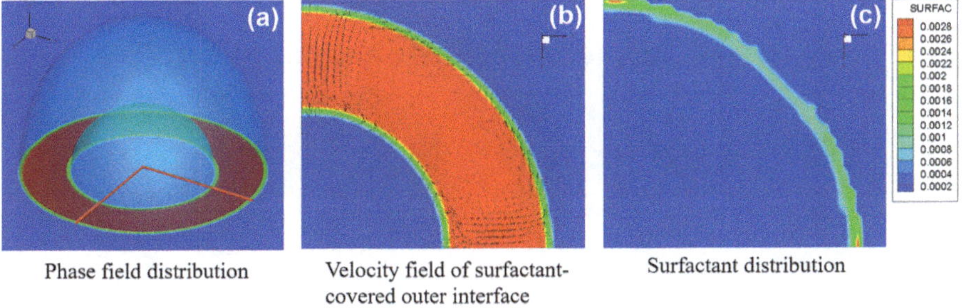


**Fig. 6.6** (**a**) Phase-field view and XY-plane cross-section of the double emulsion. (**b**) Velocity field when surfactant is present only on the outer interface. (**c**) Corresponding surfactant concentration distribution for the case shown in panel (**b**)

concentration. The resulting Marangoni force drives a vortex in the annular gap. The Marangoni stress produced by the surface tension gradient is

$$-\nabla_s \sigma_s = -\partial_\Gamma \sigma_s \cdot \nabla_s \Gamma. \tag{6.17}$$

Because $\partial \sigma_s / \partial \Gamma$ grows with the Péclet number, higher Pe values intensify the Marangoni stress and thus reduce the inner droplet's deformation. The full stability mechanism is summarized in Fig. 6.7.

Figure 6.8 plots the absolute deformation index of the inner droplet, $|DI_{in}|$, for four surfactant configurations: (I) surfactant on the inner interface only, (II) clean interfaces, (III) surfactant on both interfaces, and (IV) surfactant on the outer interface only. Applying surfactant to the outer interface (case IV) lowers $|DI_{in}|$, whereas placing surfactant on the inner interface (case I) raises it. Hence, the

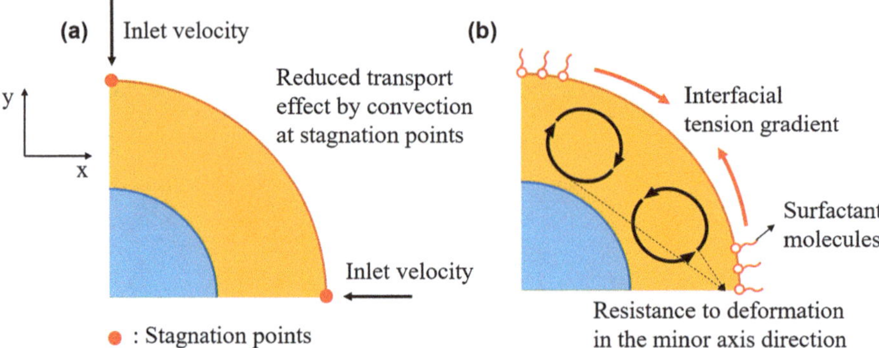


**Fig. 6.7** Illustration of how surfactant distribution affects double-emulsion deformation. (**a**) An imposed extensional flow creates stagnation points along the outer interface, where convective surfactant transport is minimal. (**b**) Uneven surfactant coverage sets up an surface tension gradient, driving a Marangoni-induced vortex in the XY plane; the resulting flow counters inward deformation of the inner droplet along its minor axis

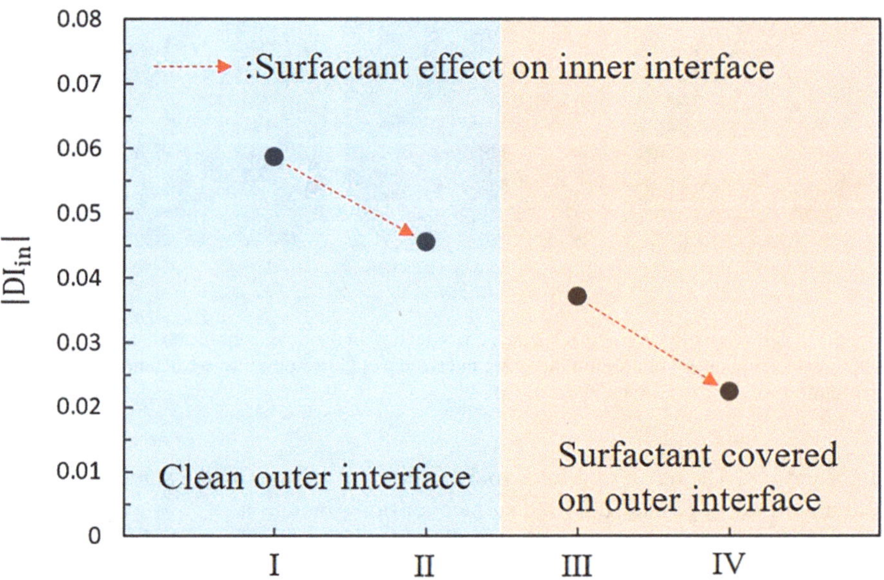


**Fig. 6.8** Variation of the absolute inner-droplet deformation index ($|DI_{in}|$) for four surfactant configurations. Applying surfactant to the outer interface (IV) induces vortices that reduce $|DI_{in}|$. In contrast, surfactant confined to the inner interface (I) concentrates near the surface, increasing $|DI_{in}|$

Marangoni force can either suppress or enhance droplet deformation, depending on which interface carries the surfactant. As discussed earlier, non-uniform coverage of the outer surface induces vortices in the annular gap, stabilizing the inner droplet. In contrast, surfactant on the inner surface produces a Marangoni force that opposes those vortices, leading to greater deformation.

**Exercise 6.3 Péclet Number Effects and Marangoni Stabilization**
At fixed $Ca = 0.06$, the $Pe$ is varied by changing the surfactant diffusivity.

(a) Explain how increasing $Pe$ changes the relative importance of convection and diffusion in the surfactant transport.
(b) According to the results, the outer droplet is only weakly affected by $Pe$, whereas the inner droplet's deformation decreases by roughly 40% at large $Pe$. Provide a mechanistic explanation for this difference.
(c) Using the Marangoni stress expression (6.17), discuss why larger $Pe$ tends to reduce inner-droplet deformation.

**Solution**

(a) Increasing $Pe$ corresponds to decreasing the surfactant diffusivity. Thus, at low $Pe$, diffusion is relatively strong compared with convection, so surfactant gradients are smoothed and the interface remains closer to uniform. At high $Pe$, diffusion is weak and convection dominates, leading to strong surfactant accumulation at certain locations (e.g., droplet tips) and pronounced nonuniform coverage.
(b) Figure 6.5 shows that the outer droplet is only weakly affected by $Pe$, whereas the inner droplet exhibits about a 40% reduction in deformation at large $Pe$. Mechanistically, surfactant resides on the outer interface, where convection and stagnation points combine to produce a nonuniform coverage. This nonuniformity generates Marangoni stresses that drive a vortex in the annular gap. The vortex primarily alters the flow near the inner interface, opposing the minor-axis compression of the inner droplet.

    Thus, even though the surfactant is applied to and transported along the outer interface, the induced flow in the gap strongly influences the inner droplet, making its deformation highly sensitive to changes in $Pe$, while the outer droplet shape changes only slightly.
(c) The Marangoni stress is proportional to the surface tension gradient. At higher $Pe$, Convection creates larger surfactant-concentration gradients along the outer interface. This leads to stronger surface tension gradients and hence larger Marangoni stresses. The Marangoni-driven vortex in the annular gap becomes more intense and more effectively counteracts the extensional flow that would otherwise compress the inner droplet.

As a result, increasing $Pe$ reduces the deformation of the inner droplet by enhancing Marangoni stabilization.

### *6.3.3 Capillary Number Effects*

To assess how capillary number (Ca) governs double-emulsion deformation and surfactant transport, the shear rate was varied while the Péclet number was held at $Pe = 100$ by adjusting the diffusion coefficient $\alpha_s$. Figure 6.9a plots the deformation

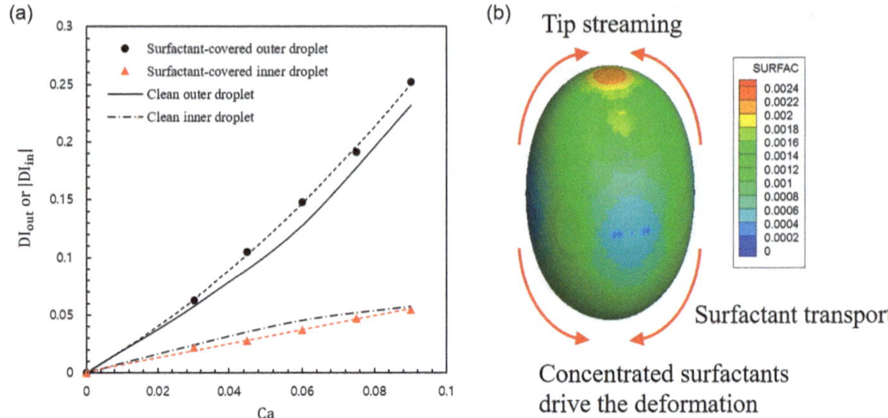


**Fig. 6.9** (**a**) Deformation indices of the inner and outer droplets in a surfactant-laden double emulsion. Relative to the clean case, outer-droplet deformation increases, whereas inner-droplet deformation decreases. (**b**) Surfactant concentration along the outer interface; convection-driven accumulation at the tips lowers local interfacial tension and enhances outer-droplet elongation

indices for a surfactant-laden system with $c_{in} = 0.4$: black circles denote the outer droplet and red triangles the inner droplet. Clean-interface results are included for reference. To isolate the effect of non-uniform coverage, the initial interfacial tension was set equal to that of the clean case.

The outer droplet exhibits greater deformation than its clean counterpart owing to tip-streaming: convection drives surfactant toward the poles (Fig. 6.9b), locally lowering interfacial tension and enhancing major-axis elongation. By contrast, although surfactant likewise accumulates at the tips of the inner droplet, its overall deformation decreases, consistent with Marangoni-induced vortices that oppose minor-axis contraction.

Figure 6.10 shows interfacial nonuniformity of the dimensionless surfactant concentration and corresponding contours. $\Delta\Gamma^* = \Gamma^*_{\text{max}} - \Gamma^*_{\text{min}}$ denotes the span of the dimensionless interfacial concentration. With Pe fixed, increasing Ca accelerates convective transport along the outer interface, enlarging $\Delta\Gamma^*$. In contrast, Ca exerts little influence on the inner interface. Surfactant accumulates at the two major-axis tips of the outer droplet, whereas four lobes appear along the minor-axis directions of the inner droplet (Fig. 6.10b), consistent with vortex-driven redistribution between the interfaces (see Fig. 6.3).

**Exercise 6.4 Capillary Number Effects and Tip Streaming**

This section investigates how increasing *Ca* at fixed *Pe* affects deformation and surfactant distribution.

(a) Explain why increasing *Ca* at constant *Pe* enhances convective transport along the outer interface.
(b) Describe how tip streaming modifies the surfactant distribution along the outer interface and how this leads to increased outer-droplet deformation.

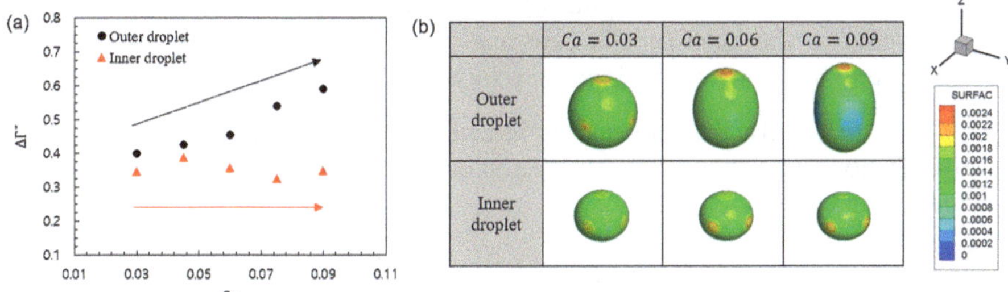


**Fig. 6.10** (**a**) Span of the dimensionless interfacial surfactant concentration, $\Delta\Gamma^* = \Gamma^*_{max} - \Gamma^*_{min}$, for the outer and inner droplets as a function of Ca. (**b**) Surfactant contours on both interfaces. With $Pe = 100$ held fixed, increasing Ca accelerates convective transport along the outer interface, enlarging $\Delta\Gamma^*$ (the inner interface is only weakly affected)

(c) Despite surfactant accumulation at the inner-droplet tips, the inner droplet deforms less as *Ca* increases. Explain this apparent paradox in terms of flow structure and Marangoni effects.

**Solution**

(a) At fixed *Pe*, increasing the capillary number *Ca* corresponds to increasing the extensional rate. A larger extensional rate strengthens the interfacial shear and extension, which in turn:

- Increases the tangential velocity along the outer interface,
- Enhances convective transport of surfactant relative to diffusion (even though *Pe* is fixed, the absolute convection level is higher),
- Promotes stronger migration of surfactant toward the droplet tips where the flow converges.

(b) Tip streaming refers to the convective accumulation of surfactant and fluid at the droplet tips under strong extensional forcing. In the outer droplet: Surfactant is swept toward the poles, leading to a local decrease in surface tension at the tips. This locally reduced surface tension reduces capillary resistance to elongation along the major axis. Consequently, the outer droplet elongates more than in the clean case, as shown in Fig. 6.9a.

Thus, tip streaming creates a positive feedback: more surfactant at the tips → lower local tension → easier elongation → further enhancement of surfactant accumulation and deformation.

(c) Although surfactant also accumulates at the inner-droplet tips, the inner deformation decreases with increasing *Ca*. This is explained by the flow structure:

- The nonuniform surfactant distribution on the outer interface sets up strong Marangoni-driven vortices in the annular gap,
- These vortices act to oppose the minor-axis compression of the inner droplet, effectively stabilizing its shape,

- The net effect of the strengthened Marangoni circulation at higher $Ca$ is a reduction in the inner-droplet deformation, even though local surface tension at certain locations on the inner interface may be lower.

In other words, the global flow induced by Marangoni stresses dominates over the local tendency for tip elongation on the inner droplet, yielding an overall stabilizing effect.

### 6.3.4 Radius Ratio Effects

Figure 6.11 examines how the radius ratio $\kappa$ influences droplet deformation and surfactant distribution. In all cases, the outer droplet has $Ca = 0.06$ and $Pe = 100$. As $\kappa$ increases, the annular gap between the two interfaces narrows; at $\kappa = 0.7$ the interfaces come into contact. Compression of the inner flow region raises viscous stresses on both interfaces, and the inner droplet behaves increasingly like a rigid sphere. Accordingly, overall deformation grows with $\kappa$ (Fig. 6.11a). The shortened circulation path also reduces the tangential interfacial velocity, slowing surfactant transport. Consistent with this, the outer interface concentration span $\Delta\Gamma^*_{out}$ decreases as the inner droplet enlarges (Fig. 6.11b). At $\kappa = 0.7$, $\Delta\Gamma^*_{out}$ increases again because the interfaces begin to touch. Although suppression of coalescence between surfactant-laden interfaces is not modeled here, the results suggest that reducing the interface spacing can facilitate inner-droplet release.

**Exercise 6.5 Radius Ratio and Interfacial Coupling**
Figure 6.11 investigates the influence of the radius ratio on deformation and surfactant transport.

(a) Describe how increasing $\kappa$ (with fixed outer radius) changes the annular gap and the viscous stresses on the interfaces.

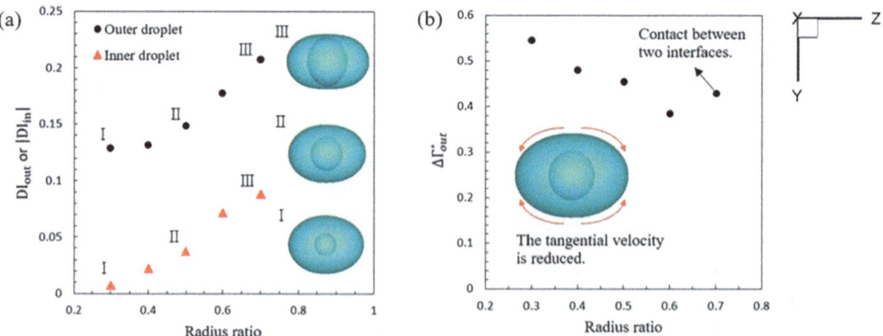


**Fig. 6.11** (**a**) Deformation indices of the inner and outer droplets; (**b**) span of the dimensionless surfactant concentration on the outer interface ($\Delta\Gamma^*_{out}$), both plotted against the radius ratio $\kappa$

(b) Explain why deformation generally increases with $\kappa$ and why the outer interface surfactant-concentration span first decreases and then increases again as $\kappa$ goes to 0.75.
(c) Discuss how these results might guide the design of double emulsions in applications where controlled inner-droplet release is desired.

**Solution**
(a) When the outer radius is fixed, increasing the radius ratio enlarges the inner droplet and consequently narrows the annular gap separating the inner and outer interfaces. As the gap becomes thinner, the fluid inside this region undergoes stronger confinement, which increases velocity gradients and thereby amplifies viscous stresses acting on both interfaces. Under these conditions, the inner droplet effectively behaves more like a rigid inclusion, transferring stronger stresses to the outer interface and intensifying the overall interfacial coupling as radius ratio increases.
(b) The deformation of the double-emulsion system tends to increase with $\kappa$ because the narrowing of the annular gap enhances viscous stresses and strengthens interfacial coupling. Meanwhile, the surfactant concentration span on the outer interface exhibits a non-monotonic response. For moderate values of $\kappa$, the reduced tangential velocity along the outer interface weakens surfactant transport, leading to a temporary decrease in concentration nonuniformity. However, as $\kappa$ approaches approximately 0.75, the inner and outer interfaces come into close proximity, and this strong confinement induces sharp gradients in surfactant distribution once again, causing the span to increase. Thus, span first decreases and then rises at high $\kappa$ due to the competing effects of weakened transport and intensified geometric confinement.
(c) These observations provide meaningful guidance for the design of double-emulsion systems requiring controlled inner-droplet release. A larger $\kappa$, corresponding to a very thin annular gap, increases viscous stresses and strengthens interfacial interactions, thereby making the system more deformable and more sensitive to flow-induced perturbations. Such configurations may facilitate situations where the inner droplet approaches the outer interface or can be released under external forcing. Consequently, adjusting $\kappa$ allows designers to balance stability and release characteristics: smaller $\kappa$ values help preserve structural stability during processing, while larger $\kappa$ values can promote release in applications such as drug delivery or encapsulation.

## 6.4 Conclusions

This chapter examined interface–interface interactions in a surfactant-covered double emulsion under uniaxial extensional flow using a three-dimensional color-gradient lattice Boltzmann framework coupled to an insoluble-surfactant convection–

diffusion model that mapped local coverage to surface tension variation. A clean double-emulsion baseline was first validated against canonical results from Stone and Leal. With surfactant present, outer-droplet deformation increased due to tip streaming and poleward accumulation that lowered local interfacial tension, whereas inner-droplet deformation decreased as Marangoni-driven vortices arose in the annular gap and resisted minor-axis compression. At fixed Péclet number, raising the capillary number intensified nonuniform coverage on the outer interface but had limited influence on the inner interface; increasing Péclet number (stronger convection) further suppressed inner-droplet deformation. The radius ratio governed both deformation and transport: a narrower gap elevated viscous stresses, increased overall deformation, reduced tangential interfacial velocity and thus surfactant transport, and could reintroduce strong gradients as the interfaces approached contact.

Taken together, the results illustrate how to simulate three-phase droplet mechanics with a color-gradient LBM, how to couple surfactant transport to interfacial tension changes, and how capillary number, Péclet number, radius ratio, and interface-selective surfactant placement operate as practical design levers to stabilize or destabilize droplets in complex emulsion systems.

**Summary**

1. **Physics of surfactant-covered double emulsions**
    - Double emulsions consist of an inner droplet encapsulated by an outer droplet (e.g., W/O/W or O/W/O), with applications in drug delivery, food encapsulation, cosmetics, and biomedical engineering.
    - Their stability depends on interfacial tension at both inner and outer interfaces; surfactants adsorb on these interfaces, lowering surface tension and preventing film rupture.
    - Nonuniform surfactant distribution under flow induces Marangoni stresses, altering droplet deformation and internal circulation.
2. **Numerical framework for three-phase surfactant-laden systems**
    - The chapter employs a three-phase color-gradient LBM to model concentric droplets with immiscible phases.
    - Surfactant transport is modeled through an insoluble-surfactant convection–diffusion equation, capturing tangential flow, curvature effects, and redistribution through a Langmuir-type surface tension law.
    - A hopscotch explicit–implicit finite-difference method solves the 3D convection–diffusion equation on the droplet interfaces.
3. **Key non-dimensional control parameters**
    - Radius ratio ($\kappa$) controls annular gap thickness and strength of interface–interface coupling.
    - Capillary number ($Ca$) characterizes the balance of viscous and capillary forces and governs deformation and tip-streaming processes.

- Péclet number ($Pe$) determines the relative strength of surfactant convection vs. diffusion, influencing gradient formation and Marangoni stresses.
- Reynolds number ($Re$) is kept $< 0.04$, ensuring creeping-flow conditions where inertia is negligible.

4. **Surfactant transport effects and Marangoni-induced flow**
   - Higher $Pe$ suppresses diffusion, producing stronger surfactant gradients along the outer interface.
   - Surfactant concentration accumulates near droplet tips due to stagnation points and convection-dominated transport.
   - The resulting surface tension gradients generate Marangoni vortices in the annular gap.
   - These vortices oppose inner-droplet minor-axis compression, reducing inner-droplet deformation significantly ($\approx$40%), while the outer droplet remains weakly affected.
5. **Capillary number influence on deformation and surfactant distribution**
   - Increasing $Ca$ increases interfacial convection and enhances tip streaming at the outer interface.
   - Outer-droplet deformation grows due to locally reduced surface tension at the poles.
   - Inner-droplet deformation decreases due to strengthening Marangoni-driven recirculation, despite having local surfactant accumulation of its own.
   - Surfactant nonuniformity increases with $Ca$ on the outer interface, while the inner interface remains only weakly sensitive.
6. **Role of radius ratio on interfacial dynamics**
   - As $\kappa$ increases, the annular gap narrows, enhancing viscous stresses and increasing the overall deformation of both droplets.
   - Reduced tangential velocity suppresses surfactant transport, resulting in smaller concentration variation on the outer interface—until the interfaces nearly touch.
   - Near $\kappa \approx 0.75$, strong confinement reintroduces large surfactant gradients, suggesting the potential for interface contact and inner-droplet release.

## References


1. A. S. Utada, E. Lorenceau, D. R. Link, P. D. Kaplan, H. A. Stone, and D. Weitz, "Monodisperse double emulsions generated from a microcapillary device," Science **308**, 537-541 (2005).
2. E. Dickinson, J. Evison, J. W. Gramshaw, and D. Schwope, "Flavour release from a protein-stabilized water-in-oil-in-water emulsion," Food Hydrocolloids **8**, 63-67 (1994).
3. C.-X. Zhao, "Multiphase flow microfluidics for the production of single or multiple emulsions for drug delivery," Advanced drug delivery reviews **65**, 1420-1446 (2013).

4. L. Cima, J. Vacanti, C. Vacanti, D. Ingber, D. Mooney, and R. Langer, "Tissue engineering by cell transplantation using degradable polymer substrates," (1991), 113, 143.
5. J.-C. Baret, "Surfactants in droplet-based microfluidics," Lab on a Chip **12**, 422-433 (2012).
6. J. Tölke, "Lattice boltzmann simulations of binary fluid flow through porous media," Philosophical Transactions of the Royal Society of London. Series A: Mathematical, Physical and Engineering Sciences **360**, 535-545 (2002).
7. W. Milliken, and L. Leal, "Deformation and breakup of viscoelastic drops in planar extensional flows," Journal of non-newtonian fluid mechanics **40**, 355-379 (1991).
8. C. D. Eggleton, T.-M. Tsai, and K. J. Stebe, "Tip streaming from a drop in the presence of surfactants," Physical review letters **87**, 048302 (2001).
9. S. B. Choi, H. M. Yoon, and J. S. Lee, "Multi-scale approach for the rheological characteristics of emulsions using molecular dynamics and lattice boltzmann method," Biomicrofluidics **8**, (2014), 052104.
10. H. Farhat, F. Celiker, T. Singh, and J. Lee, "A hybrid lattice boltzmann model for surfactant-covered droplets," Soft Matter **7**, 1968-1985 (2011).
11. H. A. Stone, and L. G. Leal, "Breakup of concentric double emulsion droplets in linear flows," Journal of Fluid Mechanics **211**, 123-156 (1990).

# Chapter 7
# Numerical Simulation of Gas–Liquid Transport in Porous Media

## 7.1 Introduction

Demand for non-aqueous Li–air batteries is rising because they are environmentally friendly, offer excellent reversibility, and provide higher theoretical gravimetric and volumetric energy densities than Li-ion systems [1]. A typical non-aqueous Li–air cell comprises a lithium anode, a porous air electrode (cathode), and a liquid electrolyte. During discharge, $Li^+$ migrates from the anode through the electrolyte while gaseous $O_2$ dissolves into the electrolyte, diffuses to the air-electrode surface, and participates in oxygen reduction reactions that generate species critical to achieving high discharge capacity. Because conventional non-aqueous electrolytes have low oxygen solubility, enhancing oxygen transport to the reactive sites at the air electrode is essential.

Carbon papers are widely employed as air-electrode substrates. Among them, carbon-nanotube (CNT) fibrous networks resist mechanical stresses arising from the deposition of discharge solids and excel at maintaining an electrically percolating network during discharge [2, 3]. These electrodes behave as porous media, and geometric characteristics—such as porosity—govern oxygen transport. Because oxygen transport is typically evaluated after the liquid electrolyte has already infiltrated the CNT scaffold, analyzing liquid-phase permeability of the air electrode is not required. In two-phase penetration simulations of the air electrode, however, the morphology of the infiltrated liquid must be specified. Reaction products that influence battery efficiency form on solid surfaces; thus, complete surface coverage by the electrolyte is desirable to enhance performance. In practice, surface and fluid properties can prevent the liquid from invading all pores, leaving voids of entrapped gas ("trapped air") [4]. Such entrapment arises naturally in porous media through multiple mechanisms, including geological $CO_2$ sequestration, water infiltration in complex soils, and liquid seepage through sediments.

 
H. M. Lee, J. S. Lee, *Lattice Boltzmann Methodology for Single-Phase and Multiphase Nanoparticle Modeling*,
https://doi.org/10.1007/978-981-95-9117-6_7

LBM is specialized for computing flow in porous media with complex geometries, such as CNT networks, owing to the bounce-back boundary scheme. Accordingly, this chapter employs the MRT color-gradient model to establish a high density ratio multiphase flow environment and investigates water transport in CNT structures under variations in viscosity ratio and wettability. Through this, the influence of trapped air within the CNT network is identified and the resulting water transport characteristics are analyzed.

**Exercise 7.1 Gas–Liquid Transport in CNT Porous Media**

(a) Explain why oxygen transport is a limiting factor in non-aqueous Li–air batteries.
(b) Describe how trapped air forms within CNT-based porous media and why it degrades oxygen transport performance.
(c) Discuss how CNT network geometry helps maintain electrochemical performance during discharge.

**Solution**

(a) In non-aqueous Li–air batteries, oxygen must be transported from the gas reservoir through the porous cathode and electrolyte to the reaction sites where the oxygen reduction reaction (ORR) occurs. If oxygen supply is insufficient, the local reaction rate is limited by mass transfer rather than by intrinsic electrochemistry. This leads to concentration polarization, reduced discharge voltage, and early capacity fade. Therefore, oxygen transport through the porous electrode—especially within complex CNT networks filled with electrolyte—is a key rate-limiting factor for overall battery performance.
(b) In non-aqueous Li–air batteries, oxygen must be transported from the gas reservoir through the porous cathode and electrolyte to the reaction sites where the oxygen reduction reaction (ORR) occurs. If oxygen supply is insufficient, the local reaction rate is limited by mass transfer rather than by intrinsic electrochemistry. This leads to concentration polarization, reduced discharge voltage, and early capacity fade. Therefore, oxygen transport through the porous electrode—especially within complex CNT networks filled with electrolyte—is a key rate-limiting factor for overall battery performance.
(c) Despite the presence of trapped air, the CNT network geometry provides continuous, interconnected pathways that help maintain electrochemical performance. The high aspect ratio and percolating structure of CNTs yield a mechanically robust and electrically conductive skeleton, which supports electron transport even when some pores are partially filled or blocked. In addition, the hierarchical porous structure (combining larger channels with smaller pores) can preserve alternative gas pathways around trapped-air zones, so that oxygen can still reach reaction sites. This combination of structural connectivity and redundancy allows the CNT network to tolerate a certain degree of saturation heterogeneity while sustaining acceptable discharge performance.

## 7.2 Numerical Methods

### 7.2.1 3D MRT Color-Gradient Model

To ensure stable simulations under high density and viscosity ratios, the 3D MRT color-gradient model described in Chap. 3, Sect. 3.3 ("Improved color-gradient LBM") is employed. The configuration of the improved color-gradient model is briefly revisited. The single-phase collision operator, $\left(\Omega_i^k\right)^{(1)}$ is given as follows [5].

$$\left(\Omega_i^l\right)^{(1)} = \mathbf{M}^{-1}\left[-\mathbf{S}\left(\mathbf{m}^l - \mathbf{m}^{l,eq}\right) + \Delta t\left(\mathbf{I} - \frac{\mathbf{S}}{2}\right)\mathbf{C}^l\right], \tag{7.1}$$

where $\mathbf{M}$ is the transformation matrix, $\mathbf{I}$ is the unit matrix, and $\mathbf{S}$ is the relaxation rate matrix, and $\mathbf{C}^l = \mathbf{M}\mathbf{G}^l$ is the error correction term. The perturbation operator, $\left(\Omega_i^l\right)^{(2)}$, can be applied in the same manner as in the SRT LBM model; however, to maintain methodological consistency with the MRT model, it is implemented as follows [5].

$$\left(\Omega_i^l\right)^{(2),MRT} = \mathbf{M}^{-1}\mathbf{S}\mathbf{M}\left(\Omega_i^l\right)^{(2)}, \tag{7.2}$$

Finally, the recoloring operator to ensure a thin fluid interface is applied as follows [6].

$$\left(\Omega_i^R\right)^{(3)} = \frac{\rho_R}{\rho}\left(f_i^R + f_i^B\right) + \beta\frac{\rho_R\rho_B}{\rho^2}\cos(\varphi_i)\sum_k f_i^{l,eq}(\rho_l, \alpha_l, \mathbf{u} = 0), \tag{7.3}$$

$$\left(\Omega_i^B\right)^{(3)} = \frac{\rho_B}{\rho}\left(f_i^R + f_i^B\right) - \beta\frac{\rho_R\rho_B}{\rho^2}\cos(\varphi_i)\sum_k f_i^{k,eq}(\rho_l, \alpha_l, \mathbf{u} = 0), \tag{7.4}$$

where $\beta$ is a parameter between 0 and 1 that controls the thickness of the interface.

## 7.3 Results

### 7.3.1 Model Validation

To validate the LBM model, a simulation was performed on a porous medium composed of a spherical bed with porosity $\varepsilon = 0.55$ and a contact angle of 180°. Periodic boundary conditions were applied, and two-phase flow was induced by a body force. Capillary and Reynolds numbers were on the order of $10^{-6}$ and $10^{-4}$, ensuring flow conditions dominated by surface tension. The body force term $F_l$, following Ba et al. [7] and Guo et al. [8], was included in the collision operator. Once

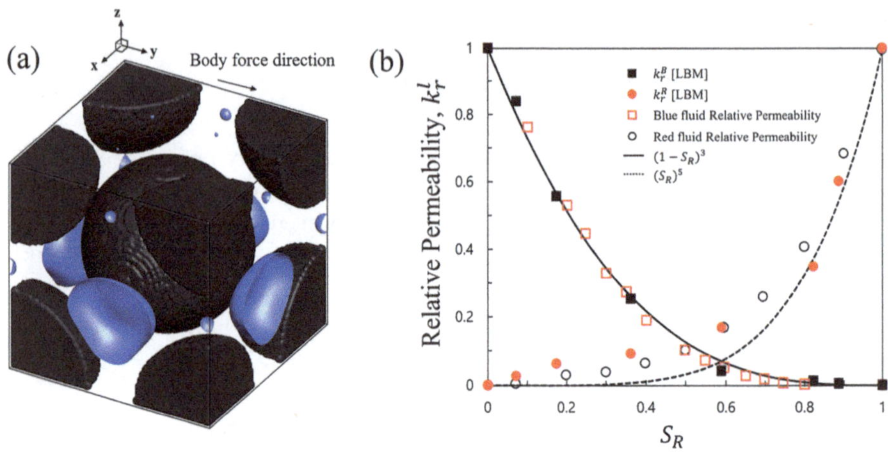


**Fig. 7.1** Validation results. (**a**) Simulation domain used for evaluating two-phase flow in porous media. (**b**) Comparison of relative permeability predicted by the color-gradient LBM with experimental data from Bryant and Blunt [9], showing good agreement across varying red fluid saturation levels

the system reached steady state, relative permeabilities of the two fluids were calculated and compared across different saturation levels. Based on the simulation results, the relative permeabilities of the red and blue fluid phases were calculated and compared across different saturation levels.

$$u^l\mathbf{u} = \frac{k^l k_r^l}{\mu^l}\left(\nabla p^l + F^l\right), \tag{7.5}$$

where $u^l$ is the fluid velocity, $k^l$ is the absolute permeability, and $k_r^l$ is the relative permeability.

As shown in Fig. 7.1b, the simulated trend of relative permeability with respect to the initial red fluid saturation shows good agreement with the experimental data reported by Bryant and Blunt [9]. Here, the red fluid saturation $S_R$ refers to the volume fraction of the porous medium occupied by the red fluid phase.

### 7.3.2 *Simulation Setup*

The CNT paper structure used in the LBM simulations is shown in Fig. 7.2. The model geometry was first generated in STL (stereolithography) format, which represents the 3D surface using triangular mesh elements. This surface geometry was created using the commercial software GeoDict (Math2Market, Germany), incorporating experimental inputs such as porosity (0.5–0.7), fiber diameter distribution (15–20 μm), and SEM imagery. Based on this method, several cases were

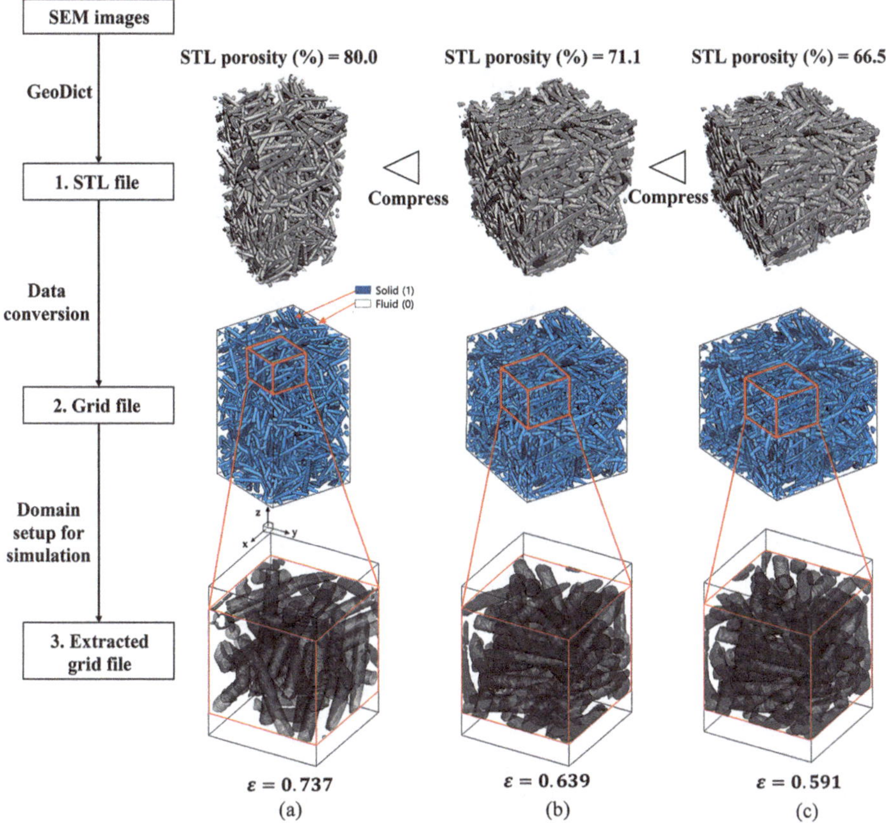


**Fig. 7.2** Data conversion workflow. (**a**) $\varepsilon = 0.737$, (**b**) $\varepsilon = 0.639$, and (**c**) $\varepsilon = 0.591$

constructed to evaluate the effects of porosity and compression along the thickness direction. The final porosity values of the modeled structures were 0.800, 0.711, and 0.665, respectively.

To incorporate CNT structures into LBM simulations, structural reconstruction is essential since the geometry must be defined on a grid. Although the STL format generated by GeoDict provides detailed surface geometry based on triangular meshes, it does not contain volumetric grid data. In contrast, LBM simulations require binary grid files that define all nodes as either fluid or solid. This enables efficient handling of complex geometries.

To address this, a data conversion process was developed to generate grid files from the STL structure by classifying each node as either fluid or solid. Subsections of these grid files were then extracted to define the simulation domain. To determine the appropriate resolution, a grid-independence test was conducted using a CNT structure with porosity $\varepsilon = 0.800$ (see Fig. 7.2a. Three grid resolutions were tested—$34^3$, $100^3$, and $200^3$—each covering the same physical domain. As shown in

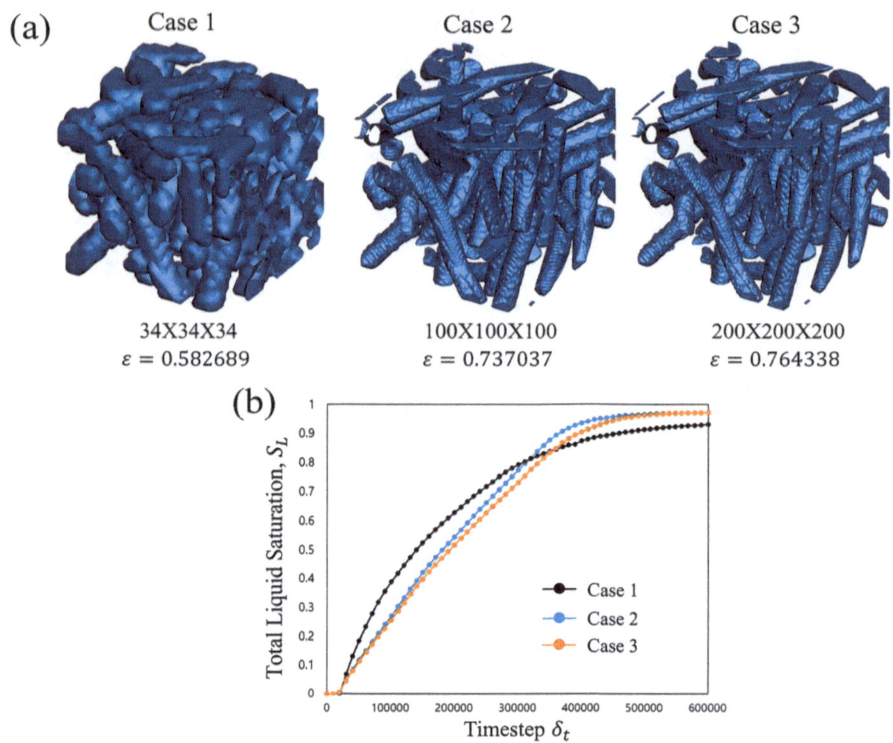


**Fig. 7.3** Grid-independence test results: (**a**) Simulation cases were constructed using three grid resolutions. (**b**) Liquid saturation progression was also monitored across the three cases

Fig. 7.3a, the porosity approached the original STL value as the grid resolution increased. The difference in porosity between Cases 2 and 3 was less than 3.5%.

Figure 7.3b presents the evolution of liquid saturation $S_L$, which quantifies the volume fraction of the liquid phase within the porous medium. The results showed minimal deviation between Cases 2 and 3. Across all cases, the error remained within 5.26%, and the steady-state liquid saturation differed by less than 0.5%. Based on this analysis, the resolution used in Case 2 ($100 \times 100 \times 100$) was selected for subsequent simulations, balancing accuracy and computational cost.

Figure 7.4 illustrates the simulation domain used for modeling transient liquid transport in porous CNT media. The domain consists of a porous structure centered vertically, with buffer (free fluid) zones above and below to enable fluid inflow and outflow. The total computational domain spans $L_x^* \times L_x^* \times L_x^* = 100 \times 100 \times 140$ lattice units.

In the LBM framework, all base lattice units (length, time, mass) are normalized to 1, i.e., $L_{LB}^* = t_{LB}^* = m_{LB}^* = 1$. Given that the physical length in the x-direction is $L_x = 300$ μm, the corresponding lattice unit scale is $L_{LB} = 3.0 \times 10^{-6}$ m. Based on this conversion, physical parameters were mapped to lattice units (as summarized in Table 7.1).

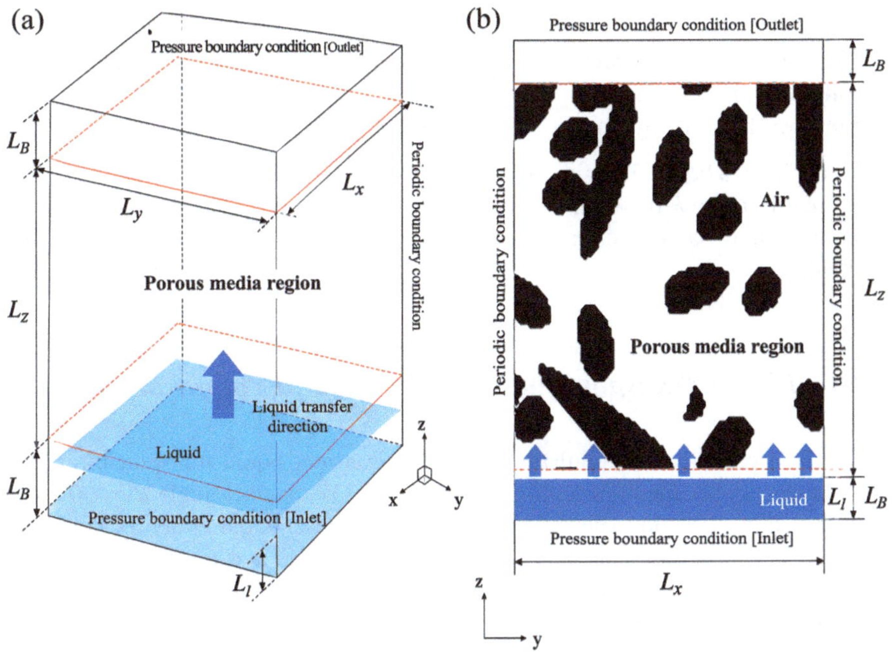


**Fig. 7.4** Simulation domain. (**a**) 3D schematic of the entire simulation domain, including the porous CNT medium positioned centrally and buffer (free fluid) zones at the top and bottom for fluid flow. (**b**) Cross-sectional view in the y–z plane at x = 0, illustrating the internal porous structure and boundary layout

**Table 7.1** The simulation variables for this research

| Variables | Physical value | Lattice value | Unit lattice scale |
|---|---|---|---|
| Length | $L_x = 300\mu m$<br>$L_y = 300\mu m$<br>$L_z = 300\mu m$<br>$L_B = 60\mu m$<br>$L_l = 45\mu m$ | $L_x^* = 100$<br>$L_y^* = 100$<br>$L_z^* = 100$<br>$L_B^* = 20$<br>$L_l^* = 15$ | $L_{LB} = 3.0 \times 10^{-6}$m |
| Mass | | | $m_{LB} = 2.7 \times 10^{-15}$kg |
| Time | | | $t_{LB} = 3.6 \times 10^{-9}$s |
| Initial density | $\rho_{in}^R = 1000\text{kg/m}^3$<br>$\rho_{in}^B = 1\text{kg/m}^3$ | $\rho_{in}^{R*} = 10$<br>$\rho_{in}^{B*} = 0.1$ | |
| Viscosity | $\mu^R = 1.0$ mPa · s<br>$\mu^B = 0.015$ mPa · s | $\mu^{R*} = 0.004$<br>$\mu^{B*} = 0.00006$ | |
| Surface tension | $\gamma = 0.072$N/m | $\gamma^* = 0.0003456$ | |
| Pressure | $P = 0.1$MPa | $P^* = 0.00144$ | |

The initial densities $\rho_R^{in}$ and $\rho_B^{in}$ represent the red (liquid) and blue (gas) fluids, respectively, with a density ratio of 1000:1. The viscosity ratio was set to 66, reflecting typical values for organic electrolytes used in lithium-air batteries. Relaxation times were $\tau_\nu^R = 1.5$ for red fluid and $\tau_\nu^B = 0.515$ for blue fluid.

Boundary conditions were as follows:

- **Top and bottom boundaries:** Dirichlet (pressure) boundary conditions
- **Side boundaries:** Periodic
- **Solid walls:** Halfway bounce-back condition

At the inlet, a liquid film was introduced to permeate the porous structure due to a pressure difference $P$ between the inlet and outlet. This setup enables observation of dynamic liquid saturation and displacement behavior within the porous CNT scaffold.

### 7.3.3 Liquid Transport and Air Saturation in Porous Media

Figure 7.5a presents the simulation results of transient liquid transport through a CNT-based porous medium with porosity $\varepsilon = 0.737$, contact angle $\theta = 90°$, and viscosity ratio $M = 66$. The light blue surface represents the evolving liquid–air interface at various time steps. The simulation confirms that the liquid gradually infiltrates the pore space within the CNT network. Due to the randomness of the CNT structure, the advancing liquid front exhibits a non-uniform profile upon reaching the bottom of the porous layer.

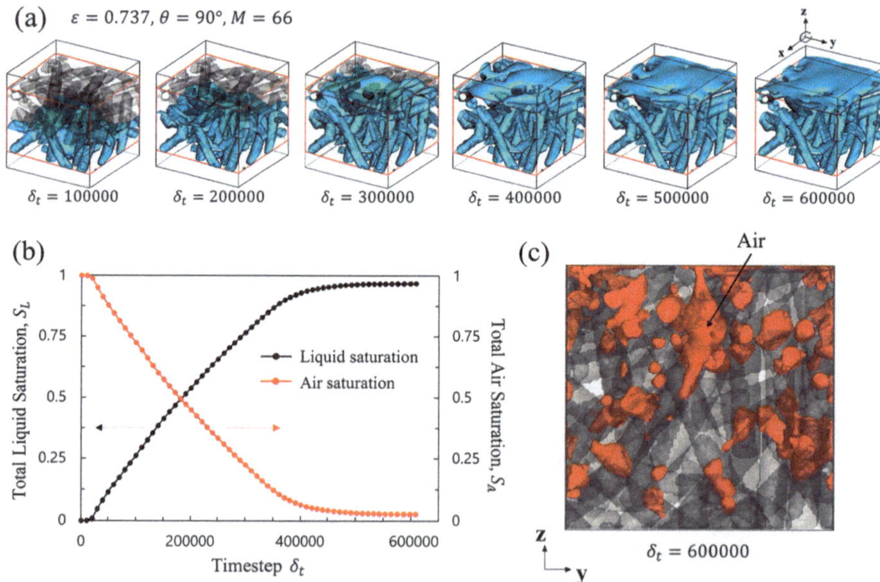


**Fig. 7.5** Liquid transport simulation in CNT porous media. (**a**) Temporal evolution of liquid transport. (**b**) Evolution of phase saturation. (**c**) Air distribution at timestep 600,000

Figure 7.5b presents the evolution of phase saturation within the CNT porous layer over time. The total liquid saturation $S_L$ and total air saturation $S_A$ are defined as:

$$S_L = \frac{\varepsilon_L}{\varepsilon}, S_A = \frac{\varepsilon_A}{\varepsilon}, \tag{7.6}$$

where $\varepsilon_L$ and $\varepsilon_A$ are the volume fraction of pores filled with liquid and the volume fraction of pores occupied by air, respectively. These saturation metrics represent the relative occupancy of each phase within the pore volume. The plot shows how liquid gradually replaces air as it infiltrates the porous structure, while the sum of $S_L$ and $S_A$ remains equal to 1 at all times.

As shown in Fig. 7.5, the total saturation—defined as the sum of liquid and air saturation—remains equal to 1 throughout the simulation. As liquid progresses through the porous CNT medium, the liquid saturation $S_L$ increases while the air saturation $S_A$ decreases. However, even after the liquid front reaches the outlet of the porous layer, full saturation is not achieved ($S_L < 1$). This indicates the presence of regions within the porous domain where liquid does not infiltrate—commonly referred to as trapped-air zones.

Figure 7.5c visualizes this residual air distribution. The transparent gray regions represent the CNT-based porous structure, while red surfaces denote liquid–air interfaces, highlighting the remaining air pockets. These pockets tend to form vertically along the direction of flow and are attributed to irregularities in local porosity and pore connectivity.

To further understand how pore structure and fluid properties affect infiltration, simulations were carried out for three different geometries. Figure 7.6 presents how total liquid saturation changes with porosity, contact angle, and viscosity ratio. In Fig. 7.6a, porosity is varied while keeping contact angle $\theta = 90°$ and viscosity ratio $M = 66$ constant. The results show that lower porosity leads to slower infiltration and lower steady-state liquid saturation. Despite these differences in infiltration rate, the final amount of trapped air remains nearly constant across all cases, indicating that the residual saturation is relatively insensitive to the detailed fiber geometry under the conditions tested.

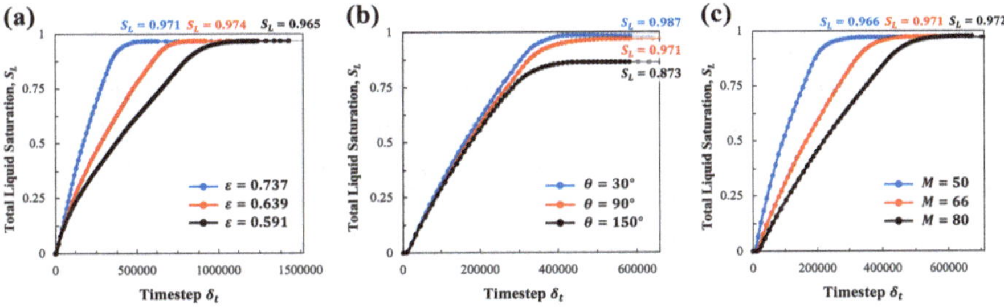


**Fig. 7.6** Effects of porosity, wettability, and viscosity ratio on liquid saturation in porous CNT media. (**a**) Variation with porosity. (**b**) Variation with wettability. (**c**) Variation with viscosity ratio

Figure 7.6b illustrates the influence of wettability on total liquid saturation in porous CNT media, under fixed conditions of porosity $\varepsilon = 0.737$ and viscosity ratio $M = 66$. Wettability, typically characterized by the contact angle, affects how easily a liquid spreads on a solid surface in the presence of another immiscible phase [10]. In this study, three contact angles were considered: hydrophilic ($\theta = 30°$), neutral ($\theta = 90°$), and hydrophobic ($\theta = 150°$).

Unlike the results observed for varying porosity, changes in contact angle did not significantly affect the overall liquid transport rate. In all three cases, the liquid reached the outlet of the porous domain in similar time frames and reached a steady saturation state. However, wettability strongly influenced the amount of residual air trapped within the medium. For example, under more hydrophilic conditions (30° and 90°), the total air saturation was lower, indicating more complete liquid infiltration. Conversely, for the hydrophobic surface, more air remained in the pores, suggesting that reduced surface wettability inhibits liquid spreading and promotes air entrapment.

Quantitative simulations were conducted to assess how the viscosity ratio affects liquid transport in porous CNT media. Three viscosity ratios—50, 66, and 80—were tested, based on a reference value of 66 from prior studies [11], while keeping porosity $\varepsilon = 0.737$ and contact angle $\theta = 90°$ constant. As shown in Fig. 7.6c, lower viscosity ratios (e.g., $M = 50$) facilitated faster infiltration, with the liquid reaching the outlet more quickly. In contrast, higher viscosity ratios (e.g., $M = 80$) slowed down the transport process. Notably, despite these differences in infiltration rates, the final air saturation remained nearly unchanged—similar to the trends observed in the porosity variation cases.

To further investigate residual air behavior, Fig. 7.7 visualizes the distribution of trapped air for the cases discussed in Fig. 7.6. Figure 7.7a shows how trapped air is distributed across different porous structures (i.e., varying porosity). While the locations of trapped-air pockets vary with geometry, the total amount remains relatively consistent. Figure 7.7b highlights the influence of wettability. In hydrophilic media, the liquid adheres more strongly to the surface and penetrates deeper into narrow pores, reducing the amount of trapped air. In contrast, hydrophobic media repel the liquid, preventing it from accessing tighter spaces and resulting in larger regions of residual air. These differences are attributed to surface interactions—hydrophilic surfaces promote liquid spreading, while hydrophobic surfaces hinder it. Figure 7.7c depicts air saturation patterns under different viscosity ratios. Although the shapes of the trapped-air regions vary slightly, their locations and the extent of liquid penetration are generally consistent across all viscosity ratios tested.

These findings underscore the importance of wettability and viscosity control in enhancing electrolyte infiltration and minimizing trapped air in porous electrodes, which is critical for optimizing Li–air battery performance.

**Exercise 7.2 Liquid and Air Saturation Dynamics**

(a) Derive the saturation definitions in (7.6) and discuss their physical meaning.

(b) Explain why total saturation remains equal to 1 in a two-phase system.

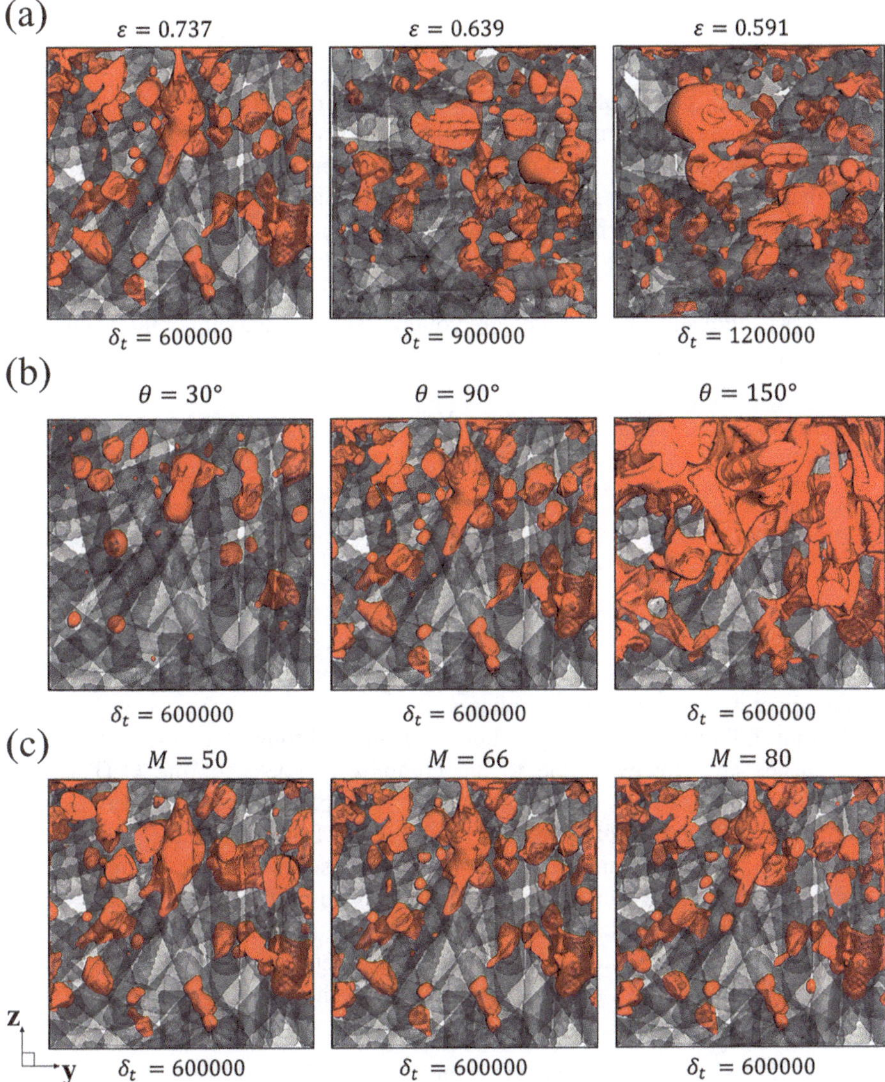


**Fig. 7.7** Trapped-air saturation in CNT porous media. (**a**) Effect of porosity. (**b**) Effect of wettability. (**c**) Effect of viscosity ratio

(c) Describe how air entrapment leads to the residual saturation patterns shown in Fig. 7.5c.

**Solution**

(a) In two-phase flow through porous media, saturation is defined as the volume fraction of each phase within the pore space. For a liquid phase $L$ and a gas phase $G$, the saturation of phase $\alpha$ is typically given by

$$S_\alpha = \frac{V_\alpha}{V_{pore}},$$

where $V_\alpha$ is the volume occupied by phase $\alpha$, and $V_{pore}$ is the total pore volume. In the LBM implementation with a color-gradient model, one can define the average saturation by integrating the phase indicator (or density) over the domain and normalizing by the pore volume, which is equivalent to (7.6). This definition has the clear physical meaning that $S_L$ and $S_G$ represent the fractions of the pore space occupied by liquid and gas, respectively.

(b) Because the pore space is assumed to be fully occupied by only two immiscible phases (liquid and air), the total pore volume is exactly the sum of the volumes of the two phases. Therefore,

$$V_{pore} = V_L + V_G,$$

and correspondingly,

$$S_L + S_G = \frac{V_L + V_G}{V_{pore}} = 1.$$

This identity simply expresses volume conservation of the pore space: at any given time, every point in the pore region is either filled with liquid or with air, and there are no voids or third phases.

(c) During infiltration, the invading liquid advances through the pore network and displaces air. However, due to the complex geometry of the CNT porous medium and capillary effects, not all air is displaced; some pockets become isolated and remain as trapped air. As the liquid front passes, these air pockets are cut off from the main flow paths and cannot escape, leading to a residual gas saturation pattern. Figure 7.5c shows that trapped air persists in certain regions even after infiltration has reached a quasi-steady state. This residual saturation arises from the combination of pore-scale geometrical constraints, unfavorable local wettability, and capillary pressure barriers that prevent complete displacement of the gas phase.

**Exercise 7.3 Effects of Porosity, Wettability, and Viscosity Ratio**

(a) Explain why lower porosity causes slower infiltration and reduced steady-state liquid saturation.

(b) Describe why wettability strongly affects the amount of trapped air, even though infiltration rates remain similar.

(c) Discuss why viscosity ratio affects infiltration rate but not the final trapped-air saturation.

**Solution**

(a) Lower porosity means that, for a given geometric volume, the total pore volume available for flow is smaller and the flow pathways are narrower and more tortuous. As a result, hydraulic resistance is higher, and capillary infiltration proceeds more slowly. In addition, reduced pore volume limits the amount of liquid that can enter the structure, leading to a lower steady-state liquid saturation. In the simulations, this manifests as a slower advance of the liquid front and a smaller final value of $S_L$ when porosity is reduced.

(b) Wettability strongly affects the contact angle and hence the capillary pressure driving infiltration, but once the liquid has entered the structure, the overall advance rate of the front can be similar for different wettabilities under the chosen conditions. However, wettability has a profound effect on how easily air can be displaced from pore corners and blind pores. Hydrophilic conditions (good wettability for the liquid) promote liquid spreading along solid surfaces and help drain air, reducing trapped-air volume. In contrast, less wetting conditions result in larger equilibrium contact angles, weaker capillary suction into certain pores, and a higher tendency for air to remain in corners or isolated pockets. Therefore, even if the infiltration rates are comparable, the amount and distribution of trapped air are strongly controlled by wettability.

(c) The viscosity ratio between liquid and gas primarily influences the rate at which the liquid front advances because it changes the viscous resistance to flow. A more viscous liquid or a larger viscosity contrast generally slows down infiltration. However, the final trapped-air saturation is mainly determined by the pore geometry and wettability-driven capillary pressures, which dictate whether the gas phase can be completely displaced from certain pores. Once a quasi-steady configuration is reached, changing the viscosity ratio does not significantly alter which pores remain filled with air, so the final trapped-air saturation is relatively insensitive to viscosity ratio in the parameter range considered.

### 7.3.4 Effective Diffusion Coefficient

Oxygen transport in a liquid electrolyte occurs through diffusion, which describes how molecules move from regions of high concentration to regions of low concentration. This process is critical for the operation of non-aqueous Li–air batteries, where oxygen and lithium ions must dissolve and move through the electrolyte to reach reaction sites.

In porous electrodes such as CNT networks, oxygen diffusion is not as straightforward as in bulk liquid due to the complex geometry of the pore space. To account for this, the effective diffusion coefficient ($D_{eff}$) is used. It adjusts the bulk diffusion value to reflect the influence of porosity and tortuosity of the medium [12]:

$$D_{eff} = \frac{\varepsilon_{eff}}{T^2} D, \tag{7.7}$$

where $D$ is the diffusion coefficient in the bulk electrolyte (depends on temperature), $\varepsilon_{eff}$ is the effective porosity—portion of the pore volume accessible to the electrolyte, and $T$ is the tortuosity—describes how convoluted the diffusion path is within the porous structure. This relation highlights how both the structure of the porous medium and fluid accessibility directly impact oxygen transport efficiency. A lower tortuosity and higher effective porosity lead to greater $D_{eff}$, which enhances battery performance.

In porous media, tortuosity is a dimensionless quantity that describes how much longer the actual path of fluid flow or diffusion is compared to the straight-line distance across the medium. It reflects the geometric complexity and connectivity of the pore network and is always greater than 1. In this study, tortuosity was calculated using velocity-based data obtained from LBM simulations. The formula is given as [13]:

$$T = \frac{\langle |\mathbf{u}(\mathbf{x})| \rangle}{\langle |u_z(\mathbf{x})| \rangle}, \tag{7.8}$$

This method enables direct evaluation of tortuosity from the simulated velocity field, making it particularly suitable for LBM-based studies of fluid or mass transport in complex porous media such as CNT networks.

The effective diffusion coefficient $D_{eff}$ quantifies how efficiently oxygen diffuses through a porous medium filled with liquid electrolyte. It is strongly influenced by the geometry of the pore structure, particularly effective porosity and tortuosity. However, in practice, trapped air—as observed in the LBM simulations—significantly impacts $D_{eff}$.

Simulation results revealed that trapped air forms predominantly on the surfaces of the CNT structures. These regions of residual air:

- Prevent the liquid electrolyte from fully wetting the CNT surfaces.
- Block access of dissolved oxygen to portions of the electrode.
- Function as dead-end zones where oxygen cannot participate in electrochemical reactions.

This effectively reduces the active transport area, lowering the efficiency of oxygen diffusion across the porous structure.

To reflect the influence of trapped air, the effective porosity $\varepsilon_{eff}$ is redefined as:

$$\varepsilon_{eff} = \varepsilon_L = \varepsilon - \varepsilon_A, \tag{7.9}$$

where $\varepsilon_A$ is the air saturation (volume fraction of trapped air) and $\varepsilon_L$ is the volume fraction accessible to liquid—i.e., the true effective porosity for oxygen transport. Once the simulation reaches steady state, the trapped air volume $\varepsilon_A$ becomes fixed, and the available pore volume for oxygen diffusion is effectively reduced.

Figure 7.8 analyzes how air porosity and the effective diffusion coefficient vary with porosity, wettability, and viscosity ratio. Each simulation fixed one variable while varying the other two to isolate their effects. Figure 7.8a, b fix the viscosity ratio at $M = 66$ and vary porosity and wettability. Figure 7.8c, d fix porosity at $\varepsilon = 0.737$, while Fig. 7.8e, f fix the contact angle at $\theta = 90°$.

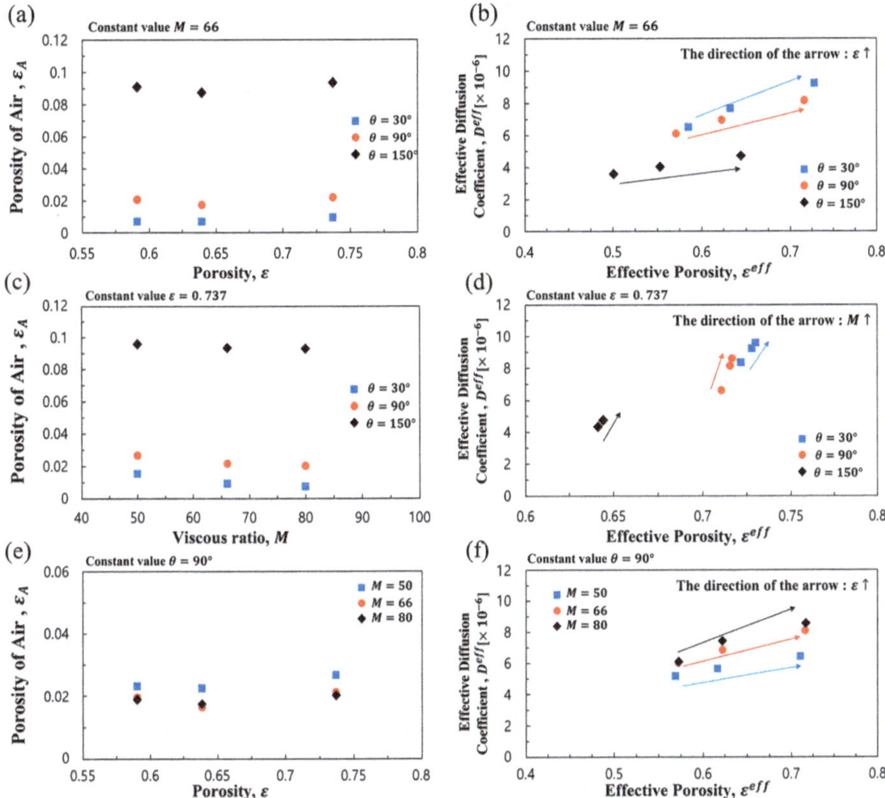


**Fig. 7.8** Effects of wettability, porosity, and viscosity ratio on air porosity and effective diffusion coefficient. (**a**, **b**) Variation with porosity and wettability at constant viscosity ratio $M = 66$. (**c**, **d**) Variation with wettability and viscosity ratio at constant porosity $\varepsilon = 0.737$. (**e**, **f**) Variation with porosity and viscosity ratio at constant contact angle $\theta = 90^{\circ}$

Among all parameters, wettability had the strongest impact on air porosity. Hydrophobic surfaces showed significantly higher air porosity, while porosity and viscosity ratio showed weaker, less consistent effects. Increased viscosity tended to slightly reduce air porosity. Complex pore structures made porosity effects irregular, but overall variation remained small.

Figure 7.8b, d, f present the effective diffusion coefficients for various cases based on the effective porosity. In all cases, the effective diffusion coefficient increases with effective porosity, showing a proportional relationship. In Fig. 7.8b, the increase in porosity leads to a higher effective diffusion coefficient, particularly influenced by wettability. Hydrophobic surfaces exhibited a significant reduction in effective porosity, resulting in lower diffusion coefficients. Conversely, hydrophilic surfaces maintained a wider distribution of diffusion coefficients due to enhanced wettability. At a fixed porosity of $\varepsilon = 0.737$, Fig. 7.8d shows that the effective diffusion coefficient is higher for hydrophilic conditions and narrower for

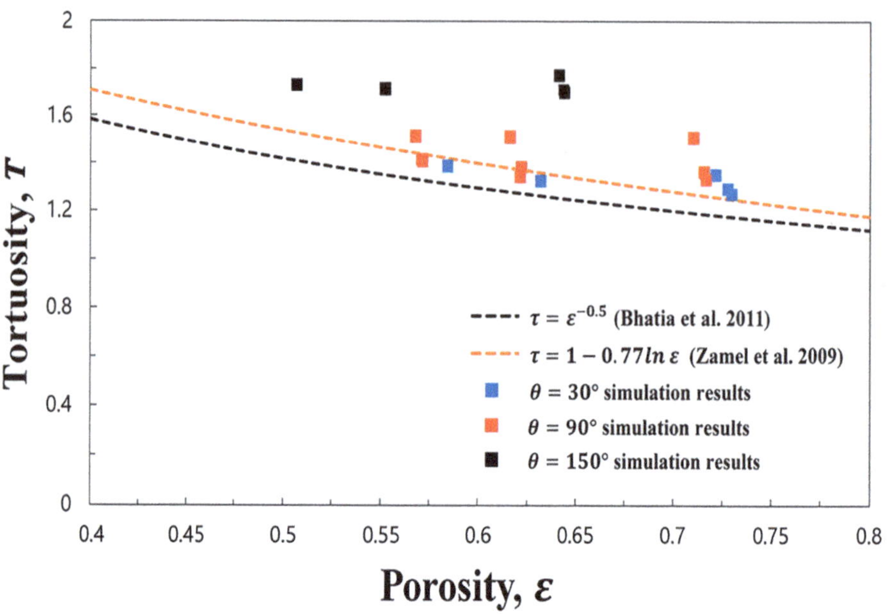


**Fig. 7.9** Tortuosity comparison between simulation results and existing correlations for all cases

hydrophobic ones. Additionally, as shown in Fig. 7.8f, increasing the viscosity ratio also leads to an increase in the effective diffusion coefficient, reinforcing the trend that both porosity and viscosity ratio positively influence diffusive transport through porous media.

Figure 7.9 illustrates the variation in tortuosity as a function of effective porosity across all simulation cases. The results show a general trend of decreasing tortuosity with increasing porosity, consistent with established correlations reported in the literature [14]. Due to the complex and irregular nature of the porous structure, direct measurement of the average path length $L_e$ is not feasible; therefore, tortuosity was estimated using the simulated fluid velocity according to Eq. (7.8). When compared with existing literature, the trend aligns in part, but higher tortuosity values were observed for surfaces with increased wettability. This overestimation indicates that wettability plays a significant role in enhancing tortuosity by increasing the complexity of the flow paths within the porous medium.

Figure 7.10 compares the normalized diffusivity $D^{eff}/D$ obtained from all simulation cases with theoretical models from the literature [15]. The black dashed line represents the Bruggeman equation, a widely used empirical correlation in electrode modeling that relates diffusivity to porosity. The orange dashed line corresponds to the diffusion correlation derived from a carbon paper structure geometry similar to that used in this study [15]. The simulation results consistently showed higher diffusivity values than those predicted by the correlation from Zamel et al. [15], indicating that the structural features and transport properties of the CNT-based porous media in this study differ from those in conventional carbon paper. These

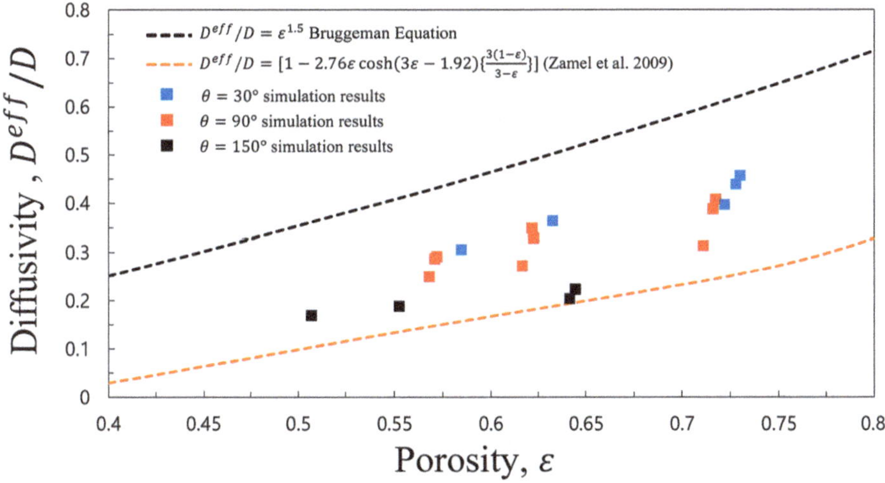


**Fig. 7.10** Comparison of diffusivity for all cases

results highlight the importance of considering the effects of trapped air, effective porosity, wettability, and viscosity ratio when evaluating the effective diffusion coefficient in such complex porous media.

**Exercise 7.4 Tortuosity and Effective Diffusion Coefficient**

(a) Define tortuosity and explain why it is always greater than 1.

(b) Describe how tortuosity is computed using the velocity-based method in (7.8).

(c) Explain how trapped air modifies effective porosity and consequently reduces the effective diffusion coefficient.

**Solution**

(a) Tortuosity is a measure of how convoluted the transport pathways are compared with a straight line. It is often defined as

$$T = \frac{L_{eff}}{L},$$

where $L_{eff}$ is the effective path length that a diffusing species or flowing fluid must travel through the porous medium, and $L$ is the macroscopic (straight) length of the sample. Because real pore paths are always longer and more winding than the straight-line distance, $L_{eff} > L$, and therefore tortuosity $T$ is always greater than 1.

(b) In the velocity-based method used in (7.8), tortuosity is obtained by analyzing the pore-scale velocity field computed from LBM. Conceptually, one compares the average magnitude of the fluid velocity vectors to their average projection in the main transport direction. If the flow channels are highly tortuous, a significant fraction of the local velocity is not aligned with the macroscopic direction, so the effective path length is increased. Mathematically, this leads to an expression of the form

$$T = \frac{\langle |\mathbf{u}| \rangle}{\langle u_{\parallel} \rangle},$$

or an equivalent ratio that reflects how much the streamlines deviate from a straight path. In this way, the complex 3D geometry of the CNT porous medium is encoded into a scalar tortuosity factor via the simulated velocity field.

(c) Trapped air reduces the pore volume that is actually available for gas-phase oxygen transport. In the diffusion model, this is expressed as a reduction in the effective porosity $\varepsilon_{eff}$, because some regions of the nominal pore space are blocked by liquid-filled zones or isolated air pockets that do not contribute to continuous transport pathways. The effective diffusion coefficient is commonly modeled as

$$D_{eff} = D_0 \frac{\varepsilon_{eff}}{T},$$

where $D_0$ is the bulk diffusion coefficient. Trapped air both decreases $\varepsilon_{eff}$ (by removing parts of the pore space from the transport network) and often increases $T$ (by forcing oxygen to follow more convoluted paths), leading to a significant reduction in $D_{eff}$. This is why accurate modeling of trapped air and tortuosity is crucial for predicting oxygen diffusion in CNT electrodes.

## 7.4 Conclusions

In this chapter, we investigate multiphase flow and oxygen diffusion through CNT-based porous media using a 3D MRT color-gradient lattice Boltzmann framework. The simulations quantify the effects of porosity, wettability, and viscosity ratio on liquid and air saturation, trapped-air distribution, tortuosity, and the effective diffusion coefficient. As liquid propagates through the porous structure, air becomes trapped primarily near CNT surfaces, reducing effective porosity and forming dead-end zones that hinder oxygen transport. Wettability was found to be the most influential factor, with hydrophobic surfaces producing higher air saturation and lower diffusivity. While porosity and viscosity ratio also affected transport, their impact was secondary. A velocity-based calculation of tortuosity showed inverse correlation with effective porosity and highlighted how surface interactions intensify flow complexity. Comparisons with theoretical models confirmed the predictive accuracy of the simulation, with deviations attributed to air entrapment.

Collectively, these results demonstrate how interface topology and material properties co-govern electrolyte transport and offer practical design insights for optimizing electrode architectures in Li–air battery systems.

**Summary**

1. **Motivation: Oxygen transport limits Li–air battery performance**
   - Non-aqueous Li–air batteries offer high theoretical energy density but suffer from low oxygen solubility in organic electrolytes.
   - Oxygen must diffuse through the electrolyte-filled porous CNT electrode to reach reaction sites, making oxygen transport the dominant rate-limiting step.
   - Liquid infiltration of CNT electrodes often leaves void regions of trapped air, blocking transport pathways and reducing accessible reaction surface area.
2. **CNT fibrous networks as porous electrodes**
   - CNT sheets provide high mechanical robustness, good conductivity, and continuous percolation pathways during discharge.
   - Their hierarchical pore structure enhances gas accessibility but can also promote complex liquid–gas displacement patterns.
   - Understanding fluid displacement and trapped-air formation in CNT structures is essential for predicting battery performance.
3. **Numerical framework: 3D MRT color-gradient LBM**
   - The chapter employs the high-density-ratio MRT color-gradient model, for stable simulation of two-phase flows in porous CNT media.
   - Components include:
     - MRT collision operator with stabilization matrix
     - Perturbation operator for interfacial tension
     - Recoloring operator for sharp interfaces
4. **Liquid infiltration dynamics and phase saturation**
   - Liquid introduced from the inlet gradually infiltrates the CNT domain, forming a non-uniform moving front due to pore randomness.
   - Liquid saturation $S_L$ increases while air saturation $S_A$ decreases, with $S_L + S_A = 1$ maintained at all times.
   - Even after breakthrough, significant trapped-air pockets remain—indicating incomplete infiltration.
5. **Influence of porosity, wettability, and viscosity ratio**
   - Porosity
     - Lower porosity → slower infiltration and lower final liquid saturation.
     - However, the amount of residual trapped air is largely insensitive to porosity.
   - Wettability
     - Wettability has the strongest effect on trapped-air volume.
     - Hydrophilic surfaces reduce gas entrapment; hydrophobic surfaces retain significantly more trapped air.

- Viscosity ratio
  - Higher liquid viscosity slows infiltration but has little impact on final trapped-air saturation.

6. **Trapped-air distribution and pore-scale mechanisms**
   - Trapped air forms mostly along fiber surfaces and in dead-end pore regions.
   - These pockets remain isolated after the liquid front passes due to capillary barriers and geometric confinement.
   - Spatial patterns depend on fiber alignment and local pore connectivity but total trapped-air volume is robust across cases.

## References


1. W. Xu, J. Xiao, D. Wang, J. Zhang, and J.-G. Zhang, "Effects of nonaqueous electrolytes on the performance of lithium/air batteries," Journal of the Electrochemical Society **157**, A219 (2009).
2. A. Nomura, K. Ito, and Y. Kubo, "Cnt sheet air electrode for the development of ultra-high cell capacity in lithium-air batteries," Sci. Rep. **7**, 45596 (2017).
3. K. Ushijima, S. Iwamura, and S. R. Mukai, "Simple and cost-effective method to increase the capacity of carbon nanotube sheet cathodes for lithium–air batteries," ACS applied energy materials **3**, 6915-6921 (2020).
4. S. Iglauer, A. Paluszny, and M. Blunt, "Simultaneous oil recovery and residual gas storage: A pore-level analysis using in situ x-ray micro-tomography," Fuel **103**, 905-914 (2013).
5. Z. X. Wen, Q. Li, Y. Yu, and K. H. Luo, "Improved three-dimensional color-gradient lattice boltzmann model for immiscible two-phase flows," Phys. Rev. E **100**, 023301 (2019).
6. M. Latva-Kokko, and D. H. Rothman, "Diffusion properties of gradient-based lattice boltzmann models of immiscible fluids," Phys. Rev. E **71**, 056702 (2005).
7. Y. Ba, H. H. Liu, Q. Li, Q. J. Kang, and J. J. Sun, "Multiple-relaxation-time color-gradient lattice boltzmann model for simulating two-phase flows with high density ratio," Physical Review E **94**, (2016).
8. Z. Guo, C. Zheng, and B. Shi, "Discrete lattice effects on the forcing term in the lattice boltzmann method," Physical review E **65**, 046308 (2002).
9. S. Bryant, and M. Blunt, "Prediction of relative permeability in simple porous media," Physical review A **46**, 2004 (1992).
10. R. Aziz, V. Niasar, H. Erfani, and P. J. Martínez-Ferrer, "Impact of pore morphology on two-phase flow dynamics under wettability alteration," Fuel **268**, 117315 (2020).
11. J. Read, "Ether-based electrolytes for the lithium/oxygen organic electrolyte battery," Journal of The Electrochemical Society **153**, A96 (2005).
12. J. Yuan, and B. Sundén, "On mechanisms and models of multi-component gas diffusion in porous structures of fuel cell electrodes," Int. J. Heat Mass Transf. **69**, 358-374 (2014).
13. P. Wang, "Lattice boltzmann simulation of permeability and tortuosity for flow through dense porous media," Mathematical Problems in Engineering **2014**, 694350 (2014).
14. S. K. Bhatia, M. R. Bonilla, and D. Nicholson, "Molecular transport in nanopores: A theoretical perspective," Physical Chemistry Chemical Physics **13**, 15350-15383 (2011).
15. N. Zamel, X. Li, and J. Shen, "Correlation for the effective gas diffusion coefficient in carbon paper diffusion media," Energy & Fuels **23**, 6070-6078 (2009).

# Chapter 8
# Particle Suspended Drop-on-Demand Inkjet Printing

## 8.1 Introduction

Drop-on-Demand (DoD) inkjet printing has emerged as a promising technology for display manufacturing due to its high material utilization, scalability to large-area substrates, and compatibility with thin and flexible displays. Unlike conventional contact-based fabrication methods, DoD printing enables precise, non-contact deposition of functional inks, making it particularly attractive for next-generation display applications.

Despite these advantages, achieving reliable printing performance requires a detailed understanding of droplet ejection behavior. Stable jetting is not determined solely by whether a droplet is generated, but also by how accurately and cleanly it is delivered to the target substrate. The formation of satellite droplets, deviations in droplet trajectory, and variations in ejection velocity can lead to non-uniform patterns and reduced printing reliability. In addition, thermal effects near the nozzle or induced by the actuation process may alter ink properties such as viscosity and surface tension, thereby affecting ligament breakup, droplet volume, and jetting stability. Therefore, a detailed understanding of droplet ejection dynamics is essential for predicting and controlling printing accuracy in DoD inkjet systems.

In the preceding Chapter 5, we covered the applications and utilization of DoD inkjet printing. In precision applications, DoD inkjet printing must ensure both positional accuracy of droplet deposition and reliable droplet formation at high frequencies. Key dimensionless numbers governing the precision and stability of the DoD process are the Reynolds number (*Re*), Weber number (*We*), and Ohnesorge number (*Oh*), defined as follows:

$$Re = \frac{\rho UR}{\mu}, We = \frac{\rho U^2 R}{\sigma}, Oh = \frac{\mu}{\sqrt{\rho\sigma R}} = \frac{We^{\frac{1}{2}}}{Re}. \tag{8.1}$$



H. M. Lee, J. S. Lee, *Lattice Boltzmann Methodology for Single-Phase and Multiphase Nanoparticle Modeling*,
https://doi.org/10.1007/978-981-95-9117-6_8

Here, $\rho$ is the ink density, $\mu$ is the dynamic viscosity, $\sigma$ is the surface tension, $U$ is the droplet velocity, and $R$ is the droplet radius.

Numerous studies have examined the influence of fluid properties—such as viscosity and surface tension—on droplet velocity and satellite formation during DoD inkjet printing. Castrejón-Pita performed large-scale experiments and compared results with Lagrangian simulations [1]. Liu and Derby investigated stable droplet formation across a wide range of $Z$ numbers by varying ink properties [2]. Zhang used simulation to analyze clogging within the printhead, confirming its impact on jetting accuracy [3], while Zhang and Cheng studied thermal effects on repeated jetting from a heated reservoir [4].

Despite its advantages, inkjet printing faces challenges in real-world manufacturing. Functional inks often contain pigments or conductive particles, introducing issues not present in pure fluids. These particles can cause asymmetric jetting, reducing placement accuracy. They also alter the ink's rheology, leading to uneven filament evolution and misaligned droplet deposition. In addition, particle-induced satellite droplets reduce process reliability. Particles trapped in the filament accelerate droplet breakup and promote satellite formation [5]. The formation varies with particle distribution, leading to inconsistent ejection behavior. Thermal effects further complicate the process. In piezoelectric DoD systems, repeated actuation generates heat, altering viscosity and surface tension. These changes affect droplet formation and particle distribution. Post-deposition, the coffee-ring effect—where particles migrate to the contact line—can occur during evaporation, degrading print quality.

In this chapter, the DoD inkjet injection process is analyzed by incorporating particle dynamics and thermal effects. A multiphase flow was modeled using the MRT color-gradient LBM, while particle motion was computed via the SPM. Heat transfer was captured through the convection–diffusion equation solved with the LBM scheme. All models were implemented with GPU-based parallelization to ensure computational efficiency. The overall jetting behavior was investigated through a unified simulation framework integrating multiphase flow, particle dynamics, and thermal effects.

**Exercise 8.1 Governing Dimensionless Numbers in Inkjet Printing**

(a) Define the Reynolds (*Re*), Weber (*We*), and Ohnesorge (*Oh*) numbers used in DoD inkjet printing and describe the physical meaning of each.

(b) Explain why maintaining an appropriate $Z$ number ($= 1/Oh$) is essential for stable droplet formation.

**Solution**

(a) Reynolds number (*Re*):

- Ratio of inertial to viscous forces.
- Large *Re*: inertia-dominated flow; small *Re*: viscous-dominated, more damped dynamics.

Weber number (*We*):

- Ratio of inertial to capillary forces.
- Large *We*: inertia easily overcomes surface tension $\rightarrow$ strong jetting, more propensity to break up.

Ohnesorge number ($Oh$):

- Ratio of viscous forces to the combined inertial–capillary scale.
- High $Oh$: strong viscous damping, jet tends to be sluggish and may not break off cleanly; low $Oh$: weak viscous damping, jet can be too "violent," leading to excessive ligaments and satellites.

These three numbers together characterize the relative importance of inertia, viscosity, and surface tension during droplet formation and breakup.

(b) The printability parameter $Z$ is widely used in inkjet printing. A moderate Z (typically $1 \leq Z \leq 10$) is associated with stable, single-droplet jetting.

If $Z$ is too small:

- Viscous dissipation is too strong.
- The meniscus may deform but retract without forming a free droplet, or a very long, strongly damped filament forms that does not pinch-off cleanly.

If $Z$ is too large:

- Inertia dominates.
- The liquid column is easily overstretched, producing long ligaments and multiple satellite drops.

Maintaining an appropriate Z ensures a balance between inertia, viscosity, and surface tension so that a main droplet forms and separates cleanly without excessive satellites.

## 8.2 Numerical Methods

This study analyzes the effect of heat transfer in an inkjet printing system containing particles. Accordingly, a multiphysics simulation model that integrates multiphase flow, particle dynamics, and heat transfer is required. However, as the number of coupled physical models increases, computational efficiency may decrease. To address this issue, this study employs a GPU-based parallel computing model. As illustrated in Fig. 8.1, all physical computations—excluding the initialization and output stages—are performed on the GPU. The following sections describe the numerical methodologies applied for each of the integrated models.

### *8.2.1 GPU-Implemented MRT Color-Gradient Model*

To simulate droplet jetting in inkjet printing, the multiphase flow model proposed by Wen et al. [6], based on the MRT color-gradient method, was employed. All computations are carried out on the GPU. Most of the methodological components are consistent with those discussed in the previous chapter; however, for parallel

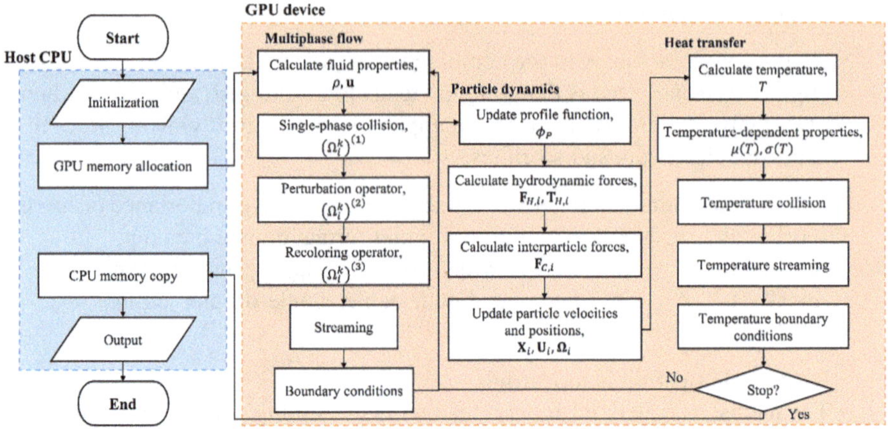


**Fig. 8.1** Flowchart of a multiphysics LBM simulation model coupling multiphase flow, particle dynamics, and heat transfer

computing optimization, the Structure of Array (SoA) scheme and pull scheme introduced in Chap. 2 were applied.

If the LBM model is implemented using CUDA, each computational component requires a dedicated kernel function. For instance, consider the following equation for computing the fluid density:

$$\rho = \sum_i f_i. \tag{8.2}$$

In this case, a CUDA kernel must be written to perform this summation for each lattice node in parallel. The CUDA kernel for computing fluid density is shown below.

```
__global__ void Calc_Density(double* f, double* rho)
{
  int x = threadIdx.x + blockIdx.x * blockDim.x;
  int y = threadIdx.y + blockIdx.y * blockDim.y;
  int z = threadIdx.z + blockIdx.z * blockDim.z;
 int tid = z + ZL * (y + YL * x);
 int SIZE = XL * YL * ZL;

  rho[tid] = f[tid+SIZE*0] + f[tid+SIZE*1] + f[tid+SIZE*2] + f[tid
+SIZE*3] + f[tid+SIZE*4] + f[tid+SIZE*5] + f[tid+SIZE*6] + f[tid
+SIZE*7] + f[tid+SIZE*8] + f[tid+SIZE*9] + f[tid+SIZE*10] + f[tid
+SIZE*11] + f[tid+SIZE*12] + f[tid+SIZE*13] + f[tid+SIZE*14] + f[tid
+SIZE*15] + f[tid+SIZE*16] + f[tid+SIZE*17] + f[tid+SIZE*18];
}
```

Here, XL, YL, and ZL represent the number of lattice nodes in the *x*-, *y*-, and *z*-directions, respectively. The remaining components of the LBM model can also be implemented using similar methodologies. Each physical or numerical operation—

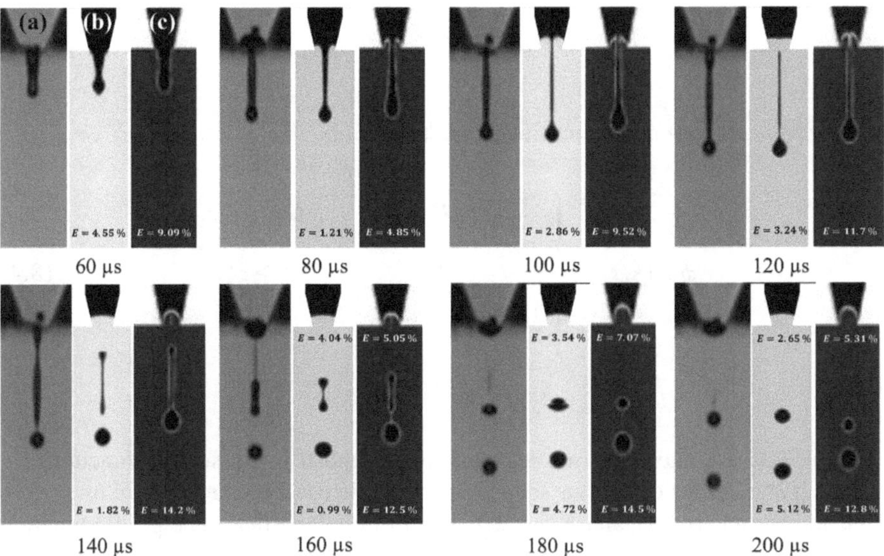


**Fig. 8.2** Validation of the DoD inkjet printing process (**a**) experimental observation [7], (**b**) results from the proposed LBM model, and (**c**) VOF-based simulation [7]

**Table 8.1** Physical properties of ink for validation

| Density (kg/m$^3$) | Surface tension (mN/m) | Viscosity (mPa s) | Z number |
|---|---|---|---|
| 975 | 25.144 | 7.68 | 4.56 |

such as collision, streaming, force calculation, or boundary treatment—can be expressed as a dedicated CUDA kernel, with parallel execution over the entire 3D lattice domain. This structure allows for scalable and efficient computation on the GPU.

The validation of the GPU-implemented LBM model was subsequently performed. Figure 8.2 presents the inkjet printing results obtained from experiments [7], the LBM model, and the volume of fluid (VOF) model [7]. The position errors of the main droplet and the satellite droplet are indicated at the bottom and top of each panel, respectively. The nozzle diameter is 50 μm, and the physical properties of the ink are listed in Table 8.1. The droplet flight velocity is approximately 1.67 m/s, corresponding to a Reynolds number of 5.27. The simulation results demonstrate that the LBM model accurately captures the droplet detachment and satellite droplet formation, achieving a maximum position error of 5.12%, which is significantly lower than the maximum error of 14.5% observed in the VOF model.

### 8.2.2 *Smoothed Profile Method*

In SPM, particle boundaries are smoothed profiles, and fluid–particle forces are computed accordingly. For $N_P$ dispersed particles, the overall profile $\phi_P$ is the sum of individual profiles.

$$\phi_P(\mathbf{x}, t) = \sum_{i=1}^{N_P} \phi_{P,i}(\mathbf{x}, t). \tag{8.3}$$

Here, $\phi_{P,\,i}$ represents the profile of the $i$th particle.

$$\phi_{P,i}(\mathbf{x}, t) = \begin{cases} 0 & h < -\frac{1}{2}\xi, \\ 1 & h > \frac{1}{2}\xi, \\ \frac{1}{2}\left[\sin\left(\frac{\pi h}{\xi}\right) + 1\right], & \text{else} \end{cases} \tag{8.4}$$

$$h(\mathbf{x}) = R_P - |\mathbf{X}_i - \mathbf{x}| \tag{8.5}$$

Here, $\xi$ is the interface thickness, set to $\Delta x$, and $h$ is the surface function. By choosing $h$, particles of various shapes can be modeled; for spheres, $h$ follows (8.5). $R_P$ is the particle radius, and $\mathbf{X}_i$ is the center of the $i$th particle. The profile function $\phi_P$ ranges from 0 to 1: 0 for fluid, 1 for solid, and intermediate values for the interface.

Once the total force and torque are known, the particle's linear and angular accelerations follow from the momentum equations.

$$M_P \frac{d\mathbf{U}_i}{dt} = \mathbf{F}_{H,i} + \mathbf{F}_{C,i}, \tag{8.6}$$

$$I_P \frac{d\mathbf{\Omega}_i}{dt} = \mathbf{T}_{H,i}, \tag{8.7}$$

where $M_P$ and $I_P$ are the particle's mass and moment of inertia, while $\mathbf{U}_i$ and $\mathbf{\Omega}_i$ are its linear and angular velocities. $\mathbf{F}_{H,\,i}$ and $\mathbf{T}_{H,\,i}$ denote the hydrodynamic force and torque, and $\mathbf{F}_{C,\,i}$ is the collision force.

For a rigid particle, the velocity at position $\mathbf{x}$ within the particle, $\mathbf{u}_P$, is given by its linear and angular motion.

$$\phi_P(\mathbf{x}, t)\mathbf{u}_P(\mathbf{x}, t) = \sum_{i=1}^{N_P} \phi_{P,i}(\mathbf{x}, t)[\mathbf{U}_i + \mathbf{\Omega}_i \times (\mathbf{x} - \mathbf{X}_i)]. \tag{8.8}$$

The fluid velocity $\mathbf{u}$ is computed via LBM, and $\mathbf{u}_P$ from (8.8). Using these, the interfacial hydrodynamic force $\mathbf{f}_P$ is evaluated from the momentum change rate.

$$\mathbf{f}_P(\mathbf{x}, t) = \phi_P(\mathbf{x}, t)\frac{\mathbf{u}(\mathbf{x}, t) - \mathbf{u}_P(\mathbf{x}, t)}{\Delta t}. \tag{8.9}$$

The hydrodynamic force on the fluid, $\mathbf{f}_H$, acts opposite to $\mathbf{f}_P$ and is applied as a body force in LBM. The total force and torque are obtained by integrating $\mathbf{f}_P$ over the particle volume $\forall_{P,\,i}$.

$$\mathbf{F}_{H,i} = \int_{\forall_{P,i}} \rho(\mathbf{x})\mathbf{f}_P(\mathbf{x}, t) d\forall, \tag{8.10}$$

$$\mathbf{T}_{H,i} = \int_{\forall_{P,i}} (\mathbf{x} - \mathbf{X}_i) \times \rho(\mathbf{x})\mathbf{f}_P(\mathbf{x}, t) d\forall. \tag{8.11}$$

The following are simulation results conducted to validate the SPM model. A single particle was allowed to settle within a fluid-filled container, and its position and velocity were compared against experimental data [8]. Figure 8.3 illustrates the schematic of the simulation domain, while Table 8.2 provides the fluid properties used in each case. The particle has a diameter of 15 mm and a density of 1120 kg/m$^3$. Figure 8.4 presents the comparison of particle position and velocity. The particle position shows excellent agreement with the experimental data, demonstrating the high accuracy of the simulation. Although a discrepancy in the peak settling velocity is observed in Case 4, which corresponds to the highest Reynolds number, this deviation is likely attributed to particle–wall interaction effects that become more significant under higher inertial conditions.

In the inkjet simulation, particles are dispersed within the ink, and when particles are located at the fluid interface, the interaction between the particles and the fluid boundary can influence the simulation results. To validate the modeling of particle–

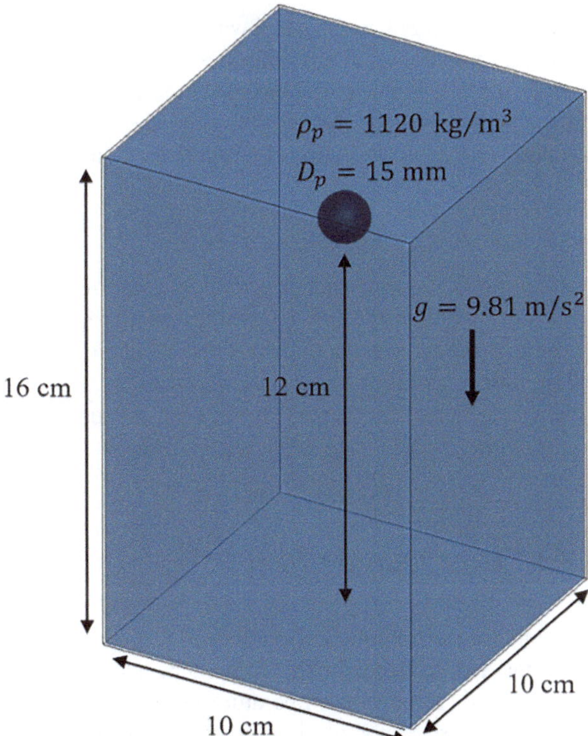


**Fig. 8.3** Simulation domain for validating the SPM model

**Table 8.2** Physical properties of ink for validation

| | Density (kg/m$^3$) | Viscosity (mPa s) | Reynolds number ($=\rho V_{fall} D_p/\mu$) |
|---|---|---|---|
| Case 1 | 970 | 373 | 1.5 |
| Case 2 | 965 | 212 | 4.1 |
| Case 3 | 962 | 113 | 11.6 |
| Case 4 | 960 | 58 | 31.9 |

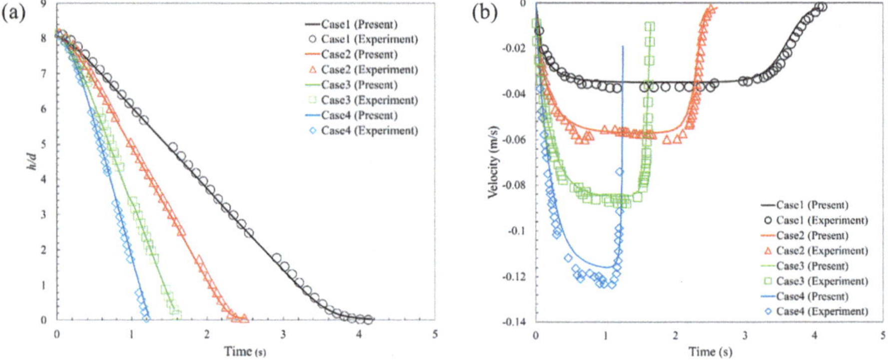


**Fig. 8.4** Validation of SPM model: (**a**) sphere–wall distance, (**b**) sedimentation velocity

fluid interface interactions, a test was conducted as shown in Fig. 8.5, where a single particle is positioned at the interface between red and blue fluids. The measured contact angle was evaluated based on the prescribed wettability of the particle. As the wettability varies, the particle's center shifts by a distance $l$ from its initial position. The measured contact angle $\theta^*$ is then calculated as follows:

$$\theta^* = \pi - \cos^{-1}(2l/D). \tag{8.12}$$

Figure 8.6 presents the measured contact angle as a function of the particle's prescribed wettability. The simulation results show excellent agreement with the theoretical predictions, demonstrating the accuracy of the contact angle implementation. The maximum error observed was 3.0%, indicating that the SPM model reliably captures the interaction between the particle and the fluid interface. This confirms the model's capability to represent wetting behavior with high fidelity.

### 8.2.3 *Heat Transfer Model*

To maintain methodological consistency, the heat transfer model was implemented using the LB scheme. In this approach, a temperature distribution function $g_i$ is introduced to solve the temperature convection–diffusion equation. This function evolves similarly to the fluid distribution functions in the standard LBM framework,

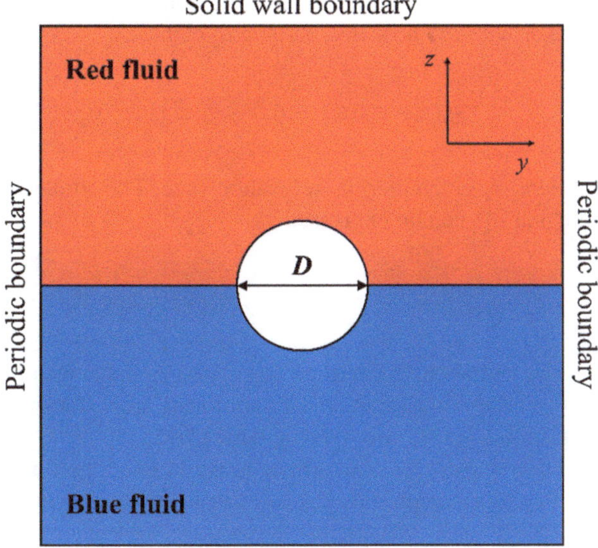


**Fig. 8.5** Schematic of the simulation for validating particle–immiscible fluid interface interactions

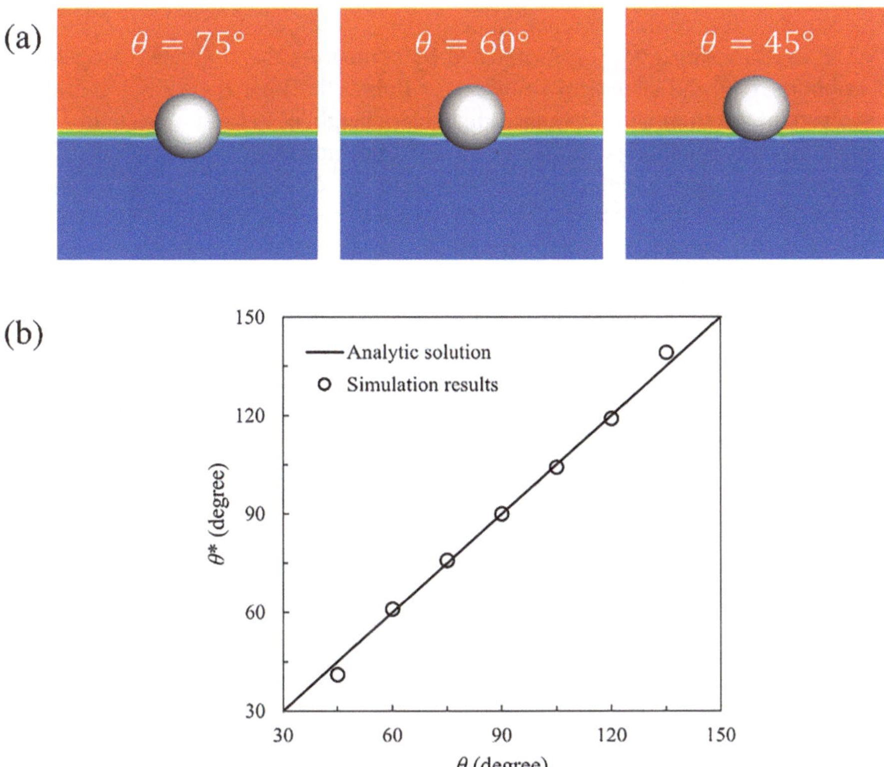


**Fig. 8.6** (**a**) Particle positions at contact angle of 75°, 60°, and 45°. (**b**) Comparison between measured and theoretical contact angles

allowing the temperature field to be computed in a mesoscopic and fully coupled manner with the flow and particle dynamics.

$$g_i(\mathbf{x} + \mathbf{e}_i\Delta t, t + \Delta t) - g_i(\mathbf{x}, t) = -\frac{1}{\tau_T}\left[g_i(\mathbf{x}, t) - g_i^{eq}(\mathbf{x}, t)\right]. \tag{8.13}$$

Here, $\tau_T$ is the thermal relaxation time. The equilibrium temperature distribution function $g_i^{eq}$ can be expressed as

$$g_i^{eq} = \omega_i T\left[1 + \frac{3}{c^2}(\mathbf{e}_i \cdot \mathbf{u}) + \frac{9}{2c^4}(\mathbf{e}_i \cdot \mathbf{u})^2 - \frac{3}{2c^2}|\mathbf{u}|^2\right]. \tag{8.14}$$

The temperature $T$ can be calculated as the zeroth-order moment of the temperature distribution function $g_i^{eq}$, similar to how density is computed from the fluid distribution function in the standard LBM.

The viscosity, $\mu$, and surface tension, $\sigma$, of the fluid vary with temperature, and the following empirical relations can be applied to model this dependence:

$$\mu(T) = \mu_{ref} + \frac{\partial\mu}{\partial T}(T - T_{ref}), \sigma(T) = \sigma_{ref} + \frac{\partial\sigma}{\partial T}(T - T_{ref}), \tag{8.15}$$

$\mu_{ref}$, $\sigma_{ref}$, and $T_{ref}$ are the reference values.

The heat transfer model was validated by comparing the simulated temperature distribution with the analytical solution for Poiseuille flow. Figure 8.7 shows the schematic of the simulation domain, which has a height of $2H$ and a length of $L$. The top and bottom boundaries are treated as walls maintained at a constant normalized temperature $T_w$. At the outlet, a zero temperature gradient condition and a constant pressure condition are applied. The inlet is maintained at a uniform temperature of $T^* = 0$, and a parabolic velocity profile, as described by the equation below, is imposed to drive the flow.

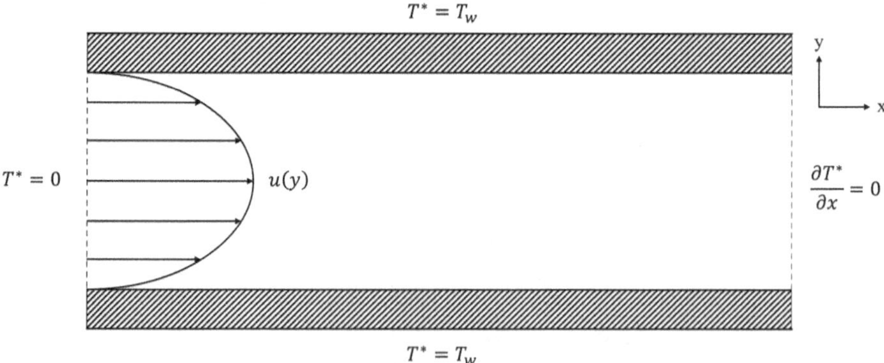


**Fig. 8.7** Schematic of the simulation for validating heat transfer model

$$u(y) = u_0\left[1 - \left(\frac{y}{H}\right)^2\right], (-H \leq y \leq H), \tag{8.16}$$

Under the conditions described above, the analytical solution for the temperature distribution as a function of position is given by [9]:

$$T^*(x,y) = T^*_w\left[1 - \sum_m C_m \exp\left(-\frac{m^4 x}{HPe} - \frac{m^2 y^2}{2H^2}\right){}_1F_1\left(\frac{1}{4} - \frac{m^2}{4}, \frac{1}{2}, m^2\frac{y^2}{H^2}\right)\right], \tag{8.17}$$

Here, $Pe = u_0 H/\alpha$ is the Péclet number, with $\alpha$ as thermal diffusivity, and ${}_1F_1$ denotes the hypergeometric function. The first ten roots of $m$ satisfying the boundary condition are: 1.2967, 2.3811, 3.1093, 3.6969, 4.2032, 4.6548, 5.0662, 5.4467, 5.8023, and 6.1373. The corresponding $C_m$ values are: 1.2008, −0.2991, 0.1608, −0.1074, 0.0796, −0.0627, 0.0515, −0.0435, 0.0375, −0.0329.

In the LBM simulation, the domain size was set to $N_x \times N_y = 1280 \times 323$, with $H = 160$, $u_0 = 0.05$, and $T^*_w = 1$. The Péclet number was varied as 25, 50, and 75 by adjusting thermal diffusivity to 0.32, 0.16, and 0.107, respectively. Figure 8.8 compares the simulation with the analytical solution, confirming strong agreement after reaching steady state.

## 8.3 Results

### 8.3.1 Simulation Setup

The simulation was designed to model the inkjet ejection process. Figure 8.9 shows the geometry of the printhead and substrate. The nozzle outlet was 25 μm in diameter with a 70° angle. Contact angles were 90° for the printhead and 10° for the substrate. Spherical particles ($D_P = 3$ μm, $\rho_P = 6000$ kg/m$^3$) were used. Physical parameters are summarized in Table 8.3.

In lattice units, the domain measured 120 × 120 × 600 lu$^3$, with nozzle inlet and outlet widths of 84 lu and 48 lu. Side boundaries were periodic. The top surface had mixed boundary conditions: constant pressure condition was applied outside the nozzle, and velocity condition was applied inside. The inlet velocity followed a sine profile, as shown in Fig. 8.10. Equation (8.18) corresponds to the waveform shown in Fig. 8.10, where the velocity amplitude $U$ and period $T$ are defined in segments as follows: $U_1 = 3.47$ m/s, $U_2 = -2.53$ m/s, $U_3 = 1.98$ m/s, $T_1 = 8$ μs, $T_2 = 4.56$ μs, $T_3 = 2.46$ μs.

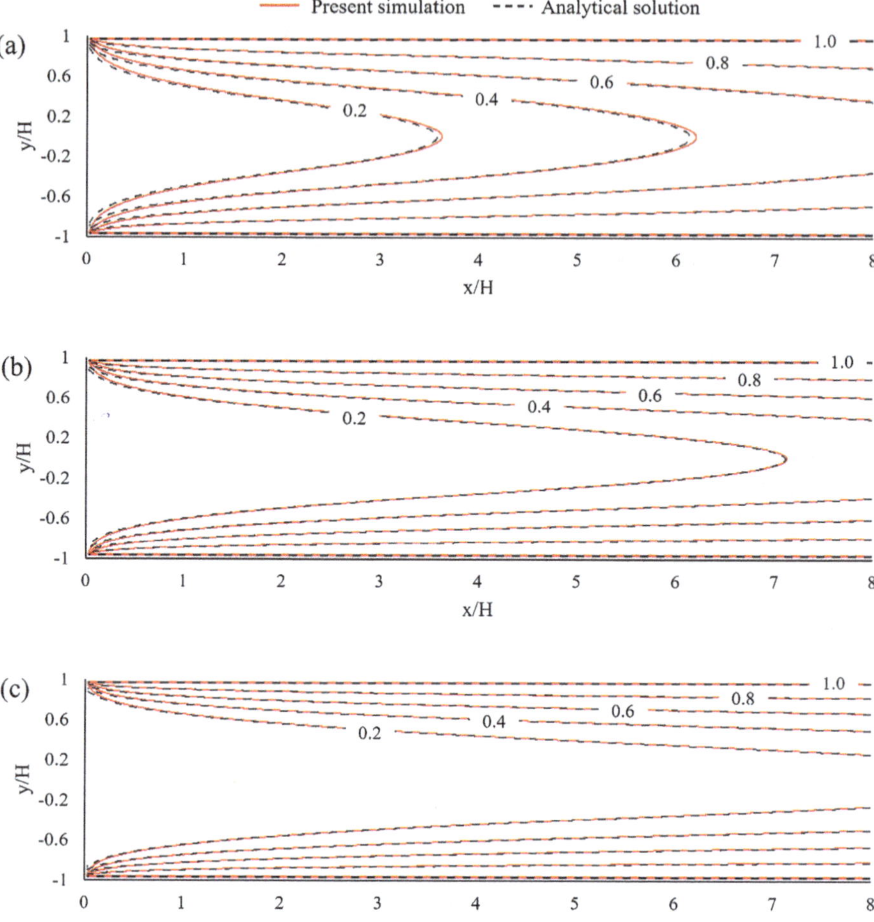


**Fig. 8.8** Normalized temperature contours: present model (red solid lines) vs. analytical solution (black dotted lines). (**a**) $Pe = 25$, (**b**) $Pe = 50$, and (**c**) $Pe = 75$

$$u_{in} = \begin{cases} U_1 \sin\left(\dfrac{\pi t}{T_1}\right) & t < T_1 \\ U_2 \sin\left(\dfrac{\pi[t - T_1]}{T_2}\right) & t < T_1 + T_2 \\ U_3 \sin\left(\dfrac{\pi[t - T_1 - T_2]}{T_3}\right) & t < T_1 + T_2 + T_3 \\ 0 & t \geq T_1 + T_2 + T_3 \end{cases}, \tag{8.18}$$

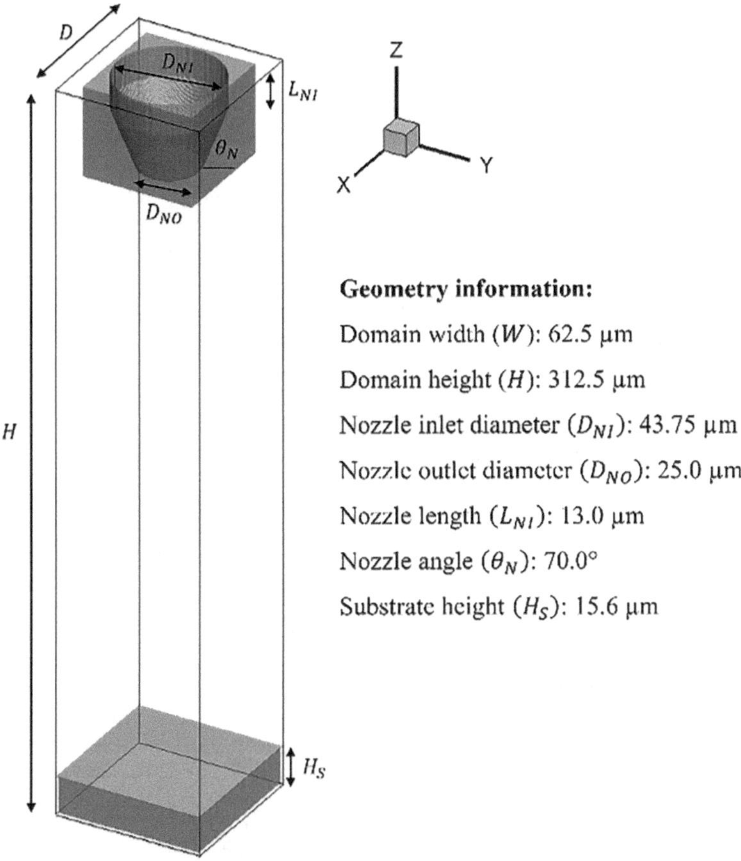


**Fig. 8.9** Schematic of the inkjet printing simulation

**Table 8.3** Inkjet printing simulation parameters

| Parameters | Symbols | Values | Units |
|---|---|---|---|
| Length conversion factor | $\delta_x$ | $5.21 \times 10^{-7}$ | m |
| Time conversion factor | $\delta_t$ | $1.63 \times 10^{-10}$ | s |
| Mass conversion factor | $\delta_m$ | $1.41 \times 10^{-17}$ | kg |
| Ink density | $\rho_I$ | 1000 | kg/m$^3$ |
| Ink dynamic viscosity | $\mu_I$ | 0.01 | Pa s |
| Air density | $\rho_A$ | 100 | kg/m$^3$ |
| Air dynamic viscosity | $\mu_A$ | 0.0002 | Pa s |
| Surface tension | $\sigma$ | 0.054 | N/m |
| Z number | $Z$ | 3.67 | – |
| Particle diameter | $D_P$ | $3.0 \times 10^{-6}$ | m |
| Particle density | $\rho_P$ | 6000 | kg/m$^3$ |

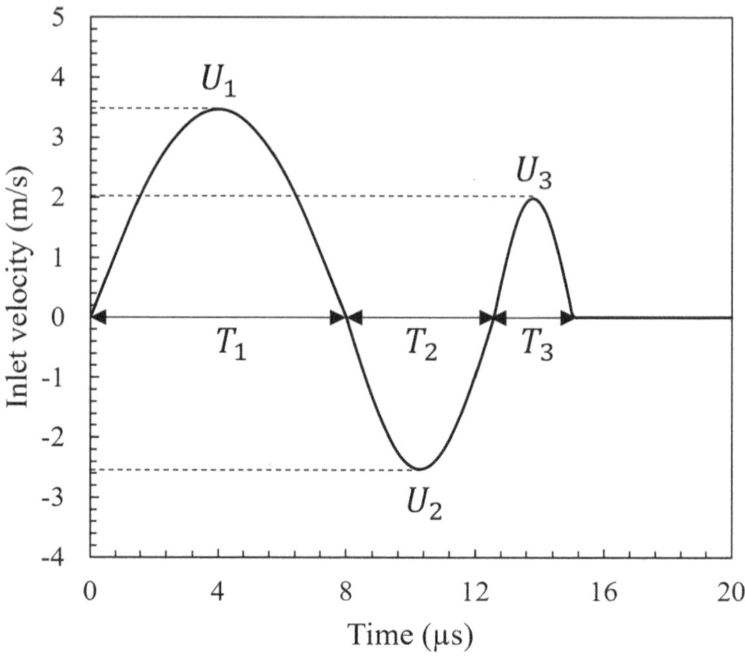


**Fig. 8.10** Inlet velocity profile applied inside the nozzle

**Table 8.4** Grouping of cases based on *Re* and *Oh*

| Case | *Re* | *Oh* |
|---|---|---|
| A | 7.7 | 0.36 |
| B | 6.93 | 0.36 |
| C | 7.7 | 0.324 |
| D | 6.93 | 0.324 |

## 8.3.2 Dimensionless Number Analysis

Droplet formation was analyzed by varying jetting speed and surface tension, which affected the Reynolds (*Re*) and Ohnesorge (*Oh*) numbers as summarized in Table 8.4. All cases fall within the printable regime defined by McKinley and Renardy [10], with simulation points shown in Fig. 8.11.

In Case A (highest *Re* and *Oh*), a long tail formed due to high speed and low surface tension; part of it merged with the main droplet after breakup. In Cases B and D (lower *Re*), insufficient momentum delayed spreading on the substrate. When *Oh* decreased (Cases C and D), the droplet broke rapidly via Rayleigh–Plateau instability. The fastest breakup occurred in Case D ($Re = 6.93$, $Oh = 0.324$). Since this study focuses on particle effects and satellite droplet control, Case A—with slower breakup and sufficient impact energy—is selected for further simulations (Fig. 8.12).

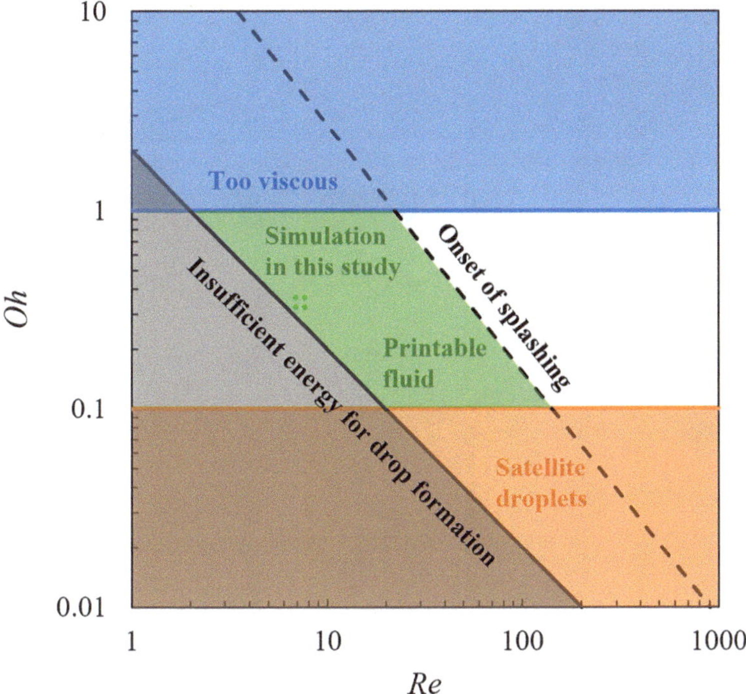


**Fig. 8.11** Operating region for stable drop-on-demand inkjet printing [10]. Simulation cases from this study are marked as green dots within the printable regime

### 8.3.3 *Effects of Particles on Asymmetric Jetting and Satellite Droplet Formation*

This section compares droplet ejection between pure ink and particle-laden ink. Figure 8.13 shows the time evolution of droplet formation. The particle volume fraction was set to 1.0%, with random initial distribution. As seen in Fig. 8.13c, this leads to an asymmetric filament, and a thinner neck appears where particles are concentrated. A satellite droplet forms at 65 μs. These results show that particles promote asymmetry and accelerate satellite droplet generation.

Figure 8.14 presents the asymmetric droplet ejection behavior at different particle concentrations. To quantify asymmetry, the "asymmetric length" is defined as the distance between the droplet's center of mass and the nozzle centerline, indicating deviation from the target point. At 2% particle concentration, asymmetry increased consistently, while at 1%, a temporary decrease occurred due to thread recovery. In the 2% case, breakup was enhanced, leading to separate motion of the main and satellite droplets. Conversely, at 1%, surface tension allowed the unbroken thread to reconnect with the main droplet. Upon substrate impact, asymmetry growth slowed, then rose again due to capillary forces. The 10° contact angle caused stronger

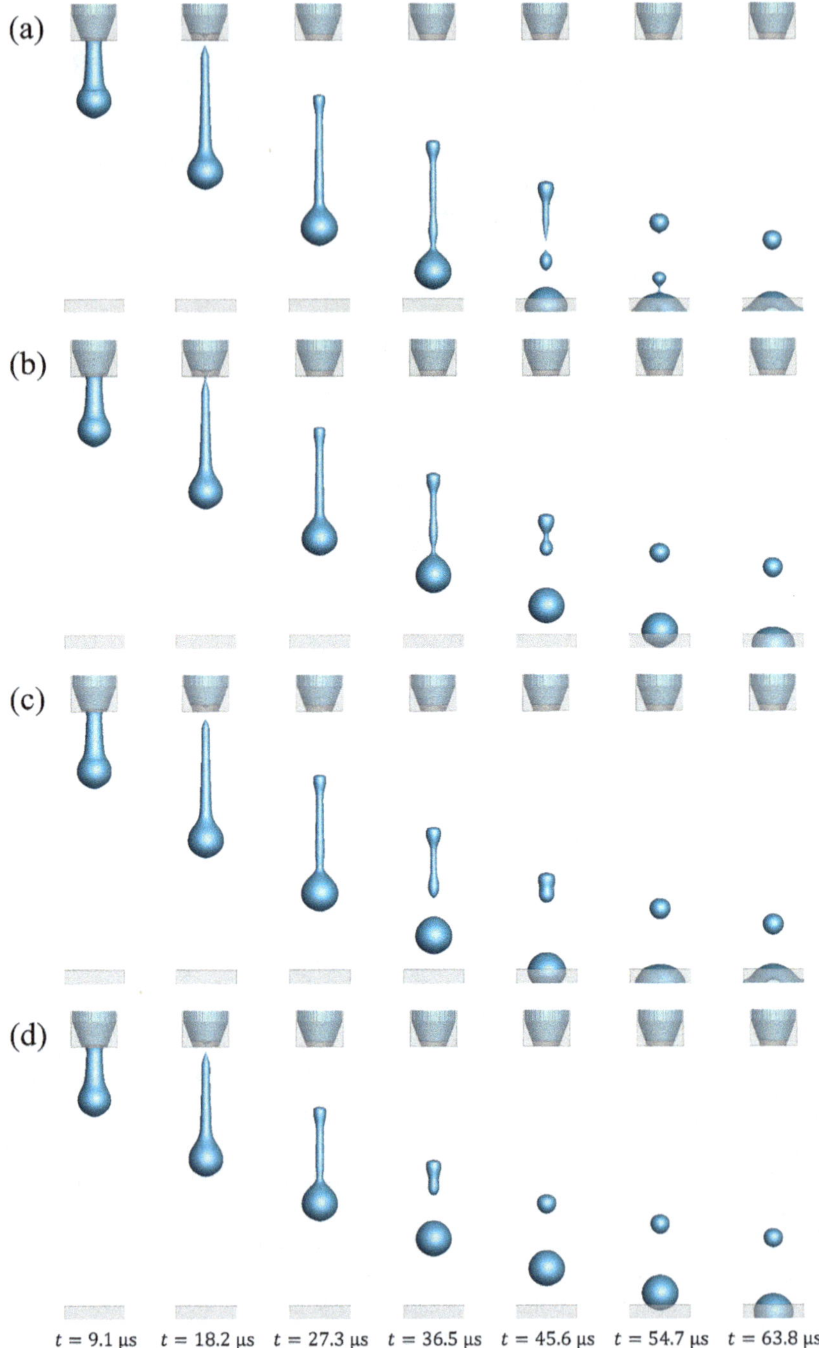


**Fig. 8.12** Time-sequenced droplet snapshots: (**a**) Case A, (**b**) Case B, (**c**) Case C, (**d**) Case D

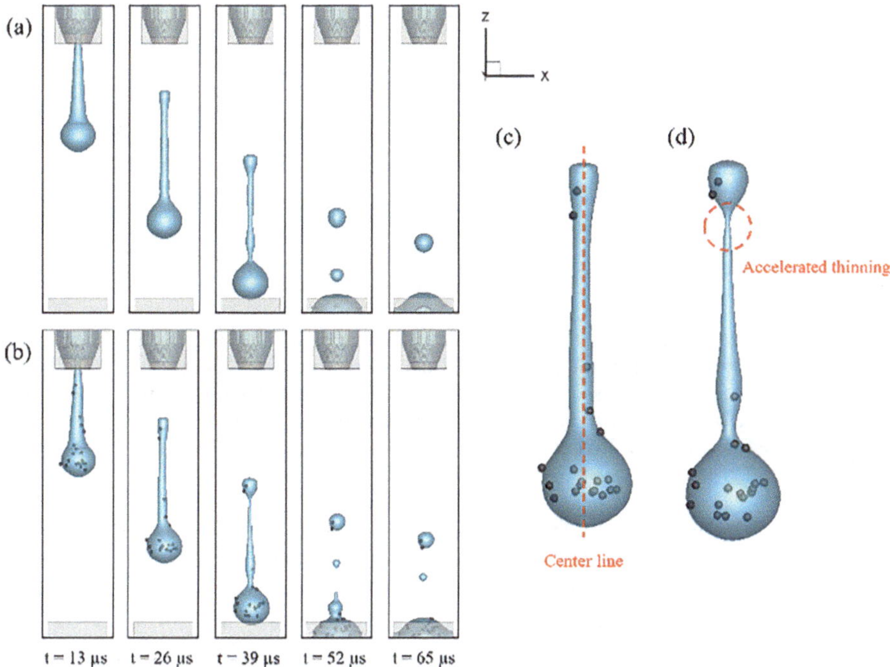


**Fig. 8.13** Time sequence of droplet ejection: (**a**) pure ink and (**b**) particle-laden ink. Time progresses from left to right: 13, 26, 39, 52, and 65 μs. (**c**) Asymmetric ejection at 26 μs; (**d**) accelerated neck thinning at 39 μs

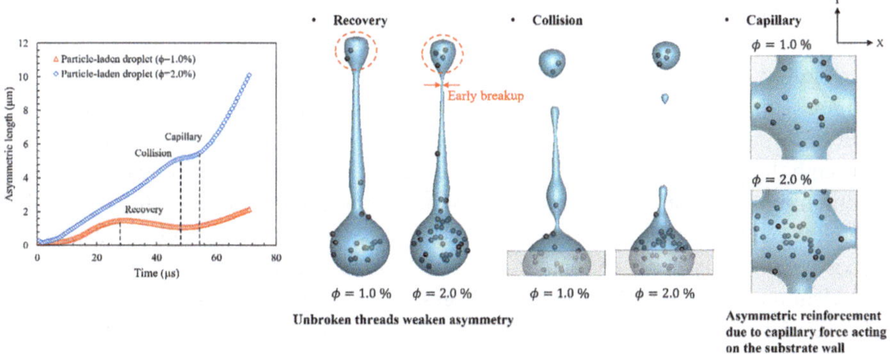


**Fig. 8.14** Asymmetric length over time. Higher particle concentration led to increased droplet asymmetry. The injection process involves three stages: recovery (thread retraction), collision (substrate impact), and capillary (motion driven by wall adhesion)

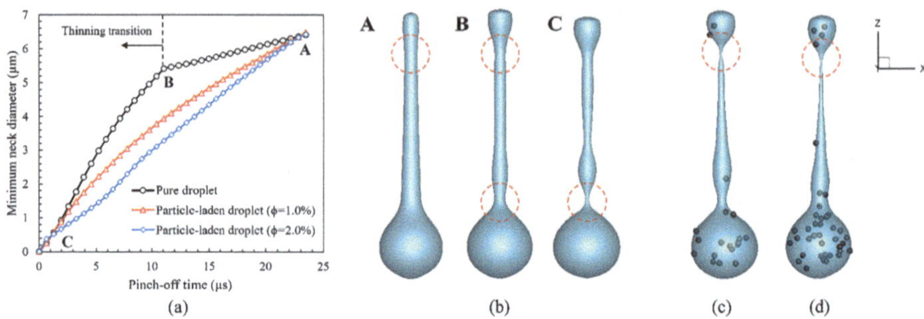


**Fig. 8.15** Minimum neck diameter and droplet shapes prior to breakup. (**a**) Minimum neck diameter vs. pinch-off time for pure ink, $\phi = 1\%$ and $\phi = 2\%$ particles concentrations. (**b**) In the pure ink case, the neck location shifts from the thread tip (A) to both the tip and main droplet (B), and finally to the main droplet neck (C) where separation occurs. (**c**) Shape at $\phi = 1\%$ and (**d**) $\phi = 2\%$ just before pinch-off. Red dotted circles mark the minimum neck locations

asymmetry toward the wall. Notably, despite identical wettability, the 2% case showed a greater asymmetry rise post-impact, influenced by the initial collision point. These asymmetry changes are governed by three main stages: recovery, collision, and capillary effects.

To investigate particle-induced satellite formation, Fig. 8.15 compares the pinch-off behavior of pure ink and inks with 1% and 2% particle concentrations. Figure 8.15a shows the minimum neck diameter versus pinch-off time $\tau = \tau_P - t$, where $\tau_P$ is the breakup time. This metric allows comparison of thread-thinning dynamics across cases. Thinning occurs at two distinct regions, as indicated by the red dotted circles in Fig. 8.15b, corresponding to the thread tip and the main droplet.

In particle-laden cases, the minimum neck diameter steadily decreased. In contrast, the pure droplet showed a different trend due to a shift in the necking location—from the thread tip to the main droplet. After the point marked by the black dotted line, rapid thinning occurred near the main droplet. This shift accelerated breakup, driven by the higher curvature at the droplet body (see Fig. 8.15b).

In both particle-laden cases (Fig. 8.15c, d), breakup occurred at the thread tip. Although the particle concentration was low and did not alter rheological properties, the thinning behavior differed from that of pure droplets. Previous studies [5, 11, 12] attributed such differences to rheological changes induced by particles. Suspension viscosity is strongly dependent on particle volume fraction $\phi$, and several semi-empirical models have been proposed to describe this relationship. In this study, the relative viscosity of the particle-laden ink was estimated using the Maron–Pierce model [13].

$$\mu_r(\phi) = \left(1 - \frac{\phi}{\phi_m}\right)^{-2}. \tag{8.19}$$

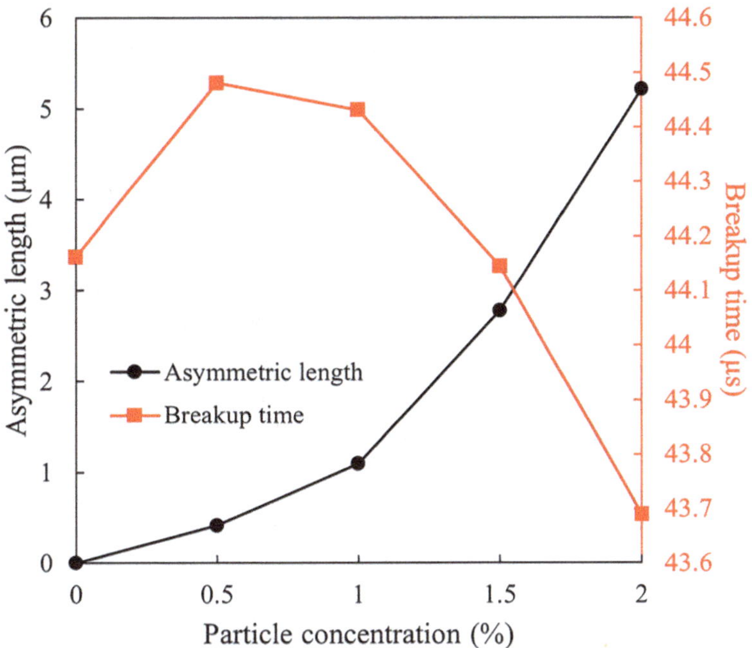


**Fig. 8.16** Asymmetric length and breakup time vs. particle concentration

The relative viscosity is defined as $\mu_r = \mu/\mu_m$, where $\mu_m$ is the medium viscosity, and the maximum packing fraction is assumed as $\phi_m = 0.64$. In this study, particle concentrations ranged from 1.0% to 2.0%, corresponding to viscosity increases from 3.2% to 6.6% based on the Maron–Pierce model. However, prior studies [14, 15] indicate that such small viscosity changes have minimal impact on filament thinning. Thus, viscosity variation alone cannot explain the observed behavior.

Initially uniform, the particles split between the thread tip and the main droplet during ejection. Some particles trapped in the filament hinder thinning beyond their size, introducing excessive local curvature that disrupts self-similar thinning [16], thereby accelerating satellite droplet formation. However, at higher concentrations, a different behavior appears near the main droplet neck (Fig. 8.15d). With enough particles, mobility is restricted [17], slowing necking and producing a cone-like shape. This dual role of particles in filament dynamics aligns with observations by Furbank and Morris [18]. Accelerated satellite formation at the thread tip results in low-velocity satellites that may remain suspended, reducing injection precision.

Figure 8.16 presents the trends of asymmetric length and breakup time with respect to particle concentration. Black circles indicate the asymmetric length at droplet impact ($t = 48.8$ μs), and red squares represent the breakup time. As particle concentration increased, asymmetry grew nonlinearly. Notably, at 0.5%, breakup was delayed compared to the pure ink due to a shift in necking location from the thread tip to the main droplet. In contrast, at higher concentrations, particles at the tip-accelerated breakup.

**Exercise 8.2 Particle-Induced Asymmetric Jetting**

(a) Using Figs. 8.13, 8.14, and 8.15, describe how particle concentration influences asymmetric jetting during DoD injection.

(b) Explain why particles cause neck thinning to accelerate and why satellite droplets form more readily at higher volume fractions.

(c) Discuss why the viscosity increase predicted by the Maron–Pierce model cannot fully explain the observed breakup behavior.

**Solution**

(a) Figures 8.13, 8.14, and 8.15 indicate that increasing particle volume fraction enhances asymmetry in the jetting and breakup process.

- At low particle concentration, the jet is nearly axisymmetric and the neck forms roughly at the geometric center of the filament.
- As particle concentration increases, particles migrate and accumulate preferentially in certain regions (often near the main droplet), causing local viscosity and microstructure to become heterogeneous.
- This heterogeneity leads to unequal thinning rates along the filament: one side (often near the tip) thins faster than the other, so necking and pinch-off occur at off-center locations.

Thus, higher particle loading promotes asymmetric jet profiles and off-axis neck positions, which in turn change the trajectory and size of satellite droplets.

(b) Particles accelerate neck thinning and promote satellite formation mainly through microstructural and local-viscosity effects:

- Particles disturb the local flow around the neck, creating regions where the liquid is effectively "squeezed out" more quickly.
- When particles accumulate on one side of the neck, they can locally increase stress concentration and encourage faster thinning in adjacent particle-free regions (e.g., at the tip).
- As the filament becomes thinner, discrete particles hinder smooth reconnection of the liquid thread, leading to multiple pinch-off events and the release of smaller ligaments that transform into satellites.

At higher particle volume fractions, these effects are stronger: more particles interact within the neck region, so more secondary pinch-off events occur, and satellite droplets become more numerous and energetically favored.

(c) The Maron–Pierce model predicts an effective viscosity which accounts for the bulk viscosity increase with particle volume fraction. However, the observed breakup behavior in the simulations cannot be explained solely by a uniform viscosity increase:

- A higher uniform viscosity would tend to slow down thinning everywhere, whereas in the simulations certain regions (e.g., near the tip) thin faster due to particle rearrangements.

- The model does not capture local clustering, collisions, or migration of particles, which produce strong spatial variations in stress and effective rheology.
- Necking is highly sensitive to these local microstructures, so a scalar bulk viscosity is insufficient to describe the asymmetric neck shift and satellite formation.

Therefore, while the Maron–Pierce model explains the global trend of viscosity versus $\phi$, the detailed breakup dynamics require fully resolved particle–fluid coupling as provided by the SPM–LBM framework.

**Exercise 8.3 Necking Location and Breakup Dynamics**

(a) Compare the neck-shift behavior between pure ink and particle-laden ink as shown in Fig. 8.15b–d.

(b) Explain why increased particle presence near the main droplet delays necking there but accelerates pinch-off at the thread tip.

**Solution**

(a) Figure 8.15b–d show that: For pure ink, the neck typically forms at a relatively symmetric location along the filament between the nozzle and the main droplet. The neck position shifts smoothly as the filament stretches and retracts. For particle-laden ink, necking becomes clearly off-centered. The neck tends to move toward the filament tip or regions with lower particle concentration, rather than remaining at the geometric center.

Thus, the presence of particles breaks the symmetry of the neck evolution, causing the pinch-off point to drift away from the center and leading to asymmetric droplet and satellite formation.

(b) Near the main droplet, particle concentration tends to be higher because particles are convected and accumulate there during jetting. This has two effects:

- High particle loading near the main droplet increases local effective viscosity and mechanical resistance, which slows neck thinning in that region.
- At the filament tip, where particle concentration is lower and the thread is thinner, there is less resistance to thinning. Capillary forces dominate and the neck can form and pinch-off more rapidly.

As a result, necking near the main droplet is delayed, while pinch-off at or near the thread tip is accelerated, shifting the breakup location toward the tip.

### *8.3.4 Effects of Heat Transfer on Injection*

This section examines the effects of heat transfer from the printhead. Thermal simulation details are shown in Fig. 8.17. The initial and substrate temperatures were set to 300 K, while the printhead was maintained at 350 K. Ink thermal

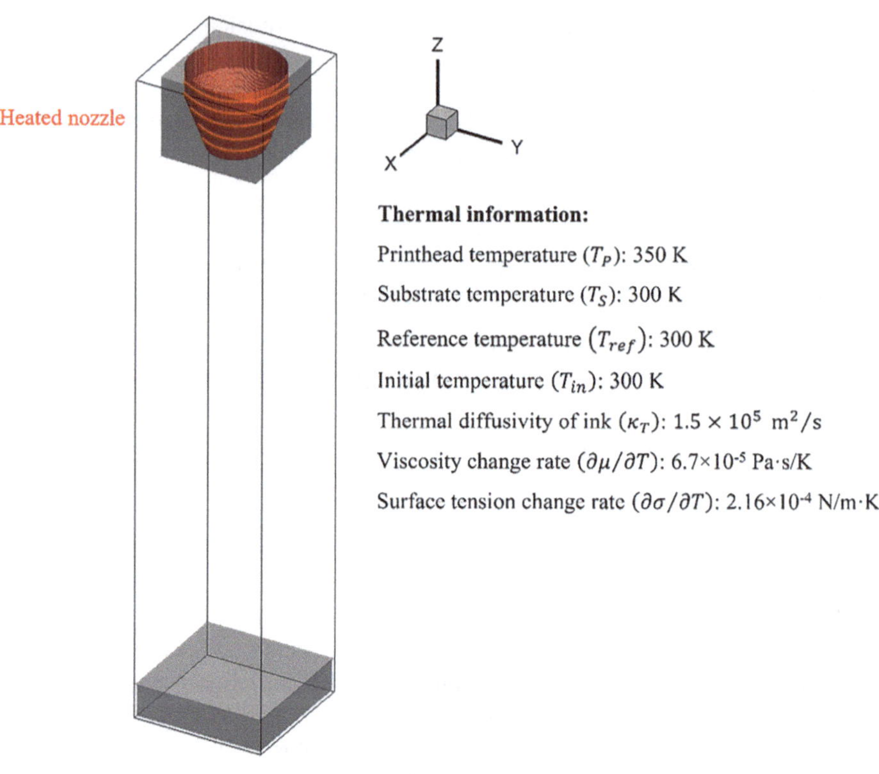


**Fig. 8.17** Thermal model configuration with nozzle heating applied

diffusivity was $1.5 \times 10^{-5}$ $m^2/s$, yielding a Péclet number of 5.13. Viscosity and surface tension vary with temperature as described in (8.15), referenced at 300 K with baseline properties from Table 8.3. Their temperature coefficients were $-6.7 \times 10^{-5}$ Pa · s/K and $-2.16 \times 10^{-4}$ N/m · K, respectively. Boundary conditions matched those of the previous section, except for thermal input. Particle distributions remained consistent across equal concentrations.

Figure 8.18 illustrates the time evolution of droplet injection with heat transfer, and Fig. 8.19 presents the corresponding temperature contours. When heat was applied, droplet viscosity decreased, leading to a slight increase in velocity and earlier separation. Figure 8.20 shows the breakup time for different printhead temperatures. In all cases, breakup occurred at the neck of the main droplet, and the breakup time decreased with increasing temperature.

These results suggest that heat buildup from repeated ejection may reduce process stability by accelerating droplet separation. A comparison between Fig. 8.18 and Fig. 8.13 reveals that, with heating, separation consistently occurs at 39 μs. Two factors explain this acceleration: (1) Lower viscosity increases injection velocity, raising the Reynolds number and promoting satellite formation; (2) A temperature gradient along the droplet tail induces thermocapillary instability [19], where surface tension gradients cause satellite droplets to coalesce post-formation.

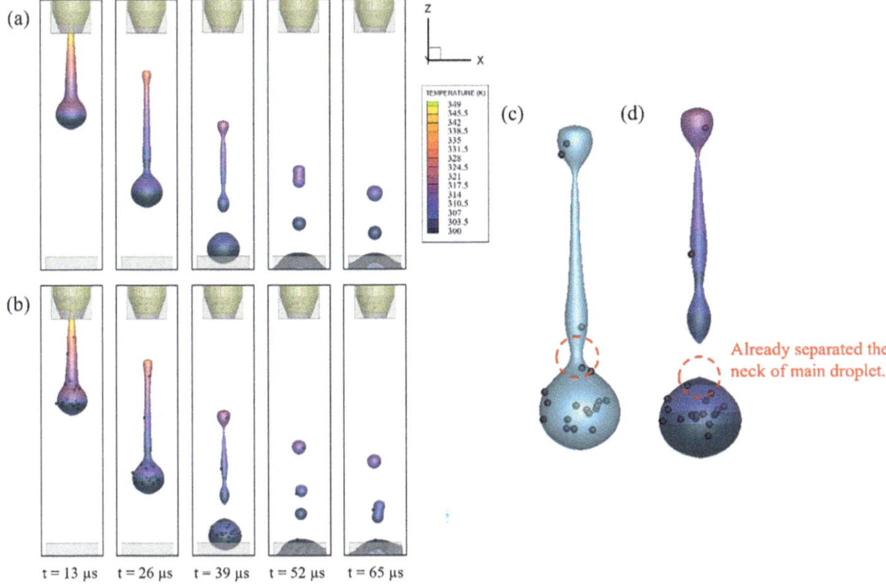


**Fig. 8.18** Time sequence of droplet ejection with heat transfer: (**a**) pure ink, (**b**) particle-laden ink ($\phi = 1.0\%$). Time progresses from left to right: 13, 26, 39, 52, and 65 μs. In both cases, main droplet separation occurred earlier compared to the non-heated condition

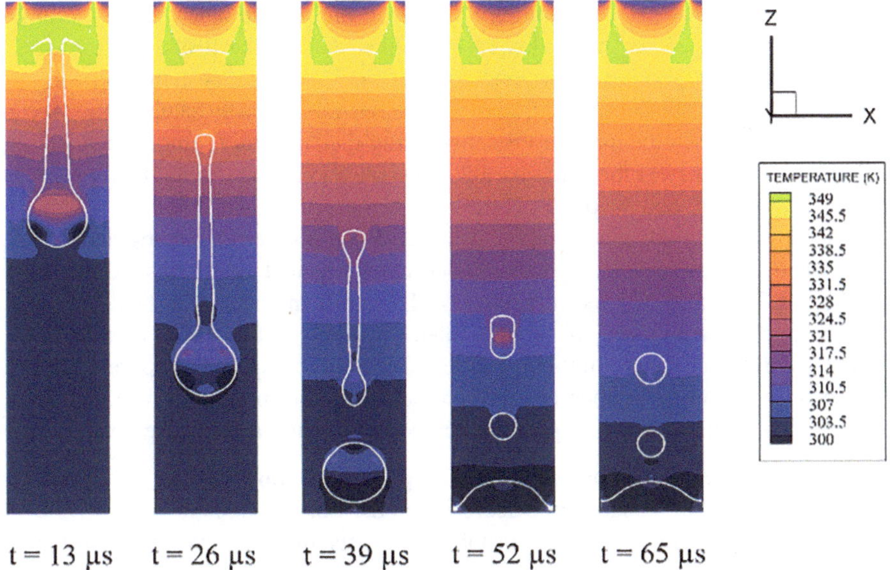


**Fig. 8.19** Temperature distribution of the pure droplet during ejection

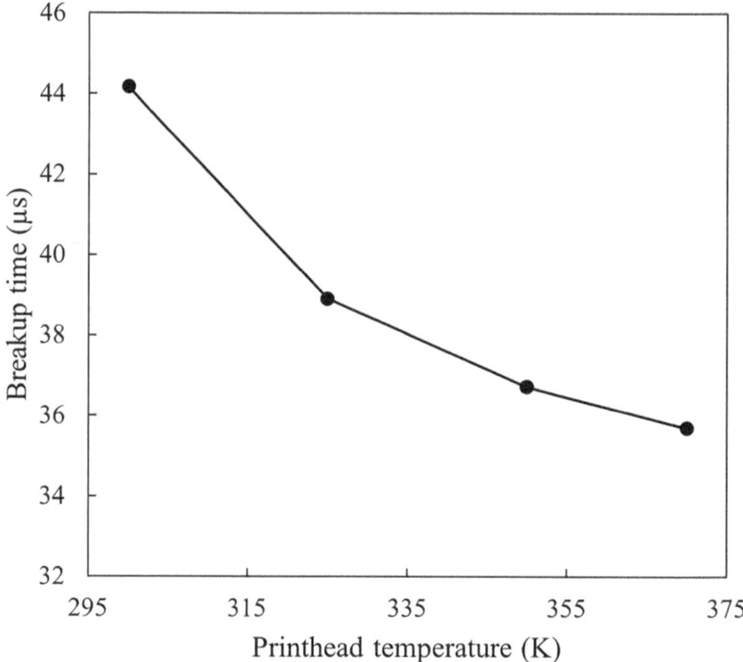


**Fig. 8.20** Breakup time according to printhead temperature

Figure 8.21 shows the asymmetry behavior of particle-laden droplets under thermal conditions. The trend closely resembles that of the 2% particle case in Fig. 8.14, as heat accelerates droplet separation. Up to ~25 μs, the asymmetric length follows a similar pattern in both cases. However, with heating, the thread detaches earlier from the main droplet, bypassing the "recovery" stage and resulting in greater asymmetry at the same particle concentration. The droplet then proceeds through "collision" and "capillary" phases.

Notably, heating advances breakup timing. While previous separations occurred at 41.0 and 40.4 μs (pure and particle-laden cases), heated conditions led to earlier breakup at 34.5 and 36.5 μs, respectively. This shift indicates that heat promotes necking and breakup even in particle-laden droplets. In Fig. 8.22, particle-laden droplets exhibited faster thinning at the thread tip compared to pure droplets. However, the elevated temperature shifted the location of minimum neck diameter, and particles near the main droplet suppressed the thread breakup rate.

**Exercise 8.4 Heat Transfer Effects on Droplet Formation**

(a) Explain why the breakup time decreases under nozzle heating.
(b) Discuss how thermocapillary instability affects filament evolution in heated jetting.
(c) Using Figs. 8.21 and 8.22, explain why heating removes the "recovery stage" and increases asymmetric ejection.

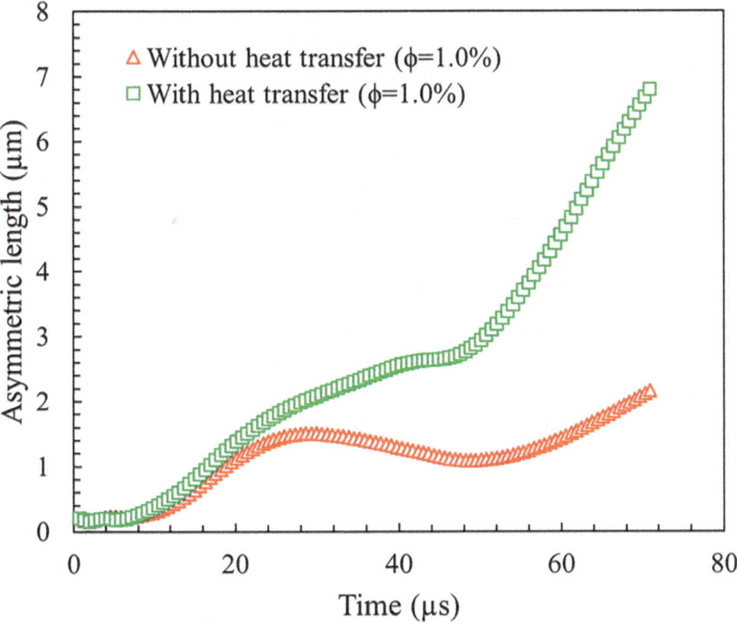


**Fig. 8.21** Asymmetric length over time with heat transfer. When heating was applied (green squares), the thread detached from the main droplet, eliminating the "recovery" stage

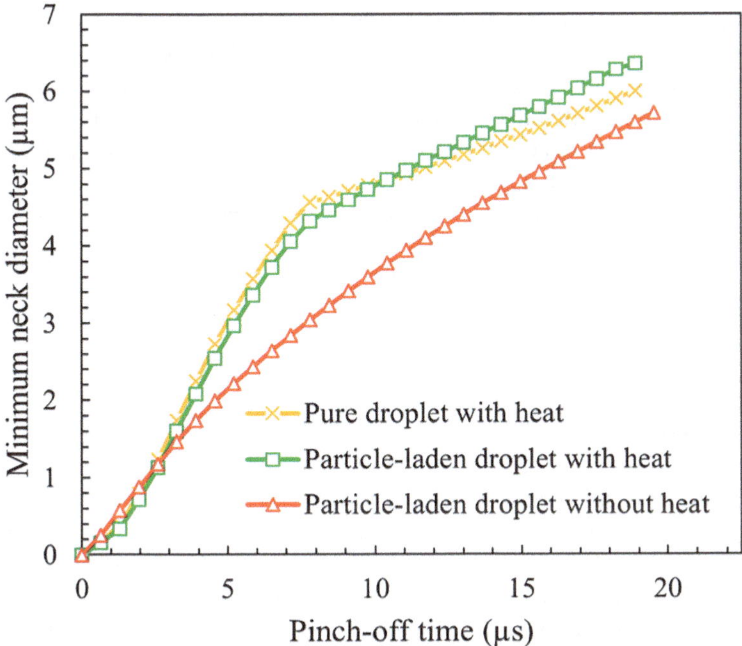


**Fig. 8.22** Comparison of minimum neck diameter. With heat applied, the particle-laden droplet exhibited a thinning transition

**Solution**

(a) Figures 8.18, 8.19, and 8.20 show that nozzle heating leads to shorter breakup times. The main mechanisms are as follows:

- Heating reduces viscosity $\mu(T)$, which lowers viscous resistance and allows the filament to stretch and thin more rapidly.
- Heating often reduces surface tension $\sigma(T)$ as well, which lowers the capillary time scale and makes pinch-off occur sooner.
- The resulting decrease in Ohnesorge number $Oh(T)$ and increase in $Z$ place the system deeper into the printable or breakup-prone regime.

Thus, the combined thermal effects accelerate filament thinning and shift the breakup time earlier compared with the isothermal case.

(b) Thermocapillary (Marangoni) effects arise from temperature-dependent surface tension. Surface tension gradients $\nabla_s\sigma(T)$ drive tangential flow along the interface from low- $\sigma$ (hot) regions toward high- $\sigma$ (cold) regions. In a heated nozzle, this can induce additional shear along the filament, enhancing stretching near certain regions and disturbing symmetry. These flows can locally thin the filament more rapidly, especially near the nozzle exit or at temperature gradients, thereby promoting earlier and more irregular pinch-off.

In short, thermocapillary instability adds a new driving mechanism for filament deformation, superimposed on inertial and capillary forces, and thus significantly modifies the filament evolution under heated conditions.

(c) Figures 8.21 and 8.22 show that, under isothermal conditions, the filament sometimes undergoes a recovery stage. With heating, Reduced viscosity and surface tension cause the filament to thin too quickly, leaving insufficient time for elastic or capillary recoil to restore symmetry. Thermocapillary flows introduce directional bias along the filament, so the neck is driven preferentially toward one side (often the tip), rather than evolving symmetrically. As a result, the system passes rapidly through the necking phase to pinch-off, without the intermediate recovery stage.

This combination of accelerated thinning and biased interfacial flow removes the recovery window and enhances asymmetric ejection, with breakup locations shifted and satellite formation patterns altered compared with the unheated case.

## 8.4 Conclusions

This chapter presented a comprehensive numerical investigation of the DoD inkjet printing process by integrating multiphase flow, particle dynamics, and thermal effects. The MRT color-gradient LBM was employed to simulate interface-resolved multiphase flow, while the SPM captured the fluid–particle interactions. Thermal transport was modeled using the convection–diffusion equation under the LBM framework. All components were implemented on a GPU-parallelized platform, enabling efficient simulation of complex droplet behaviors.

The simulation results revealed the critical roles of particle concentration and heat transfer in influencing droplet formation, asymmetric ejection, satellite droplet generation, and breakup dynamics. Even at low concentrations, particles disrupted filament thinning and altered breakup positions. Thermal effects further amplified these behaviors by reducing viscosity and surface tension, accelerating droplet separation and introducing thermocapillary instabilities.

These findings contribute to a deeper understanding of the physical mechanisms governing DoD inkjet printing and provide quantitative insights into how process parameters affect print quality. This integrated modeling approach offers practical value for optimizing high-resolution, high-reliability inkjet printing in advanced display manufacturing, functional material deposition, and micro-patterning processes. In particular, it enables predictive control of droplet behavior under varying fluid compositions and thermal environments, serving as a foundation for the design of next-generation inkjet systems.

**Summary**

1. **Motivation for particle-laden DoD inkjet printing**
    - DoD inkjet printing is a key technology for large-area, flexible, and high-resolution display manufacturing.
    - Real manufacturing environments use particle-laden functional inks, which introduce:
        - asymmetric jetting,
        - filament instability,
        - particle-induced satellite droplets,
        - clogging and placement errors,
        - thermal sensitivity due to repeated actuation.

2. **Governing physics and dimensionless parameters**
    - Droplet formation is governed by *Re*, *We*, *Oh*, and printability parameter $Z = 1/Oh$.
    - Stable DoD operation exists within a bounded region of *Re–Oh* space; satellite formation occurs when inertia dominates capillarity or viscosity is too low.
    - Realistic inkjet modeling thus requires precise control of inertial, viscous, and capillary scales, especially in particle-laden flows.

3. **Unified multiphysics numerical framework**
    - The study constructs a GPU-parallelized simulation platform integrating:
        - MRT color-gradient LBM for immiscible multiphase flow (jet formation, tailing, breakup).
        - SPM for fully resolved rigid-particle dynamics and particle–interface interactions.
        - LB-based heat transfer solver for convection–diffusion of temperature.
    - Each physical process (collision, streaming, force, particle update, thermal update) is mapped to dedicated GPU kernels.

4. **Droplet formation modes across *Re–Oh* space**
   - Four tested cases fall within printable regime.
   - Higher *Re*/lower *Oh* → faster Rayleigh–Plateau breakup, longer ligaments, more satellites.
   - Case A ($Re = 7.7$, $Oh = 0.36$) selected for particle and thermal studies due to stable but slow breakup.
5. **Particle-induced asymmetric jetting**
   - Particles disrupt axial symmetry, producing:
     - off-center filament evolution,
     - earlier or shifted necking,
     - asymmetric placement on substrate.
   - As particle concentration increases:
     - asymmetric length grows nonlinearly;
     - satellite generation increases;
     - breakup location shifts toward filament tip.
   - Three phases govern asymmetry evolution:
     - (1) recovery → (2) collision → (3) capillary-driven spreading.
6. **Particle effects on filament thinning and breakup**
   - Particle-laden inks exhibit different neck dynamics from pure ink despite small viscosity increases.
   - Particles accumulate heterogeneously, causing:
     - local curvature changes,
     - microstructural blockage of thinning,
     - accelerated pinch-off when trapped at the tip.
   - High enough concentration restricts particle mobility near the main droplet, slowing necking but promoting tip breakup.
   - Explains dual behavior: tip-accelerated breakup + delayed breakup near droplet body.
7. **Influence of heat transfer**
   - Nozzle heating (350 K → 300 K) decreases viscosity and surface tension.
   - Thermal effects lead to:
     - earlier breakup for both pure and particle-laden inks,
     - consistent pinch-off,
     - enhanced asymmetric jetting.
   - Temperature gradients induce thermocapillary (Marangoni) instability, accelerating thinning and promoting satellite formation.

- Heating removes the "recovery" stage by promoting immediate thread detachment.

## References


1. J. R. Castrejon-Pita, N. F. Morrison, O. G. Harlen, G. D. Martin, and I. M. Hutchings, "Experiments and lagrangian simulations on the formation of droplets in drop-on-demand mode," Phys. Rev. E **83**, 036306 (2011).
2. Y. Liu, and B. Derby, "Experimental study of the parameters for stable drop-on-demand inkjet performance," Phys. Fluids **31**, (2019).
3. L. Zhang, "Characteristics of drop-on-demand droplet jetting with effect of altered geometry of printhead nozzle," Sensors and Actuators A: Physical **298**, (2019), 111591.
4. L. Zhang, and X. Cheng, "Heat transfer dynamics inside the drop-on-demand inkjet printhead with a hybrid thermal lattice boltzmann model," Appl. Therm. Eng. **159**, (2019), 113789.
5. C. Bonnoit, T. Bertrand, E. Clément, and A. Lindner, "Accelerated drop detachment in granular suspensions," Phys. Fluids **24**, (2012).
6. Z. X. Wen, Q. Li, Y. Yu, and K. H. Luo, "Improved three-dimensional color-gradient lattice boltzmann model for immiscible two-phase flows," Phys. Rev. E **100**, 023301 (2019).
7. Y. Joo, S. Park, and K.-S. Kwon, "Numerical simulation of inkjet drop formation in piezo inkjet head," Journal of the Korea Academia-Industrial cooperation Society **17**, 641-647 (2016).
8. A. ten Cate, C. H. Nieuwstad, J. J. Derksen, and H. E. A. Van den Akker, "Particle imaging velocimetry experiments and lattice-boltzmann simulations on a single sphere settling under gravity," Phys. Fluids **14**, 4012-4025 (2002).
9. A. Kuzmin, M. Januszewski, D. Eskin, F. Mostowfi, and J. J. Derksen, "Lattice boltzmann study of mass transfer for two-dimensional bretherton/taylor bubble train flow," Chemical Engineering Journal **225**, 580-596 (2013).
10. G. H. McKinley, and M. Renardy, "Wolfgang von ohnesorge," Phys. Fluids **23**, (2011).
11. M. Bienia, M. Lejeune, M. Chambon, V. Baco-Carles, C. Dossou-Yovo, R. Noguera, and F. Rossignol, "Inkjet printing of ceramic colloidal suspensions: Filament growth and breakup," Chem. Eng. Sci. **149**, 1-13 (2016).
12. R. J. Furbank, and J. F. Morris, "Pendant drop thread dynamics of particle-laden liquids," International Journal of Multiphase Flow **33**, 448-468 (2007).
13. S. H. Maron, and P. E. Pierce, "Application of ree-eyring generalized flow theory to suspensions of spherical particles," Journal of colloid science **11**, 80-95 (1956).
14. A. Lindner, J. E. Fiscina, and C. Wagner, "Single particles accelerate final stages of capillary break-up," EPL (Europhysics Letters) **110**, (2015), 64002.
15. M. S. van Deen, T. Bertrand, N. Vu, D. Quéré, E. Clément, and A. Lindner, "Particles accelerate the detachment of viscous liquids," Rheologica Acta **52**, 403-412 (2013).
16. S. Chung, K. Cho, and T. Lee, "Recent progress in inkjet-printed thin-film transistors," Adv Sci (Weinh) 6, 1801445 (2019).
17. K. Y. Mitra, A. Alalawe, S. Voigt, C. Boeffel, and R. R. Baumann, "Manufacturing of all inkjetprinted organic photovoltaic cell arrays and evaluating their suitability for flexible electronics," Micromachines (Basel) 9, (2018).
18. S. P. Lin, and R. D. Reitz, "Drop and spray formation from a liquid jet.," Annual review of fluid mechanics **30**, 85-105 (1998).
19. O. A. Basaran, H. Gao, and P. P. Bhat, "Nonstandard inkjets," Annual Review of Fluid Mechanics **45**, 85-113 (2013).

# Chapter 9
# Particle Deposition Dynamics in Evaporating Droplets: Magnetic Particle Simulation

## 9.1 Introduction

When a sessile droplet is placed on a substrate, a three-phase contact line forms at the fluid–air–solid interface. The highest evaporation occurs at this contact line, inducing capillary flow toward it [1]. Suspended particles are transported along this flow and accumulate at the edge, forming a ring-like deposit after complete evaporation—commonly known as the coffee-ring effect.

Controlling this ring pattern is essential for improving the manufacturing efficiency of high-resolution, large-area displays. Conventional techniques such as spin coating and transfer printing struggle with precise pattern formation and are unsuitable for large-area production. In contrast, inkjet printing offers advantages such as high material reuse, scalability, and precise patterning capabilities [2]. However, non-uniform film formation can degrade display performance. While many studies [3] have addressed this challenge from a material and chemical perspective, inkjet printing is inherently a solution-based process. Thus, understanding the fluid dynamics of droplet evaporation is crucial for resolving the coffee-ring issue and enhancing film uniformity.

To control ring pattern formation, parameters such as temperature, contact angle, particle shape, and material properties can be modified. Among these, interparticle interactions play a critical role in determining the final deposition pattern. The use of a magnetic field offers an economical and non-invasive method for manipulating these interactions.

When an external magnetic field is applied to magnetic particles, they tend to align along the field direction, forming chain-like structures due to magnetic dipole–dipole interactions. Al-Milaji et al. [4] utilized this phenomenon to actively control particle deposition patterns. Their results showed that the structures formed under magnetic influence significantly affected the resulting ring formation.

 
H. M. Lee, J. S. Lee, *Lattice Boltzmann Methodology for Single-Phase and Multiphase Nanoparticle Modeling*,
https://doi.org/10.1007/978-981-95-9117-6_9

However, their findings also highlighted a limitation: experimental observations alone were insufficient to fully analyze how interparticle interactions contribute to the adhesion and retention mechanisms of particles during evaporation. This underscores the need for numerical simulations to provide deeper insight into the underlying physics of magnetic particle behavior and pattern formation in evaporating droplets.

This chapter aims to control ring pattern formation during the evaporation of a droplet containing nanoparticles on a substrate by applying an external magnetic field. The phase change process is modeled using the pseudopotential LBM, while the dynamics of nanoparticles are described by solving the Langevin equation in conjunction with DLVO theory. Through this framework, readers can learn the methodology for simulating both the evaporation behavior of the droplet and the active interparticle interactions among nanoparticles during the drying process.

**Exercise 9.1 Coffee-Ring Formation**

(a) Explain how non-uniform evaporation along the droplet interface generates the coffee-ring effect.

(b) Describe how contact line pinning alters internal flow during evaporation.

**Solution**

(a) Along an evaporating sessile droplet, the local evaporation flux is largest near the contact line because vapor can diffuse out more easily there. This non-uniform evaporation means that more liquid is removed at the edge than at the center. To conserve mass, liquid from the interior is driven radially outward toward the contact line. This outward capillary flow transports suspended particles from the droplet interior to the edge, where they accumulate as the solvent evaporates completely. The result is a ring-like deposit with a high particle concentration near the perimeter—the classical coffee-ring effect.

(b) When the contact line is pinned, the droplet footprint radius remains nearly constant while the height decreases. Evaporation near the pinned line continues to be strong, so radial outward flow persists throughout much of the evaporation process. Because the edge cannot retract, the only way to supply liquid to the intense evaporation region is via a sustained capillary flow from the interior to the rim. This enhances particle transport toward the edge and strengthens the coffee-ring deposition.

## 9.2 Numerical Methods

This section presents a simulation of the evaporation process of a sessile droplet containing nanoparticles and analyzes the resulting particle distribution. To achieve this, an LBM framework is proposed that couples a phase change model, a nanoparticle dynamics model, and a heat transfer model. The phase change behavior is modeled using the pseudopotential approach, while the random motion and interparticle cohesion of the nanoparticles are represented through Brownian forces and van der Waals interactions, respectively. To enable active control of particle distribution, a magnetic field-induced particle rearrangement model is incorporated. The heat transfer model follows the methodology described in Chap. 8.

The following sections provide detailed descriptions of the pseudopotential model and the nanoparticle dynamics model employed in this simulation.

### 9.2.1 Pseudopotential LBM

The pseudopotential LBM shares the same governing equation structure as the conventional LBM, with the exception of the introduction of a body force term $S_i$, which is incorporated to represent the interaction force between two fluid phases. This additional force term enables phase separation and interfacial dynamics by inducing attractive or repulsive forces based on a pseudopotential function. Aside from this modification, the core formulation—such as collision, streaming, and macroscopic variable extraction—remains consistent with that of standard LBM.

$$f_i(\mathbf{x} + \mathbf{e}_i\Delta t, t + \Delta t) - f_i(\mathbf{x}, t) = -\frac{1}{\tau}[f_i(\mathbf{x}, t) - f_i^{eq}(\mathbf{x}, t)] + S_i. \tag{9.1}$$

The body force term is defined as

$$S_i = f_i^{eq}(\rho, \mathbf{u} + \Delta\mathbf{u}) - f_i^{eq}(\rho, \mathbf{u}). \tag{9.2}$$

Here, $\Delta\mathbf{u} = \mathbf{F}\Delta t/\rho$ is the velocity increment due to the external force.

Phase separation was achieved using interparticle interaction forces. The discretized form of the interaction force is given as:

$$\mathbf{F}_{int} = -3c_0 g\left[\beta\psi(\mathbf{x})\sum_i \omega_i\psi(\mathbf{x} + \mathbf{e}_i\Delta t)\mathbf{e}_i + 0.5(1-\beta)\sum_i \omega_i\psi^2(\mathbf{x} + \mathbf{e}_i\Delta t)\mathbf{e}_i\right], \tag{9.3}$$

where $\beta$ is the interparticle weighting factor, $g$ is the parameter that controls the interaction force, and $c_0$ is a constant determined by the lattice structure. The effective mass, $\psi$, is defined as

$$\psi(\mathbf{x}) = \sqrt{\frac{2(p - \rho c_s^2)}{c_0 G}}, \tag{9.4}$$

The Peng–Robinson equation of state (EOS) can be used to calculate pressure in the pseudopotential LBM framework to improve thermodynamic consistency. The equation is given by:

$$p = \frac{\rho RT}{1 - b\rho} - \frac{a\varepsilon(T)\rho^2}{1 + 2b\rho - b^2\rho^2}, \tag{9.5}$$

The temperature convection–diffusion equation is solved using the temperature distribution function $g_i$ within the LBM framework. However, to accurately simulate phase change phenomena, a phase transformation term, $\Phi$ is incorporated into the evolution equation. This term accounts for the latent heat exchange associated with the liquid–vapor transition and modifies the local energy balance accordingly. By introducing this additional source term, the model can capture the temperature drop near the interface and the associated mass loss due to evaporation, enabling realistic simulation of droplet evaporation dynamics.

$$g_i(\mathbf{x} + \mathbf{e}_i\Delta t, t + \Delta t) - g_i(\mathbf{x}, t) = -\frac{1}{\tau_T}\left[g_i(\mathbf{x}, t) - g_i^{eq}(\mathbf{x}, t)\right] + \Delta t\omega_i\Phi. \tag{9.6}$$

The phase transformation term can be expressed as

$$\Phi = T\left[1 - \frac{1}{\rho c_v}\left(\frac{\partial \rho}{\partial T}\right)_v\right]\nabla \cdot \mathbf{U}, \tag{9.7}$$

where $c_v$ is the specific heat capacity and $\mathbf{U}$ is the real fluid velocity.

### 9.2.2 Nanoparticle Dynamics

Nanoparticle distribution within a sessile droplet is influenced by hydrodynamic, capillary, and interparticle forces. The total force acting on the $m^{th}$ particle, $\mathbf{F}_P^m$, is given as:

$$\mathbf{F}_P^m = \mathbf{F}_D^m + \mathbf{F}_C^m + \mathbf{F}_B^m + \mathbf{F}_{vdW}^m + \mathbf{F}_R^m + \mathbf{F}_M^m + \mathbf{F}_{ev}^m, \tag{9.8}$$

Here, $\mathbf{F}_D^m$ is the drag force, and $\mathbf{F}_C^m$ is the capillary force arising from droplet–particle interactions. $\mathbf{F}_B^m$ and $\mathbf{F}_{vdW}^m$ represent Brownian and van der Waals forces, capturing random motion and interparticle attraction. $\mathbf{F}_R^m$ is the repulsive collision force preventing particle overlap. $\mathbf{F}_M^m$ and $\mathbf{F}_{ev}^m$ denote magnetic forces applied for particle manipulation.

The drag force is expressed as

$$\mathbf{F}_D^m = -\gamma\left[\mathbf{U}_P^m(t) - \mathbf{U}^*\left(\mathbf{X}_P^m, t\right)\right], \tag{9.9}$$

where $\gamma = 6\pi\mu R_P$ is the Stokes drag coefficient and $R_P$ is the radius of particle. The particle's linear velocity is denoted as $\mathbf{U}_P^m$, while $\mathbf{U}^*$ represents the fluid velocity evaluated at the particle center position $\mathbf{X}_P^m$.

Particle motion was constrained by surface tension, which suppressed their escape from the droplet. The capillary force at each fluid lattice node was computed as:

$$\mathbf{F}_C^m = -\xi \sum_i G_P(\mathbf{x} + \mathbf{e}_i \Delta t) \psi(\mathbf{x} + \mathbf{e}_i \Delta t), \tag{9.10}$$

where $\xi$ is the weighting factor, and $G_P$ is the fluid–particle interaction constant.

The Brownian force is applied to model the random motion of nanoparticles induced by thermal fluctuations. To compute this force, a stochastic differential equation (SDE) [5] is solved, typically in the form of a Langevin equation:

$$d\mathbf{r}^m(t) = \mathbf{U}_P^m(t)\Delta t + \sqrt{2D_P} d\mathbf{W}(t) + \frac{\mathbf{F}_P^m}{\gamma} \Delta t, \tag{9.11}$$

where $D_P = k_B T/\gamma$ is the diffusion coefficient. $k_B$ is the Boltzmann constant. $\mathbf{W}(t)$ is a Gaussian random variable with variance $\langle |\mathbf{W}(t + dt) - \mathbf{W}(t)|^2 \rangle = 3dt$. The van der Waals force, representing mutual attraction between nanoparticles, induces their aggregation. The attractive force between two identical spherical particles is given by [6]:

$$\mathbf{F}_{vdW} = \frac{A_H}{6} \left[ \frac{2(r_P + 1)}{r_P^2 + 2r_P} - \frac{r_P + 1}{\left(r_P^2 + 2r_P\right)^2} - \frac{2}{r_P + 1} - \frac{1}{\left(r_P + 1\right)^3} \right] \mathbf{n}^{mn}, \tag{9.12}$$

$A_H$ is the Hamaker constant, and $\mathbf{n}^{mn}$ is the unit vector between $m$th and $n$th particles. $r_P$ is defined as $r_P = |\mathbf{R}^{mn}|/2R_P - 1$ ($\mathbf{R}^{mn} = \mathbf{X}^m - \mathbf{X}^n$).

Finally, when an external magnetic field is applied to magnetic particles, the interaction force between the particles is computed using the dipole–dipole interaction formula. The magnetic interaction force $\mathbf{F}_m^P$ is given by [7]:

$$\mathbf{F}_m^P = \frac{3\mu_0 \mu_s m^2}{4\pi r^4} \left\{ \left[ 1 - 5\left(\widehat{\mathbf{m}} \cdot \widehat{\mathbf{j}}\right)^2 \right] \widehat{\mathbf{j}} + 2\left(\widehat{\mathbf{m}} \cdot \widehat{\mathbf{j}}\right) \widehat{\mathbf{m}} \right\}, \tag{9.13}$$

where $\mu_0$ and $\mu_s$ are the vacuum permeability and the magnetic permeability of medium. $\widehat{\mathbf{m}}$ and $\widehat{\mathbf{j}}$ are the unit vectors of the magnetic dipole moment and the center point between the particles. To prevent particle overlap due to finite volume, a repulsive force is introduced [7].

$$\mathbf{F}_{ev}^P = A \frac{3\mu_0 \mu_s m^2}{4\pi \left(2R^P\right)^4} \exp\left[ -B\left( \frac{r}{2R^P} - 1 \right) \right] \widehat{\mathbf{r}}, \tag{9.14}$$

where $A = 2$ and $B = 10$.

## 9.3 Results

### 9.3.1 *Simulation Setup*

Figure 9.1 shows a schematic of the simulation domain. A sessile droplet containing randomly distributed nanoparticles is initially placed at the center of the substrate. The droplet is set to a uniform initial temperature and undergoes evaporation due to heat from the substrate. After complete evaporation, the final particle deposition pattern is observed. Particles are considered deposited when located near the droplet interface ($\rho(\mathbf{X}_P^m) < 0.9\rho_l$) and sufficiently close to the substrate ($|\mathbf{X}_P^m - \mathbf{X}_w| < 4R_P$) [8]. Additional simulation parameters are summarized in Table 9.1.

The analysis was performed using the following dimensionless variables.

$$\tau^* = \frac{\alpha t}{D_0^2}, D^* = \frac{D}{D_0}, l_P^* = \frac{\sum_m^{N_P} |\mathbf{X}_P^m - \mathbf{X}_D|/D_0}{N_P}, V_P^* = \frac{\sum_m^{N_P} \left(|\mathbf{X}_P^m - \mathbf{X}_D|/D_0 - l_{av}^*\right)^2}{N_P} \tag{9.15}$$

Here, $\tau^*$ is the nondimensional time step defined by the Fourier number, with $\alpha$ as thermal diffusivity. $D_0$ is the initial contact diameter, and $D^*$ is the normalized diameter used to evaluate the contact area during evaporation. $l_P^*$ and $V_P^*$ represent the average and variance of the particle dispersion length, respectively. These

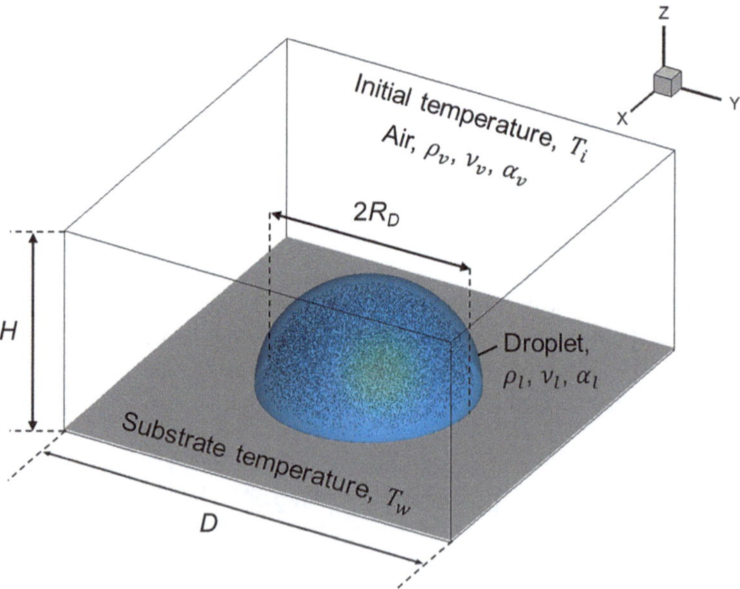


**Fig. 9.1** Schematic of the simulation domain

**Table 9.1** Domain specifications and model parameters

| Parameters | Symbols | Physical units |
|---|---|---|
| Length conversion factor | $\delta_x$ | $5.0 \times 10^{-7}$ m |
| Mass conversion factor | $\delta_m$ | $1.72 \times 10^{-17}$ kg |
| Time conversion factor | $\delta_t$ | $1.64 \times 10^{-8}$ s |
| Temperature conversion factor | $\delta_T$ | $8.88 \times 10^3$ K |
| Width of simulation domain | $D$ | 92.0 μm |
| Height of simulation domain | $H$ | 48.0 μm |
| Droplet radius | $R_D$ | 22.5 μm |
| Particle radius | $R_P$ | 50.0 nm |
| Initial temperature | $T_i$ | 517.68 K |
| Substrate temperature | $T_w$ | 528.06 K |
| Droplet density | $\rho_l$ | 991.9 kg/m$^3$ |
| Air density | $\rho_v$ | 27.13 kg/m$^3$ |
| Droplet viscosity | $\nu_l$ | $1.02 \times 10^{-6}$ m$^2$/s |
| Air viscosity | $\nu_v$ | $7.23 \times 10^{-6}$ m$^2$/s |
| Liquid thermal diffusivity | $\alpha_l$ | $1.00 \times 10^{-6}$ m$^2$/s |
| Vapor thermal diffusivity | $\alpha_v$ | $3.58 \times 10^{-6}$ m$^2$/s |
| Critical temperature | $T_c$ | 647.096 K |
| Critical density | $\rho_c$ | 322.0 kg/m$^3$ |

metrics are used to estimate the average distance from the droplet center and particle distribution density.

### 9.3.2 *Validation Results*

The numerical model was validated by comparing the simulated particle deposition pattern with the experimental results of Yunker et al. [9]. Figure 9.2 shows the final particle distribution and density from the droplet center to the initial contact line, normalized by the initial droplet radius for nondimensional analysis. Compared to a previous simulation study [8], the present model reproduced a thin ring near the contact line. The maximum particle density showed a deviation of only 5.92%, indicating high accuracy. Within the ring, particles were densely packed, and the density was influenced by particle size and shape. Since the particle diameter in this model was smaller than the lattice spacing, a higher packing density was achieved.

### 9.3.3 *Effects of Wettability on Particle Distribution Patterns*

Substrate wettability plays a crucial role in determining particle deposition patterns. Due to rapid evaporation at the contact line, an outward capillary flow is induced,

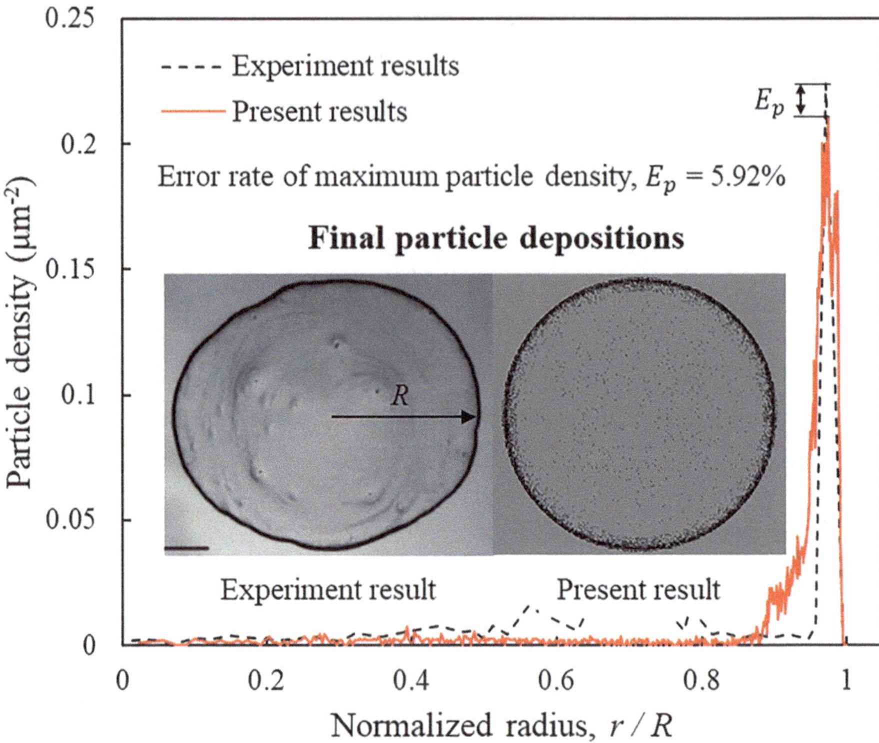


**Fig. 9.2** Validation of particle deposition. The simulation produced a thin particle ring at the contact line. The maximum particle density showed a 5.92% error, indicating high accuracy

leading to particle accumulation and the formation of a coffee-ring if the contact line remains pinned. Thus, wettability alters internal flow behavior during evaporation and enables control over particle distribution.

Sessile droplet evaporation on solid surfaces occurs in two primary modes, depending on contact line motion. In the constant contact radius (CCR) mode, the contact line remains fixed while the contact angle decreases. In contrast, the constant contact angle (CCA) mode involves a fixed contact angle with a receding contact line. These modes can be modeled using contact angle hysteresis: when the contact angle lies between the receding ($\theta_{re}$) and advancing ($\theta_{ad}$) angles ($\theta_{re} < \theta < \theta_{ad}$), it is allowed to vary. Once it reaches $\theta_{re}$, the contact angle is held constant and the contact line begins to retract.

Figures 9.3, 9.4, and 9.5 illustrate the influence of substrate wettability—specifically the receding contact angle—on droplet morphology and particle distribution. Figure 9.3 shows the morphological changes of a droplet containing 16,384 particles as a function of the receding angle. As the angle increased, the transition from CCR to CCA mode occurred earlier, resulting in a more rapid decrease in contact diameter (red dotted line). Figure 9.3d, e quantitatively show the changes in contact angle and diameter, highlighting abrupt transitions at the mode change.

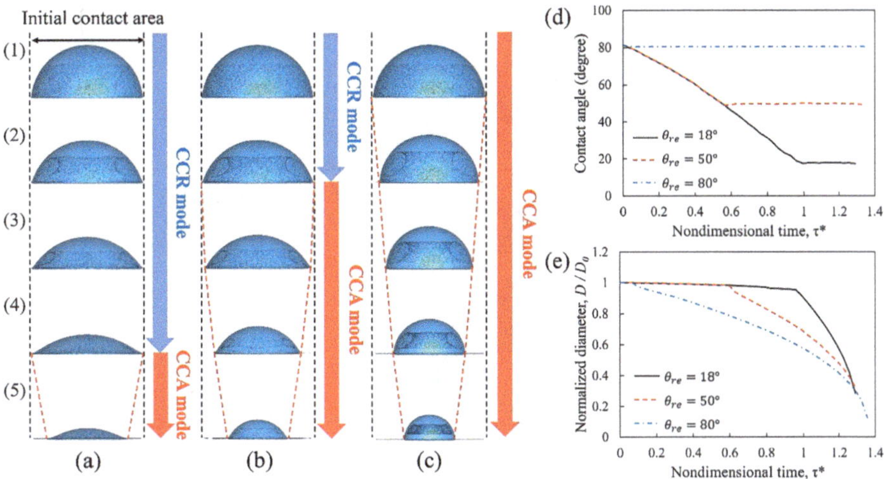


**Fig. 9.3** Snapshots of the droplet evaporation process for different receding angles: (**a**) $\theta_{re} = 18°$, (**b**) $\theta_{re} = 50°$, and (**c**) $\theta_{re} = 80°$ (non-pinning case). Frames (1)–(5) correspond to nondimensional times $\tau^* = 0.0993, 0.3640, 0.6286, 0.8933$, and 1.158, respectively. (**d**) Contact angle and (**e**) normalized contact diameter as functions of time for each receding angle

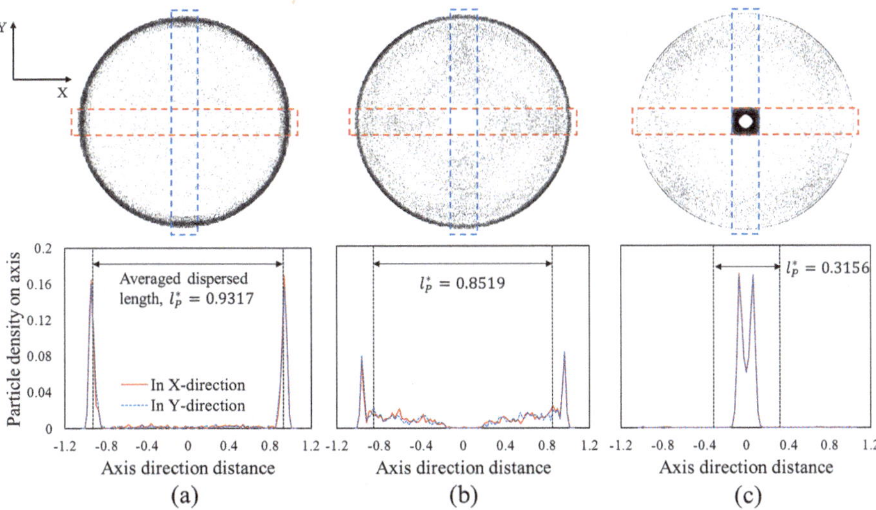


**Fig. 9.4** Particle distribution under different receding angles: (**a**) $\theta_{re} = 18°$, (**b**) $\theta_{re} = 50°$, and (**c**) $\theta_{re} = 80°$

Wettability changes not only affected droplet shape but also significantly influenced particle distribution. Figure 9.4 shows the final deposition patterns and corresponding average radial distances ($l_P^*$) of particles from the droplet center. At low receding angles, particles formed distinct coffee-ring patterns. As the angle

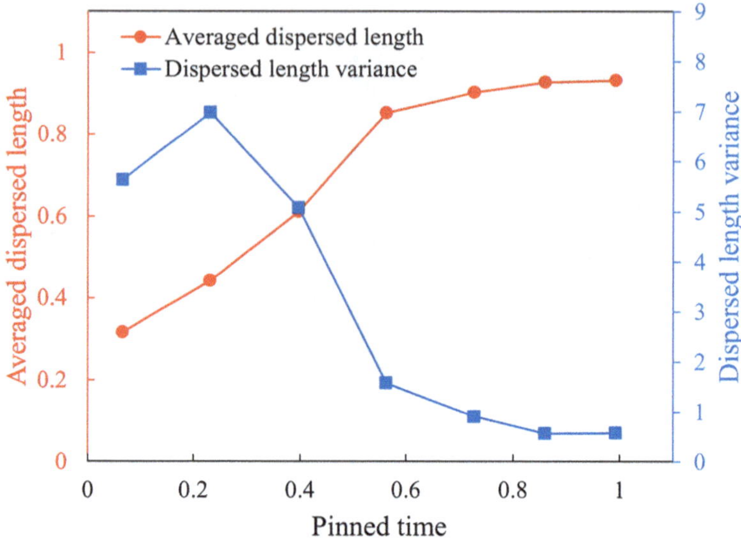


**Fig. 9.5** Average dispersed length and its variance as functions of pinning time

increased, particles shifted toward the center, eventually forming compact, ball-like deposits in unpinned droplets. This behavior was governed by the pinning duration of the contact line.

Figure 9.5 further quantifies the relationship between pinning time and particle dispersion. Longer pinning resulted in greater average dispersion and lower variance, indicating ring formation. In contrast, shorter pinning led to concentrated central deposition and reduced variance.

These differences in particle deposition arise from competing internal flows during evaporation: (1) capillary-driven radial flow caused by enhanced evaporation at the contact line, and (2) Marangoni recirculation induced by surface tension gradients. The dominance of either flow—and thus the resulting deposition pattern—depends on whether the contact line remains pinned.

To elucidate the effect of contact angle hysteresis on particle distribution, time-dependent streamlines are shown in Fig. 9.6. Initially, strong vortices were observed in both pinned and non-pinned droplets. However, their internal flow patterns diverged as evaporation progressed. In the pinned droplet, vortex strength diminished over time, and capillary-driven radial flow became dominant at $\tau^* = 0.6286$ and 0.8933. In contrast, the non-pinned droplet maintained Marangoni-driven vortices throughout evaporation.

This competition between capillary and Marangoni flows can be characterized using the Péclet number, $Pe = R_D U_{ev}/\nu$, representing the ratio of convective to diffusive transport, and the Marangoni number, $Ma = -(\partial\sigma/\partial T)(H_D \Delta T/\mu\alpha)$, representing the ratio of Marangoni to diffusive transport. Here, $R_D$ is the contact radius, $U_{ev}$ is the evaporative velocity estimated from the evaporation flux and contact area, $H_D$ is the droplet height, and $\Delta T$ is the surface temperature difference.

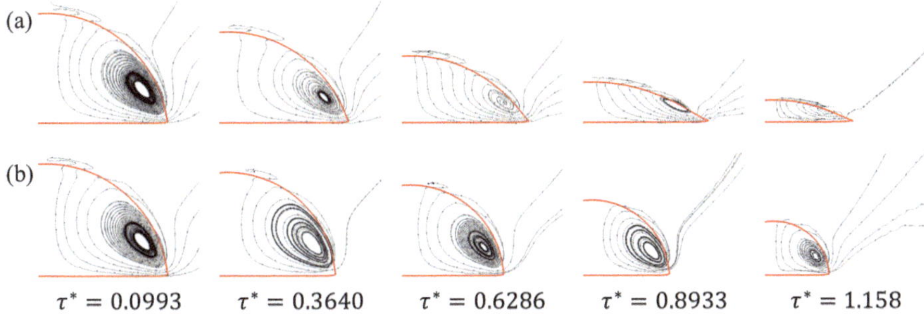


**Fig. 9.6** Streamline of evaporating droplets. (**a**) $\theta_{re} = 18°$, (**b**) $\theta_{re} = 80°$

Figure 9.7 presents the time evolution of *Pe* and *Ma* for the cases in Fig. 9.6. Both values decrease with time, but *Ma* drops more rapidly in the pinned droplet, indicating diminished Marangoni influence. This is attributed to the stronger interfacial temperature gradient in the non-pinned droplet.

Simulations revealed that capillary flow transports particles toward the contact line, promoting deposition, while Marangoni flow traps particles within the droplet, delaying their settlement. The persistence of Marangoni circulation in the non-pinned case inhibited particle adhesion and led to central accumulation.

**Exercise 9.2 Internal Flow Mechanisms**

(a) Compare the dominant flow mechanisms between pinned and non-pinned droplets using Figs. 9.3 and 9.6.

(b) Discuss how wettability (receding angle) modifies particle deposition patterns.

**Solution**

(a) For a pinned droplet (Fig. 9.3), the dominant flow is a robust, persistent outward capillary circulation. The velocity field shows strong radial motion near the substrate, driving particles toward the rim. Any Marangoni recirculation that might arise is relatively weak compared with the capillary-driven flux.

For a non-pinned droplet (Fig. 9.6), the contact line can recede as the droplet volume shrinks. In this case, the outward capillary flow is weaker and often transient. Instead, Marangoni-driven vortical structures can become dominant, especially when surface tension gradients exist along the interface. The flow becomes more recirculatory, transporting particles both outward and inward and enabling more uniform deposition or even central accumulation, rather than a pure ring.

(b) Wettability, expressed via the receding contact angle, determines how easily the contact line can move during evaporation:

- A small receding angle (strongly wetting surface) favors pinning; the contact line tends to remain fixed until the droplet becomes very thin. This maintains a large footprint and strong outward flow, promoting a pronounced coffee-ring.

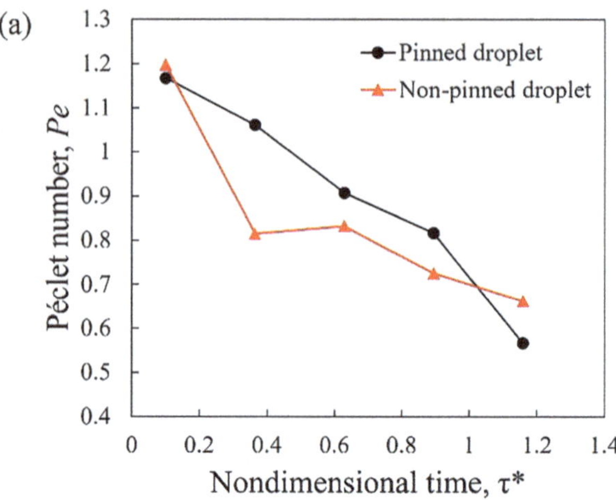


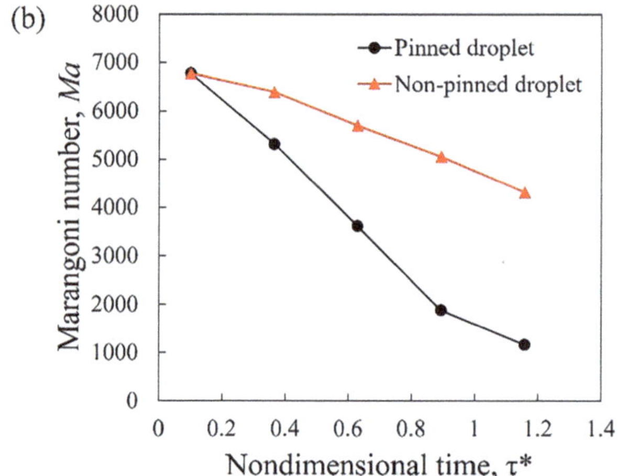


**Fig. 9.7** Variations in (**a**) Péclet number and (**b**) Marangoni number

- A larger receding angle allows the contact line to retract more readily. As the edge recedes, the radial outward flux is reduced and can be balanced by Marangoni or other recirculating flows. Particles are then less confined to the rim and can deposit in the interior, producing narrower rings or even more uniform coatings.

Thus, by tuning the receding angle, the relative importance of capillary versus Marangoni flow is adjusted, and the final particle pattern can be continuously varied from strong ring-like to nearly uniform.

### 9.3.4 *Effect of External Magnetic Field on Particle Distribution*

In the previous section, it was shown that contact angle hysteresis alters the internal flow of the droplet and, consequently, affects particle distribution. Here, we examine how particle interactions under magnetic fields influence deposition behavior. When subjected to a magnetic field, the magnetic nanoparticles tend to align and aggregate along the field direction. By adjusting the direction of the external magnetic field, the ring-like deposition pattern can be modified, enabling active control over the final particle distribution.

In this study, 16,384 particles were used, with the receding angle fixed at 18°. A uniform magnetic field was applied in the vertical (z) direction, perpendicular to the substrate.

Figure 9.8 presents the particle density profile as a function of magnetic field strength. As the field increased, the maximum particle density decreased and its peak shifted toward the droplet center, indicating progressive formation of a ring pattern influenced by the magnetic field.

Figure 9.9 shows the corresponding deposition patterns for each field strength. The ring thickness $\delta$, defined as the width where the particle density exceeds 20% of the peak value, is also indicated. A comparison between the qualitative results in Fig. 9.9 and the density profiles in Fig. 9.8 confirms the same trend.

To examine the influence of magnetic force on deposition behavior, the average velocity of all particles is shown in Fig. 9.10. Across all cases, particle velocity followed a similar trend, allowing the process to be divided into three distinct stages based on velocity variation. The right panels of the figure display particle distributions corresponding to each stage. Stage I (blue dotted line): Heat transfer to the droplet was insufficient, and particles initially rearranged along the droplet interface

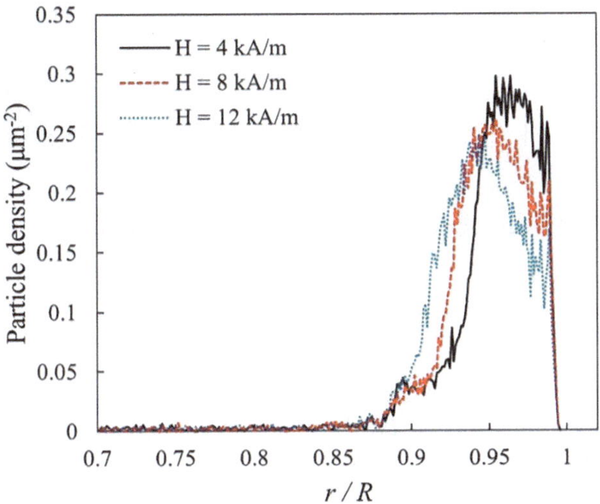


**Fig. 9.8** Particle distribution under varying magnetic field strength

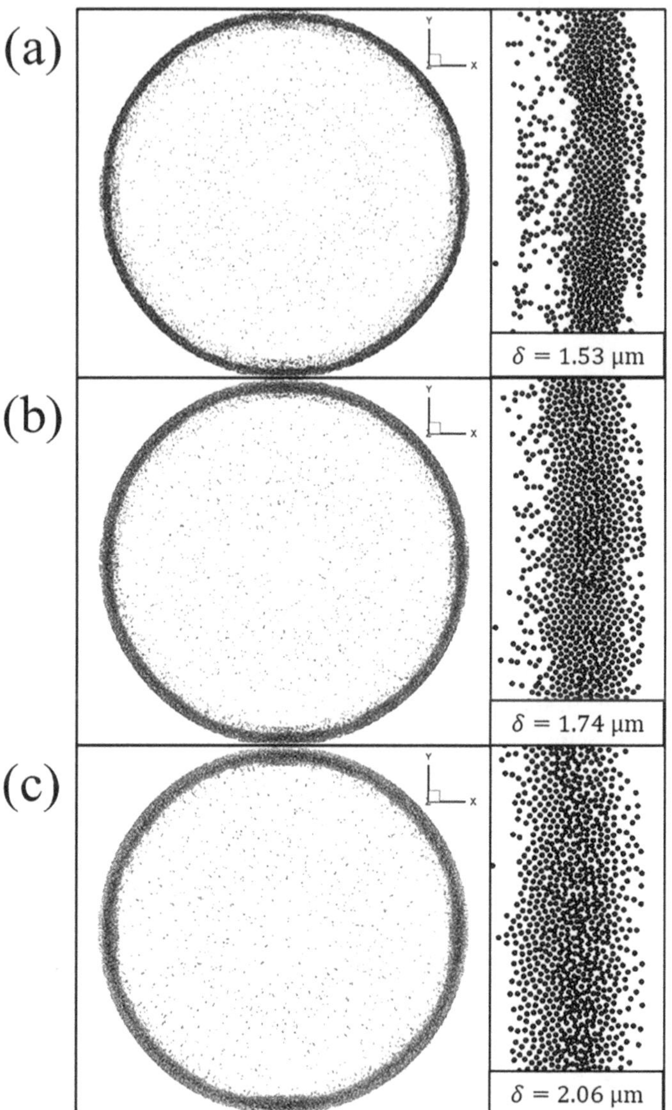


**Fig. 9.9** Ring pattern under different magnetic field strengths. (**a**) H = 4 kA/m, (**b**) H = 8 kA/m, and (**c**) H = 12 kA/m

due to interfacial tension. During this stage, particles near the center migrated toward the droplet boundary, and the average velocity slightly declined as rearrangement progressed. Stage II (red dashed line): The average velocity initially decreased, then sharply increased as particles became trapped in the Marangoni-induced vortex.

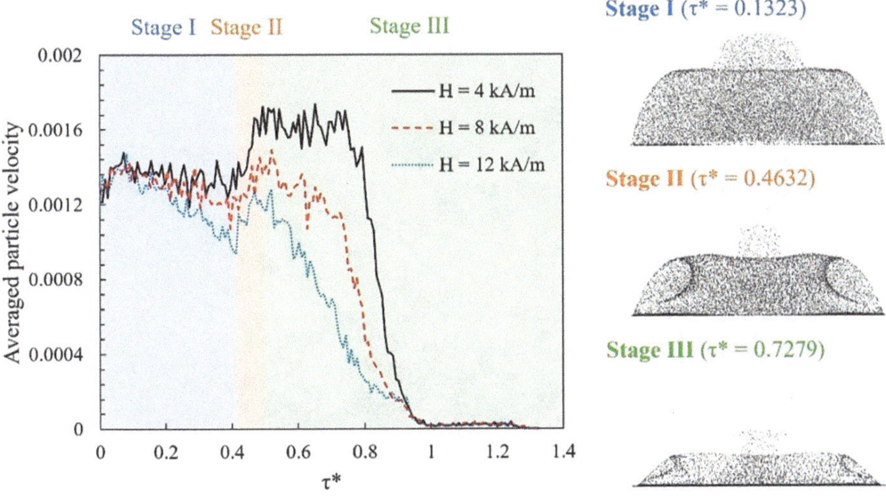


**Fig. 9.10** Average particle velocity over time. In all cases, the velocity profile exhibits three distinct stages. Right panels show the corresponding particle distributions at each stage for H = 12 kA/m

Ring-like particle structures formed during this phase, as shown in the distribution. Stage III: Particle adhesion to the substrate increased, leading to a gradual decline in average velocity.

Throughout all stages, the magnetic field reduced the rate of change in average particle velocity, as magnetically induced particle bundles resisted internal flow. Figure 9.11 compares particle distributions at $\tau^* = 0.5956$ with and without magnetic force. Distinct particle behaviors were observed in regions associated with Marangoni entrapment and substrate adhesion. These areas are highlighted with red and blue dotted circles, respectively.

In the red region, the applied magnetic field aligned particles into bundles along the field direction. These bundles acted as obstacles to the internal flow, contributing to the reduced particle velocity. In the blue region, the ring became thicker under the magnetic field, and particle adhesion was enhanced. This was due to magnetically induced attraction of suspended particles toward those already adhered to the substrate. Consequently, stronger magnetic fields led to a faster decline in particle velocity during Stage III.

Figure 9.12 presents the particle deposition rate over time. As described earlier, particle adhesion primarily occurs during Stage III, where the deposition rate increased sharply. Interestingly, the adhesion rate decreased with increasing magnetic field strength, contrary to the expectation that magnetically attracted particles would adhere more rapidly—as suggested in Fig. 9.11.

Two factors explain this behavior. First, stronger magnetic fields increased resistance to flow along the contact line, slowing particle transport. Second, particles formed vertically aligned chain structures under the magnetic field, delaying

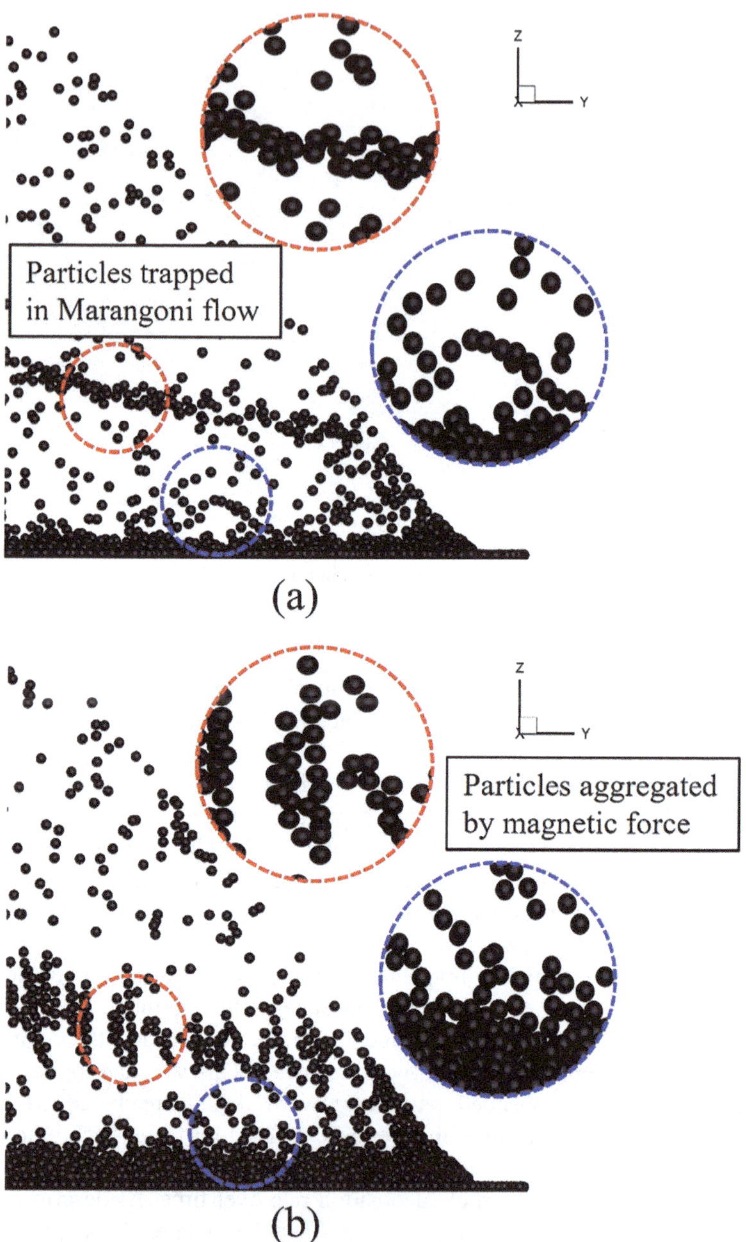


**Fig. 9.11** Particle structures near the contact line. (**a**) Without magnetic field and (**b**) with H = 12 kA/m. The red dotted circle highlights particles aligned by the Marangoni flow, while the blue dotted circle shows particles adhered to the substrate

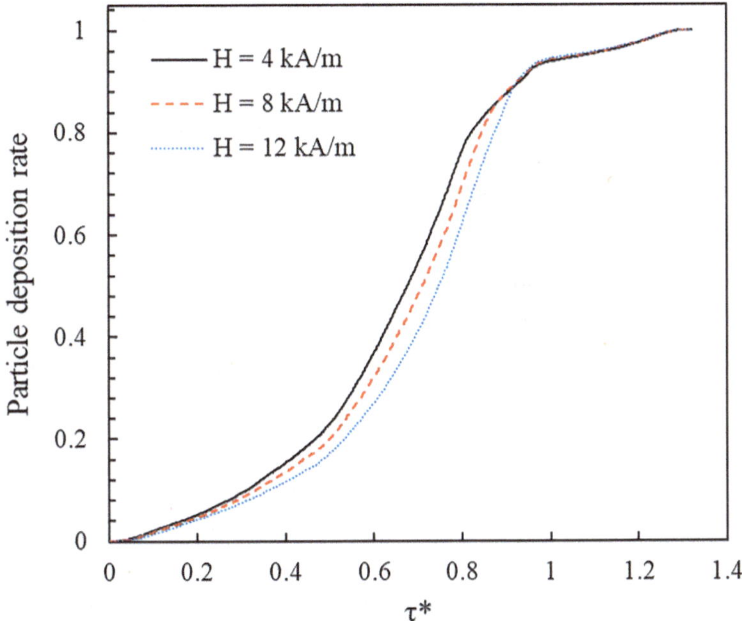


**Fig. 9.12** Comparison of particle deposition rates. In Stage III, the deposition rate began to rise. Higher magnetic field strength led to delayed particle adhesion

adhesion of particles located farther from the substrate. As the contact line receded, entire chain bundles were eventually deposited, resulting in similar overall deposition rates across all cases.

**Exercise 9.3 Dipole–Dipole Magnetic Interaction**

(a) Explain how magnetic dipole–dipole interactions in (9.13) cause particle alignment.
(b) Discuss why magnetic chains create anisotropic resistance to internal circulation.
(c) Predict deposition behavior when the magnetic field is applied horizontally (parallel to the substrate).

**Solution**

(a) Magnetic nanoparticles acquire dipole moments under an external magnetic field. The dipole–dipole interaction (9.13) favors a head-to-tail arrangement of neighboring dipoles along the field direction because this configuration minimizes the magnetic interaction energy. Particles thus experience attractive forces when aligned with the field and repulsive components when misaligned. Over time, these interactions cause particles to rotate and translate so that their dipoles become collinear with the applied field, forming chain-like structures.
(b) Magnetic chains are elongated aggregates of particles aligned along the field direction. Hydrodynamically, such chains behave like slender rigid bodies:

- They present a large cross-sectional area to flow perpendicular to the chain axis, increasing drag and resisting motion in that direction.
- Along the chain axis, they are more streamlined and offer less resistance.

As a result, internal Marangoni or capillary-driven circulation is more strongly hindered in directions transverse to the chains than along them. This produces anisotropic resistance: the droplet interior is more "blocked" for flows that cut across chains, while flows parallel to the chains remain relatively less affected.

(c) If the magnetic field is applied horizontally (parallel to the substrate), the favored dipole alignment is also horizontal. In that case:

- Particles tend to form chains lying along the substrate plane, rather than vertical columns.
- These horizontal chains can stretch from near the contact line toward the interior or along azimuthal directions.
- Internal circulation that tries to move particles radially outward or inward will interact differently with these horizontal chains, possibly guiding particles into streak-like or banded patterns along the field direction.

Overall, one would expect a less axisymmetric ring: the deposit may become elongated along the field direction, with anisotropic ring thickness or even broken ring segments aligned with the horizontal field.

**Exercise 9.4 Particle Velocity Evolution and Deposition Stages**

(a) Explain why particle velocity increases during Stage II despite evaporation-driven thinning.
(b) Using Fig. 9.11, relate particle alignment to the velocity trends.

**Solution**

(a) Even though the droplet is thinning during Stage II, particle velocity increases because the driving forces for flow grow faster than viscous dissipation:

- Evaporation rate and resulting concentration gradients rise, intensifying capillary flow from the interior to the edge.
- Temperature or concentration gradients along the interface can strengthen Marangoni stresses, enhancing recirculation.
- The decreasing characteristic length (smaller droplet height) can also amplify velocity for a given pressure or surface tension gradient.

Thus, Stage II corresponds to a regime where evaporation-induced forcing dominates, leading to stronger internal flow and higher particle speeds, even as the liquid layer becomes thinner.

(b) Figure 9.11 shows that as the magnetic field increases, particles become more strongly aligned into chains. This alignment correlates with the velocity evolution:

- During Stage II, chain formation is still ongoing; partial alignment allows significant mobility along the chain axis, contributing to relatively high velocities as particles slide and reorganize.
- As chains become more complete and rigid (toward Stage III), particles are increasingly constrained to move collectively with the chains. Local rearrangements are suppressed, and average velocities decrease.
- Regions with poorly formed or broken chains maintain higher residual motion, explaining spatial variations in velocity.

Therefore, the observed velocity peaks and subsequent decay can be interpreted as a transition from a highly mobile, weakly aligned particle ensemble to a mechanically constrained, well-aligned chain network.

## 9.4 Conclusions

This chapter presented a comprehensive numerical investigation of particle deposition dynamics in evaporating sessile droplets containing magnetic nanoparticles. A pseudopotential LBM model was employed to simulate liquid–vapor phase change under thermal gradients, while particle motion was resolved using a stochastic Langevin approach incorporating hydrodynamic, capillary, Brownian, van der Waals, and magnetic forces. The computational framework was implemented on a GPU-parallelized platform, enabling efficient simulation of coupled evaporation and particle dynamics in large-scale droplet systems.

The simulation results demonstrated that contact angle hysteresis and thermal gradients induce complex internal flows, including capillary and Marangoni convection, which critically influence the spatial distribution of particles. In the absence of external fields, particle deposition was governed by flow-pinning behavior, yielding characteristic ring-shaped patterns. When magnetic fields were applied, magnetic dipole–dipole interactions led to directional aggregation and bundling of nanoparticles, modulating the ring thickness and shifting deposition toward the droplet center. The magnetic bundles also resisted internal vortex flows, altering particle rearrangement and adhesion timing during the final evaporation stage.

These findings offer new physical insight into the interplay between interfacial flows and magnetic forces in evaporating nanoparticle-laden droplets. The results provide design guidelines for manipulating particle distributions and suppressing undesired coffee-ring effects through substrate wettability and magnetic field control. This integrated modeling approach is applicable to a wide range of solution-based printing techniques, including display fabrication, bio-patterning, and printed electronics, where precise control over nanoparticle deposition is essential for functional performance.

**Summary**

1. **Coffee-ring mechanism and motivation**
   - Sessile droplets evaporate most rapidly near the contact line, generating strong outward capillary flow that transports suspended nanoparticles toward the rim.
   - Resulting particle accumulation produces a ring-like deposit (coffee-ring effect), which degrades film uniformity in display manufacturing.
   - Magnetic nanoparticles enable external-field control of interparticle forces, offering a non-invasive method to manipulate pattern formation during droplet evaporation.

2. **Numerical framework for coupled evaporation and particle dynamics**
   - The chapter integrates three physical models:
     - Pseudopotential LBM for liquid–vapor phase separation and phase change.
     - Nanoparticle motion model combining hydrodynamic drag, Brownian motion, capillary interactions, van der Waals attraction, DLVO-based cohesion, repulsive collision, and magnetic dipole–dipole forces.
     - Heat transfer LBM with a phase transformation term to model latent heat and interface cooling.
   - This unified framework enables simultaneous simulation of temperature evolution, droplet shrinkage, and nanoparticle redistribution.

3. **Influence of wettability (receding angle)**
   - Wettability governs whether evaporation occurs in CCR (constant contact radius) or CCA (constant contact angle) mode.
   - Low receding angles → strong pinning → long CCR duration → robust outward capillary flow → clear coffee-rings.
   - High receding angles → earlier contact line recession → weakened radial flow → particles shift toward droplet center → compact central deposits.
   - Particle dispersion metrics (mean radial distance, variance) confirm the strong dependence on pinning duration.

4. **Flow-field evolution during evaporation**
   - Internal flow is governed by competition between:
     - Capillary-driven radial flow, dominant when contact line is pinned.
     - Marangoni recirculation, dominant when contact line recedes or surface tension gradients increase.
   - Pinned droplets exhibit diminishing vortices and increasingly outward flow.

- Non-pinned droplets maintain strong Marangoni vortices that trap particles and delay deposition.
- Time evolution of Péclet (*Pe*) and Marangoni (*Ma*) numbers quantifies this shift in flow dominance.

5. **Effects of external magnetic fields**

- Vertical magnetic fields cause nanoparticles to align into chain structures due to dipole–dipole forces.
- Increasing field strength leads to:
  - lower peak particle density,
  - peak location shifting toward the droplet center,
  - thicker rings and enhanced particle adhesion near the substrate,
  - slower velocity evolution due to magnetic bundle resistance.
- Particle dynamics exhibit three stages:
  - Stage I—Interfacial rearrangement,
  - Stage II—Marangoni entrapment and ring formation,
  - Stage III—Substrate adhesion.

6. **Magnetic-field-induced anisotropy**

- Chain bundles hinder internal circulation, especially in directions perpendicular to the field.
- Near the contact line, magnetic attraction enhances localized clustering and thickens the deposition ring.
- Despite stronger adhesion in local regions, overall deposition rate decreases with magnetic field due to hindered particle transport.

## References

1. R. D. Deegan, O. Bakajin, T. F. Dupont, G. Huber, S. R. Nagel, and T. A. Witten, "Capillary flow as the cause of ring stains from dried liquid drops," Nature **389**, 827-829 (1997).
2. S. Chung, K. Cho, and T. Lee, "Recent progress in inkjet-printed thin-film transistors," Adv Sci (Weinh) **6**, 1801445 (2019).
3. M. Yu, M. H. Saeed, S. Zhang, H. Wei, Y. Gao, C. Zou, L. Zhang, and H. Yang, "Luminescence enhancement, encapsulation, and patterning of quantum dots toward display applications," Advanced Functional Materials **32**, (2021).
4. K. N. Al-Milaji, R. L. Hadimani, S. Gupta, V. K. Pecharsky, and H. Zhao, "Inkjet printing of magnetic particles toward anisotropic magnetic properties," Sci Rep **9**, 16261 (2019).
5. R. Verberg, J. M. Yeomans, and A. C. Balazs, "Modeling the flow of fluid/particle mixtures in microchannels: Encapsulating nanoparticles within monodisperse droplets," J. Chem. Phys. **123**, 224706 (2005).
6. H. C. Hamaker, "The london—van der waals attraction between spherical particles," Physica **4**, 1058-1072 (1937).
7. S. Melle, O. G. Calderon, M. A. Rubio, and G. G. Fuller, "Microstructure evolution in magnetorheological suspensions governed by mason number," Phys. Rev. E **68**, 041503 (2003).

8. C. Zhang, H. Zhang, Y. Zhao, and C. Yang, “An immersed boundary-lattice boltzmann model for simulation of deposited particle patterns in an evaporating sessile droplet with dispersed particles,” Int. J. Heat Mass Transf. **181**, (2021), 121905.
9. P. J. Yunker, T. Still, M. A. Lohr, and A. Yodh, “Suppression of the coffee-ring effect by shape-dependent capillary interactions,” Nature **476**, 308-311 (2011).

# Chapter 10
# Particle Deposition Dynamics in Evaporating Droplets: Patterned Wettability Substrate Simulation

## 10.1 Introduction

In Chap. 9, we examined strategies to alter particle adhesion patterns by directly controlling interparticle interactions, such as applying magnetic fields to induce directional aggregation. However, this chapter shifts the focus to modulating the internal flow fields within the droplet—particularly the interplay between capillary and Marangoni convection—as a means to achieve uniform particle distribution.

A key parameter affecting these flows is the behavior of the contact line. A pinned contact line enhances capillary-driven outward transport, promoting edge deposition, while a moving contact line gives rise to Marangoni-induced vortices that can trap particles and suppress ring formation. By engineering the surface wettability through a radial pattern of receding angles, the contact line can be made to alternately pin and depin during evaporation. This oscillatory behavior introduces multiple flow regimes within the droplet, resulting in multi-ring deposition patterns and improved particle uniformity.

This chapter presents a simulation-based approach using the pseudopotential LBM coupled with heat transfer and nanoparticle transport modeling. The internal flow structure and nanoparticle dynamics are resolved in detail, and 51 substrate designs are evaluated to identify optimal configurations. Through this study, readers will gain insight into how patterned wettability—rather than direct particle manipulation—can be used to control deposition and suppress the coffee-ring effect in solution-based processes such as inkjet printing.



H. M. Lee, J. S. Lee, *Lattice Boltzmann Methodology for Single-Phase and Multiphase Nanoparticle Modeling*,
https://doi.org/10.1007/978-981-95-9117-6_10

## 10.2 Numerical Methods

In Chap. 9, we investigated the particle deposition pattern resulting from the direct control of interparticle interactions. In the present study, we aim to achieve uniform particle deposition by manipulating the droplet's contact line through the use of a substrate with patterned wettability. The following presents the methodology for modeling such a substrate within the pseudopotential LBM framework.

### 10.2.1 Patterned Wettability Modeling

To model contact angle hysteresis, we employed the method proposed by Ding and Spelt [1]. In this approach, a ghost density is assigned to the wall nodes based on the prescribed contact angle, allowing the enforcement of static or dynamic contact angles within the pseudopotential LBM framework. The ghost density $\rho_{i,\,j,\,0}$ at the wall is defined as

$$\rho_{i,j,0} = \rho_{i,j,2} + \tan\left(\frac{\pi}{2} - \theta_e\right)\varphi. \tag{10.1}$$

where $\varphi = \sqrt{\left(\rho_{i+1,j,1} - \rho_{i-1,j,1}\right)^2 + \left(\rho_{i,j+1,1} - \rho_{i,j-1,1}\right)^2}$, and $\theta_e$ is the equilibrium contact angle.

## 10.3 Results

This section investigates nanoparticle (NP) deposition during droplet evaporation on a substrate with spatially varying receding angles. As shown in Fig. 10.1b, the substrate is designed with a radial gradient in receding angle. During evaporation,

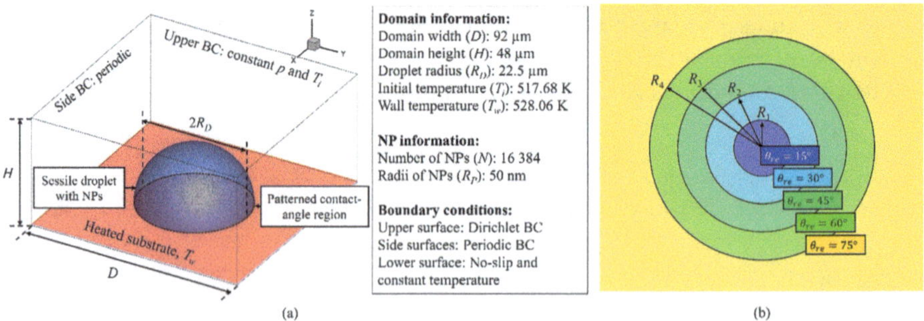


**Fig. 10.1** Simulation domain configuration. (**a**) A sessile droplet containing nanoparticles is positioned at the center of a heated patterned substrate. (**b**) Radial variation in receding angles on the patterned substrate, increasing in 15° intervals from the center

the droplet's contact line retreats inward, remaining in the constant contact radius (CCR) mode until the local contact angle matches the patterned receding angle. At this point, the contact line transitions to the constant contact angle (CCA) mode, and this CCR–CCA switching repeats. We analyze how these transitions influence NP deposition and identify an optimized pattern geometry that promotes uniform particle distribution. The underlying deposition mechanism is also discussed.

### 10.3.1 Simulation Setup

Figure 10.1a illustrates the simulation setup, where a nanoparticle-laden droplet is placed at the center of a patterned substrate, depicted in Fig. 10.1b. The substrate features circular zones with varying receding angles, each spaced 15° apart and labeled $R_1$–$R_4$ from the center outward. The droplet initially contains 16,384 randomly distributed nanoparticles, with a contact angle of 90°. The domain starts at a uniform initial temperature $T_i$, while the substrate is maintained at a constant temperature $T_w$. The bottom surface applies a no-slip and Dirichlet condition for velocity and temperature, respectively, while the top applies constant temperature and pressure boundaries. Periodic conditions are imposed on the lateral boundaries. Simulation details are provided in Table 10.1.

**Table 10.1** Domain specifications and model parameters

| Variable | Symbol | Lattice value | Physical value | Unit |
|---|---|---|---|---|
| Length conversion factor | $\delta_x$ | – | $5.0 \times 10^{-7}$ | m |
| Mass conversion factor | $\delta_m$ | – | $1.72 \times 10^{-17}$ | kg |
| Time conversion factor | $\delta_t$ | – | $1.64 \times 10^{-8}$ | s |
| Temperature conversion factor | $\delta_T$ | – | $8.88 \times 10^{3}$ | K |
| Width of simulation domain | $D$ | 184 | 92.0 | μm |
| Height of simulation domain | $H$ | 96 | 48.0 | μm |
| Droplet radius | $R_D$ | 45.0 | 22.5 | μm |
| Particle radius | $R_P$ | 0.1 | 50.0 | nm |
| Initial temperature | $T_i$ | 0.05832 | 517.68 | K |
| Substrate temperature | $T_w$ | 0.06065 | 528.06 | K |
| Droplet density | $\rho_l$ | 7.203 | 991.9 | kg/m$^3$ |
| Air density | $\rho_v$ | 0.197 | 27.13 | kg/m$^3$ |
| Droplet dynamic viscosity | $\mu_l$ | 0.4802 | $1.01 \times 10^{-3}$ | Pa s |
| Air dynamic viscosity | $\mu_v$ | 0.09358 | $1.96 \times 10^{-4}$ | Pa s |
| Droplet kinematic viscosity | $\nu_l$ | 0.067 | $1.02 \times 10^{-6}$ | m$^2$/s |
| Air kinematic viscosity | $\nu_v$ | 0.475 | $7.23 \times 10^{-6}$ | m$^2$/s |
| Liquid thermal diffusivity | $\alpha_l$ | 0.06567 | $1.00 \times 10^{-6}$ | m$^2$/s |
| Vapor thermal diffusivity | $\alpha_v$ | 0.2350 | $3.58 \times 10^{-6}$ | m$^2$/s |
| Critical temperature | $T_c$ | 0.0729 | 647.096 | K |
| Critical density | $\rho_c$ | 2.3382 | 322.0 | kg/m$^3$ |

### 10.3.2 Wettability Effects: Equally Spaced Patterned Substrates

The behavior of particle deposition is highly affected by the pinning of the droplet's contact line. On rough surfaces with low receding angles, the contact line tends to remain pinned, promoting capillary-driven outward flow and resulting in a coffee-ring deposition. In contrast, on smooth or superhydrophobic surfaces where the contact line readily recedes or the contact area is minimized, particles are often retained within internal vortices [2], producing a central accumulation known as the coffee-eye pattern. This study aims to achieve uniform particle deposition by adjusting the pinning duration through a substrate with spatially varied receding angles. As a preliminary step, we examined how the simplest patterned substrate—with uniform spacing—affects particle distribution compared to a non-patterned substrate, thereby isolating the impact of wettability contrast on deposition behavior.

Figure 10.2a presents the droplet evaporation process on the patterned substrate, while Fig. 10.2b shows the evolution of the contact diameter and angle. Both the contact diameter $D^* = D_D/D_0$ and time $\tau^* = \alpha t/D_0^2$ are normalized. As evaporation progresses, the contact line advances inward, with each color segment representing entry into a distinct receding-angle region. The transition points, labeled V, mark the locations where the contact line reaches a pattern boundary. The results indicate that the droplet cyclically alternates between constant contact radius (CCR) and constant contact angle (CCA) modes based on the receding-angle pattern, demonstrating that contact line mobility can be effectively tuned. This mobility plays a key role in determining the final particle deposition profile.

Figure 10.3 compares particle deposition on unpatterned substrates with receding angles of 15°, 45°, and 75° (Fig. 10.3a–c) to that on a patterned substrate (Fig. 10.3d). On the unpatterned surface with a low receding angle of 15°, the contact line remains pinned until the contact angle reduces from 90° to 15°, resulting in a pronounced coffee-ring pattern. As the receding angle increases, particles increasingly deposit near the center, and at 75°, the distribution inverts relative to the coffee-ring. These results highlight that while receding angle strongly influences deposition, a single uniform distribution is difficult to achieve with one fixed angle. In contrast, the patterned substrate in Fig. 10.3d generates multiple ring-like depositions, as further quantified in Fig. 10.4. This figure presents particle adhesion versus radial distance, color-coded by patterned region (as in Fig. 10.2), clearly illustrating multiple deposition peaks in the 45° and 75° zones, rather than a single dominant ring.

These multiple rings indicate enhanced uniformity in particle deposition compared to unpatterned surfaces. To evaluate this quantitatively, we introduce a uniformity index, $U_f$ which accounts for both the spatial distribution and packing density of particles. Specifically, $U_f$ is defined as the ratio of the distribution spread to the peak particle density, providing a comprehensive measure of uniformity.

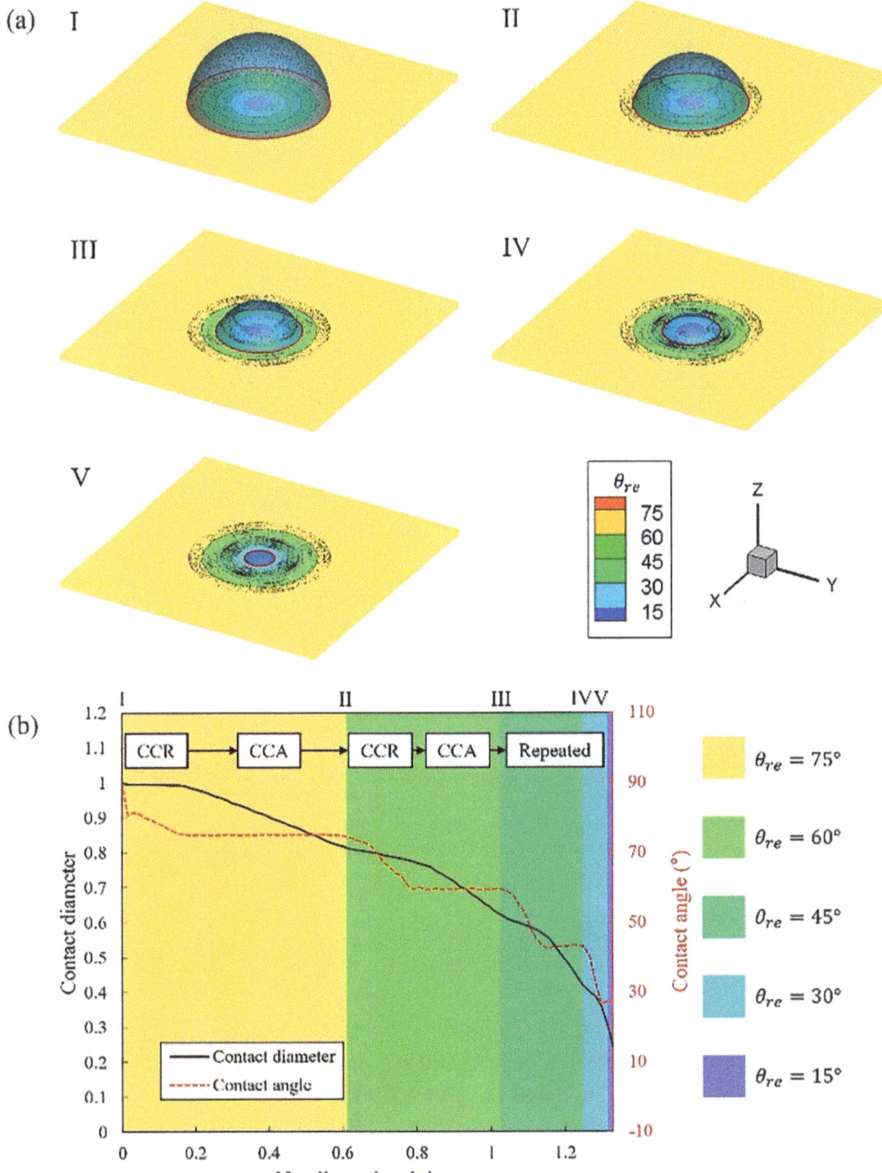


**Fig. 10.2** Droplet evaporation behavior on a patterned substrate. (**a**) Evolution of droplet shape at labeled time points (I–V), where V marks the instance when the contact line reaches a pattern boundary. (**b**) Time evolution of the contact diameter and contact angle, with color segments indicating the corresponding receding-angle region

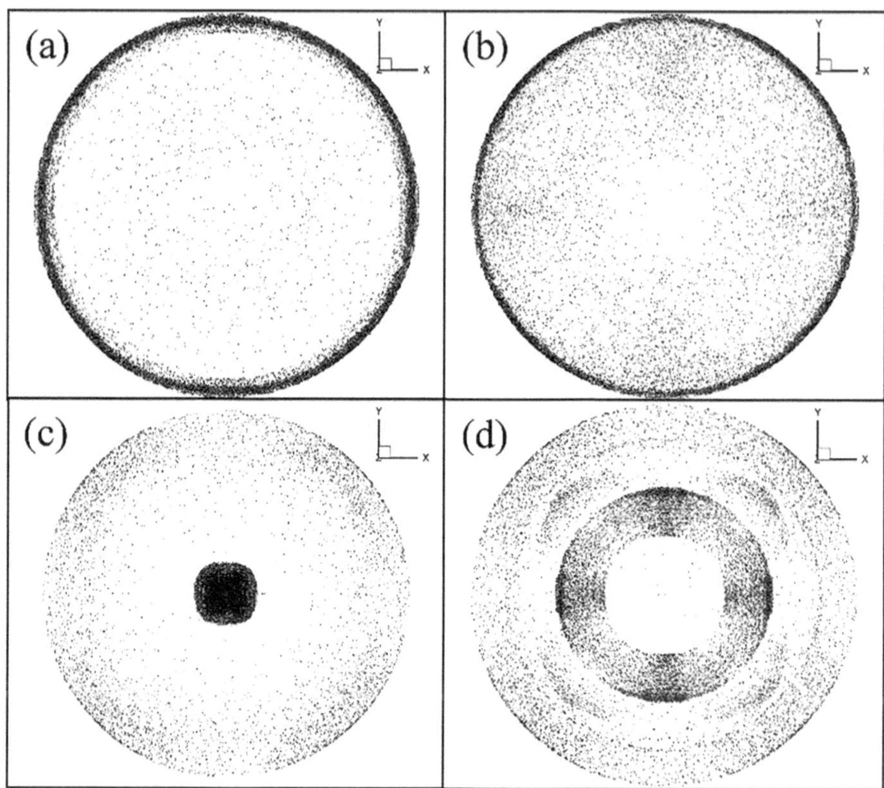


**Fig. 10.3** Particle deposition patterns on substrates with uniform receding angles. (**a**) $\theta_{re} = 15°$, (**b**) $\theta_{re} = 45°$, (**c**) $\theta_{re} = 75°$, and (**d**) on a patterned substrate with spatially varying receding angles

$$U_f = \frac{\text{Particle spreadibility}}{\text{Particle density}} = \frac{\sigma / R_D}{\rho^P_{max} / \rho^*}. \tag{10.2}$$

Here, $\sigma$ denotes the standard deviation of particle distances from the droplet center, $\rho^P_{max}$ is the peak particle density, and $\rho^*$ represents the density in an ideally uniform distribution. The calculated $U_f$ values for the cases in Fig. 10.3 are summarized in Table 10.2. Neither $\sigma$ nor the density ratio alone can accurately capture uniformity. For instance, although the case with $\theta_{re} = 75°$ shows the highest $\sigma$, it yields the lowest $U_f$ (0.00601) due to a central concentration with a peak density of 62.38. In contrast, the patterned substrate achieves the highest uniformity index of 0.04345—approximately 4.24 times greater than that of $\theta_{re} = 15°$, which exhibits a typical coffee-ring. This result highlights the potential for substantially enhancing uniformity through even simple substrate patterning.

The results indicate that equally spaced patterned substrates improve particle distribution uniformity. To further investigate this effect, simulations were

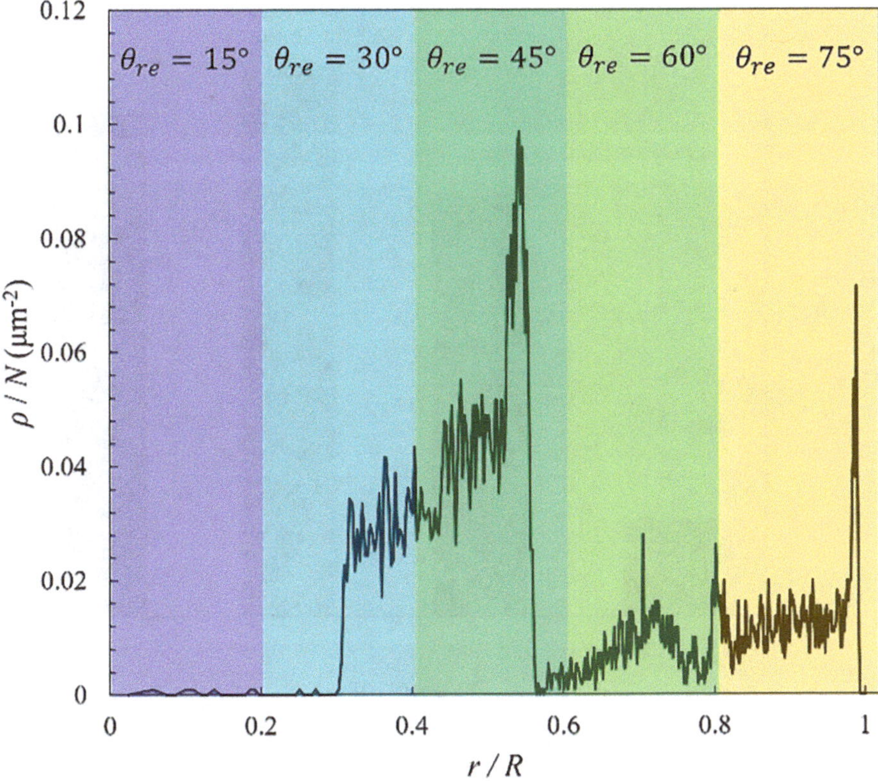


**Fig. 10.4** Radial particle density profile on the patterned substrate. Colors correspond to different regions defined by varying receding angles in the substrate design

**Table 10.2** Uniformity parameters for various receding angles and patterned substrates

| Receding angle, $\theta_{re}$ (°) | Standard deviation, $\sigma/R_D$ | Density ratio, $\rho^P_{max}/\rho^*$ | Uniformity, $U_f$ |
|---|---|---|---|
| 15 | 0.095022 | 9.273 | 0.01025 |
| 45 | 0.142767 | 7.746 | 0.01901 |
| 75 | 0.374889 | 62.38 | 0.00601 |
| Patterned | 0.203933 | 4.694 | 0.04345 |

performed on substrates with two to five-ring patterns, keeping the ring width and wettability gradient constant. As shown in Fig. 10.5, the four-ring configuration achieved the highest uniformity. Compared to this, the two-, three-, and five-ring cases exhibited uniformity reductions of 10.3%, 2.90%, and 1.07%, respectively. The lowest uniformity occurred with two rings, and a gradual improvement was observed with an increasing number of rings, reaching a peak at four. This trend is attributed to the more frequent pinning–depinning events, which facilitate multi-ring formation when sufficient ring count is present.

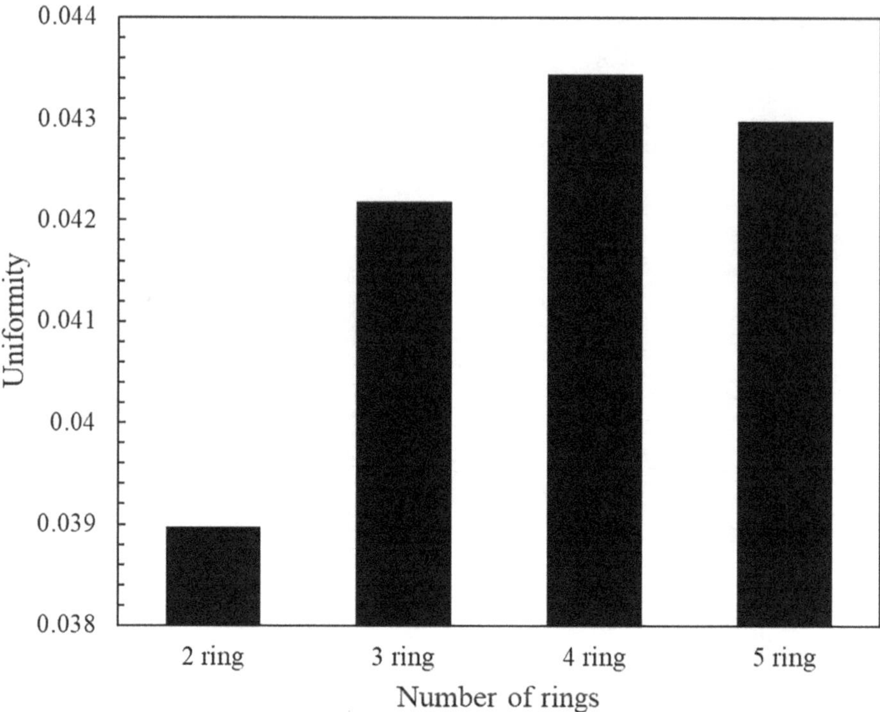


**Fig. 10.5** Uniformity of particle distribution as a function of the number of equally spaced ring patterns

Although the five-ring configuration includes more rings, it exhibited lower uniformity than the four-ring case. This indicates that pinning frequency alone does not determine multi-ring formation. Instead, uniform particle distribution also depends on the evolution of internal flow within the droplet. Adequate time between pinning events is essential for flow development that enables uniform deposition. These results underscore the need to optimize ring patterns, with the four-ring configuration serving as the most effective design based on its superior uniformity.

**Exercise 10.1 CCR–CCA Transitions and Contact Line Dynamics**

(a) Explain how spatially varying receding angles induce CCR–CCA transitions in an evaporating droplet.

(b) Using Fig. 10.2, describe how these transitions influence the overall evaporation dynamics.

**Solution**

(a) On a patterned substrate, the receding contact angle is not uniform but changes stepwise (e.g., 15° → 30° → 45° → 75°) in the radial direction. As the droplet evaporates and the apparent contact angle decreases:

- In a region where the local receding angle is small, the contact line tends to remain pinned (CCR mode). The base radius is fixed while the height decreases.
- Once the apparent angle falls below the next larger receding angle of an inner region, the outer pinned line can no longer be sustained and the contact line begins to move inward (CCA mode).
- When the contact line reaches a new region with a different receding angle, it may pin again, re-entering a local CCR state.

Thus, the spatial variation of receding angle provides a series of "pinning thresholds." As the droplet evaporates and the apparent contact angle evolves, it crosses these thresholds and naturally switches between CCR and CCA modes multiple times.

(b) Figure 10.2 shows that evaporation dynamics are segmented by the CCR–CCA transitions:

- During CCR segments, the footprint radius is fixed and the evaporative flux is highest near the pinned contact line. The droplet height decreases, but the exposed area is relatively large, maintaining a high overall evaporation rate.
- During CCA segments, the contact line recedes inward, reducing the wetted area. The flux still peaks near the moving edge, but the total evaporation rate can decrease as the footprint shrinks.

**Exercise 10.2 Multi-ring Formation Mechanism**

(a) Explain why patterned substrates generate multiple deposition peaks in Fig. 10.4.
(b) Describe how pinning at each receding-angle boundary influences ring spacing and intensity.

**Solution**

(a) Figure 10.4 shows that a patterned substrate with several wettability zones produces multiple radial peaks in the final particle density. This occurs because:

- Each wettability boundary (change in receding angle) can serve as a new pinning location for the contact line.
- During pinning at each radius, strong evaporation near that line drives capillary flow that transports particles to that radius.
- When the droplet depins and moves inward, a new ring-forming episode begins at the next boundary.

Thus, the droplet experiences multiple CCR segments at different radii, and each segment leaves behind one deposition peak, resulting in a multi-ring profile.

(b) The spacing and intensity of rings are determined by:

- Spacing: The radial positions of the wettability boundaries directly set where pinning can occur. Therefore, the ring spacing reflects the designed radial widths of each wettability region.

- Intensity: Longer pinning at a given radius means more time for particles to be carried there, increasing ring intensity. Stronger local evaporation flux and higher capillary flow during pinning also enhance accumulation. If a region pinning is short-lived or the flux is weak, the corresponding ring becomes faint or may not form clearly.

Thus, careful placement of wettability boundaries and control of pinning duration at each boundary determine how far apart the rings are and how strong each ring appears.

### *10.3.3 Wettability Effects: Pattern Optimization*

This section focuses on optimizing the patterned substrate to achieve uniform particle distribution by varying the widths of the receding-angle regions. A total of 51 configurations, detailed in Table 10.3, were simulated with the same receding angles as in the previous section. Figure 10.6 presents the uniformity results, where Cases 34, 16, and 42 showed the highest uniformities—0.05982, 0.06191, and 0.06738, respectively. Case 42 exhibited the most uniform particle distribution, with a uniformity index 6.57 times higher than that of the uniform 15° receding angle case. All three high-performing cases shared a common feature: the outer regions at 45°, 60°, and 75° had widths of 30%, 20%, and 10% of the droplet diameter, respectively. These findings imply two key points: (1) particle uniformity is largely governed by the width of the outermost region where evaporation initiates, and (2) increasing region width toward the center enhances uniformity. Since most particles deposit at early evaporation stages, outer regions contribute more to final patterns. Moreover, wider inner regions help maintain consistent pinning time despite reduced droplet volume during later stages of evaporation.

As illustrated in Fig. 10.7, particle deposition patterns are compared between Case 30, which exhibits the lowest uniformity, and Case 42, which shows the highest. In Case 30 (Fig. 10.7a), a dense single ring forms close to the center, resulting in poor uniformity. In contrast, Case 42 (Fig. 10.7b) produces three distinct ring structures aligned with the pattern boundaries, with significantly reduced peak density. The particle density profiles in Fig. 10.7c confirm this difference, highlighting that multi-ring formation plays a crucial role in achieving a uniform distribution. Moreover, cases like Case 30—with wide outermost regions—tend to suppress uniformity, reinforcing the observation that the outermost region's width critically affects deposition outcomes.

Controlling the contact line pinning time plays a key role in achieving uniform particle deposition. Thus, we evaluated the pinning durations in different regions of Case 42, known for its high uniformity, and compared them with theoretical predictions (Fig. 10.8a). Regions A1–A5 denote the patterned areas from center to edge. According to Shanahan et al. [3], the pinning time $\Delta t_{pinning}$ for a droplet to reach its receding angle is given by the following expression:

**Table 10.3** Radii defining the receding-angle zones for various patterned substrates

| Case | $R_1/R_D$ (%) | $R_2/R_D$ (%) | $R_3/R_D$ (%) | $R_4/R_D$ (%) |
|---|---|---|---|---|
| 1 | 10 | 20 | 40 | 70 |
| 2 | 10 | 20 | 50 | 70 |
| 3 | 10 | 20 | 50 | 80 |
| 4 | 10 | 30 | 40 | 70 |
| 5 | 10 | 30 | 50 | 70 |
| 6 | 10 | 30 | 50 | 80 |
| 7 | 10 | 30 | 60 | 70 |
| 8 | 10 | 30 | 60 | 80 |
| 9 | 10 | 30 | 60 | 90 |
| 10 | 10 | 40 | 50 | 70 |
| 11 | 10 | 40 | 50 | 80 |
| 12 | 10 | 40 | 60 | 70 |
| 13 | 10 | 40 | 60 | 80 |
| 14 | 10 | 40 | 60 | 90 |
| 15 | 10 | 40 | 70 | 80 |
| 16 | 10 | 40 | 70 | 90 |
| 17 | 20 | 30 | 40 | 70 |
| 18 | 20 | 30 | 50 | 70 |
| 19 | 20 | 30 | 50 | 80 |
| 20 | 20 | 30 | 60 | 70 |
| 21 | 20 | 30 | 60 | 80 |
| 22 | 20 | 30 | 60 | 90 |
| 23 | 20 | 40 | 50 | 70 |
| 24 | 20 | 40 | 50 | 80 |
| 25 | 20 | 40 | 60 | 70 |
| 26 | 20 | 40 | 60 | 80 |
| 27 | 20 | 40 | 60 | 90 |
| 28 | 20 | 40 | 70 | 80 |
| 29 | 20 | 40 | 70 | 90 |
| 30 | 20 | 50 | 60 | 70 |
| 31 | 20 | 50 | 60 | 80 |
| 32 | 20 | 50 | 60 | 90 |
| 33 | 20 | 50 | 70 | 80 |
| 34 | 20 | 50 | 70 | 90 |
| 35 | 20 | 50 | 80 | 90 |
| 36 | 30 | 40 | 50 | 70 |
| 37 | 30 | 40 | 50 | 80 |
| 38 | 30 | 40 | 60 | 70 |
| 39 | 30 | 40 | 60 | 80 |
| 40 | 30 | 40 | 60 | 90 |
| 41 | 30 | 40 | 70 | 80 |
| 42 | 30 | 40 | 70 | 90 |

(continued)

**Table 10.3** (continued)

| Case | $R_1/R_D$ (%) | $R_2/R_D$ (%) | $R_3/R_D$ (%) | $R_4/R_D$ (%) |
|---|---|---|---|---|
| 43 | 30 | 50 | 60 | 70 |
| 44 | 30 | 50 | 60 | 80 |
| 45 | 30 | 50 | 60 | 90 |
| 46 | 30 | 50 | 70 | 80 |
| 47 | 30 | 50 | 70 | 90 |
| 48 | 30 | 50 | 80 | 90 |
| 49 | 30 | 60 | 70 | 80 |
| 50 | 30 | 60 | 70 | 90 |
| 51 | 30 | 60 | 80 | 90 |

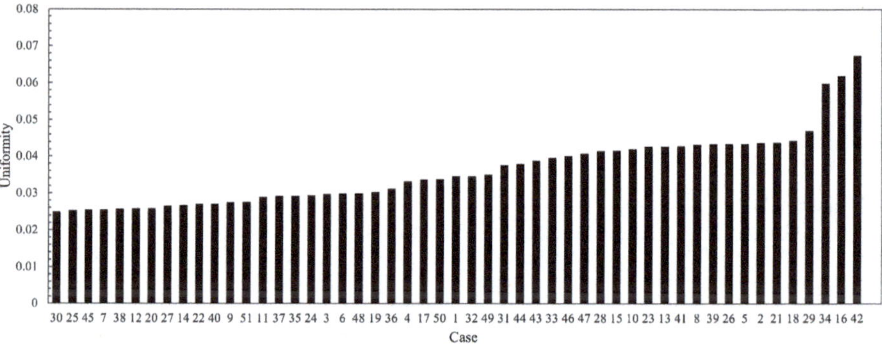


**Fig. 10.6** Uniformity results for 51 patterned substrate cases with varying region widths

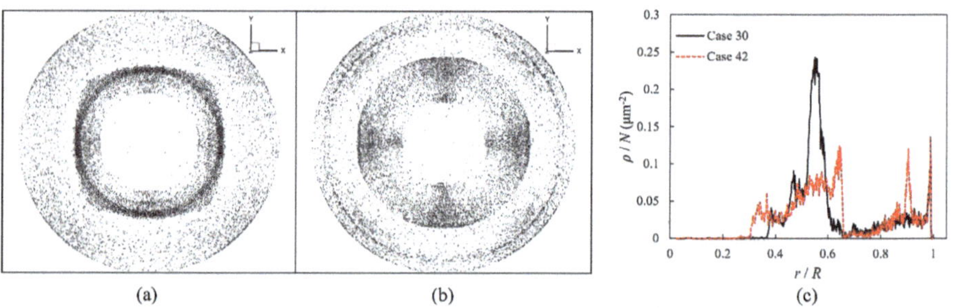


**Fig. 10.7** Particle distributions for (**a**) Case 30 with the lowest final uniformity ($U_f = 0.02486$) and (**b**) Case 42 with the highest ($U_f = 0.06738$). (**c**) Comparison of particle density profiles for Cases 30 and 42

$$\Delta t_{pinning} = \frac{\pi R^3(\delta\theta)}{(1 + \cos\theta_o)dV/dt}, \tag{10.3}$$

Here, $\theta_o$ is the equilibrium contact angle and $\delta\theta$ is the difference between the equilibrium contact and receding angles. The simulation results (red dotted bars)

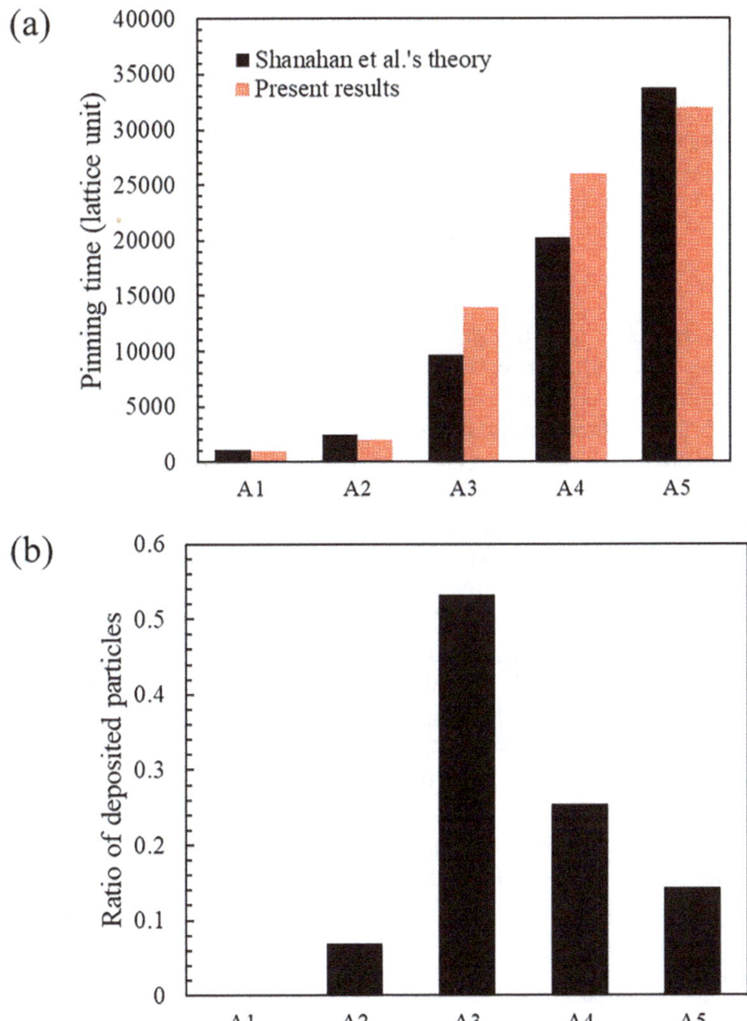


**Fig. 10.8** (**a**) Simulated versus theoretical pinning times of the contact line across regions A1–A5 in Case 42, arranged from center to edge. (**b**) Proportions of deposited particles in each corresponding region

generally align with the theoretical predictions (black bars), though minor discrepancies exist. As evaporation proceeds, the pinning time—when the contact line becomes fixed at a pattern boundary—shortens due to the shrinking contact radius. Since the theoretical model incorporates $R$ as a cubic term, it is highly responsive to changes in radius. Overall, the droplet remains longest in the outermost region (A5), with pinning time decreasing inward. Interestingly, region A3 shows the highest particle accumulation (53.2%, Fig. 10.8b), indicating that contact line pinning time alone does not fully determine particle deposition behavior.

Apart from pinning time, internal droplet flows also influence particle deposition. In an evaporating droplet, capillary flow—driven by strong evaporation near the contact line—and Marangoni flow—caused by surface tension gradients—interact and compete. These flows significantly impact how particles accumulate. To examine their relative influence, the time evolution of the Marangoni number (*Ma*) and Reynolds number (*Re*) is presented in Fig. 10.9. Their definitions are as follows:

$$Ma = -\frac{\partial \gamma}{\partial T}\frac{H_D \Delta T}{\mu \alpha}, \ Re = \frac{R_D U_{ev}}{\nu}, \tag{10.4}$$

Here, $\gamma$, $H_D$, $\Delta T$, and $U_{ev}$ denote surface tension, droplet height, maximum surface temperature difference, and evaporation velocity, respectively. All dimensionless values were computed using lattice units. As the contact line crosses each pattern region, the Marangoni number (*Ma*) exhibits oscillatory behavior over time. This is due to transitions between constant contact radius (CCR) and constant contact angle (CCA) modes—*Ma* increases during CCA and decreases during CCR, as indicated by the red and blue dashed lines in Fig. 10.9a. The negative trend in *Ma*'s rate of change suggests a continuous decline in Marangoni flow influence, highlighting its sensitivity to the evaporation mode.

Unlike *Ma*, the *Re* shows a distinct trend. As seen in Fig. 10.9b, *Re* remains relatively stable with minor oscillations before the droplet reaches the region with a 45° receding angle (marked by the black dashed line). Once this region is reached, *Re* rises sharply, accompanied by a rapid increase in particle deposition, suggesting a strong correlation between the two. This implies that capillary flow becomes dominant beyond this point. To explore the cause of the *Re* surge, Fig. 10.10 plots the contact radius and evaporation velocity. While the contact radius (red dashed line) steadily decreases, both *Re* and evaporation velocity show abrupt increases. This trend aligns with predictions from the theoretical model of local evaporative flux $J$ [4, 5].

$$J = \frac{D_f(c_{sat} - c_\infty)}{R} \times \left[\frac{1}{2}\sin\theta + \sqrt{2}(\cosh\Theta + \cos\theta)^{\frac{3}{2}}\int_0^\infty \frac{\cosh\theta\tilde{\tau}}{\cosh\pi\tilde{\tau}}\tanh[(\pi-\theta)\tilde{\tau}]P_{-1/2+i\tilde{\tau}}(\cosh\Theta)d\tilde{\tau}\right], \tag{10.5}$$

Here, $D_f$ is the diffusion coefficient, while $c_{sat}$, and $c_\infty$ represent the saturation and ambient vapor concentrations, respectively. The function $P_{-1/2+i\tilde{\tau}}(\cosh\Theta)$ denotes the Legendre function of the first kind, where the degree $-1/2 + i\tilde{\tau}$ is defined in terms of the toroidal coordinate $\Theta$.

As shown in Fig. 10.11, the rise in evaporation velocity near the contact line influences particle deposition. The figure captures the internal flow and particle movement near the black dashed line in Figs. 10.9b and 10.10. Prior to the onset of CCR mode, many particles remain trapped in Marangoni-induced vortices.

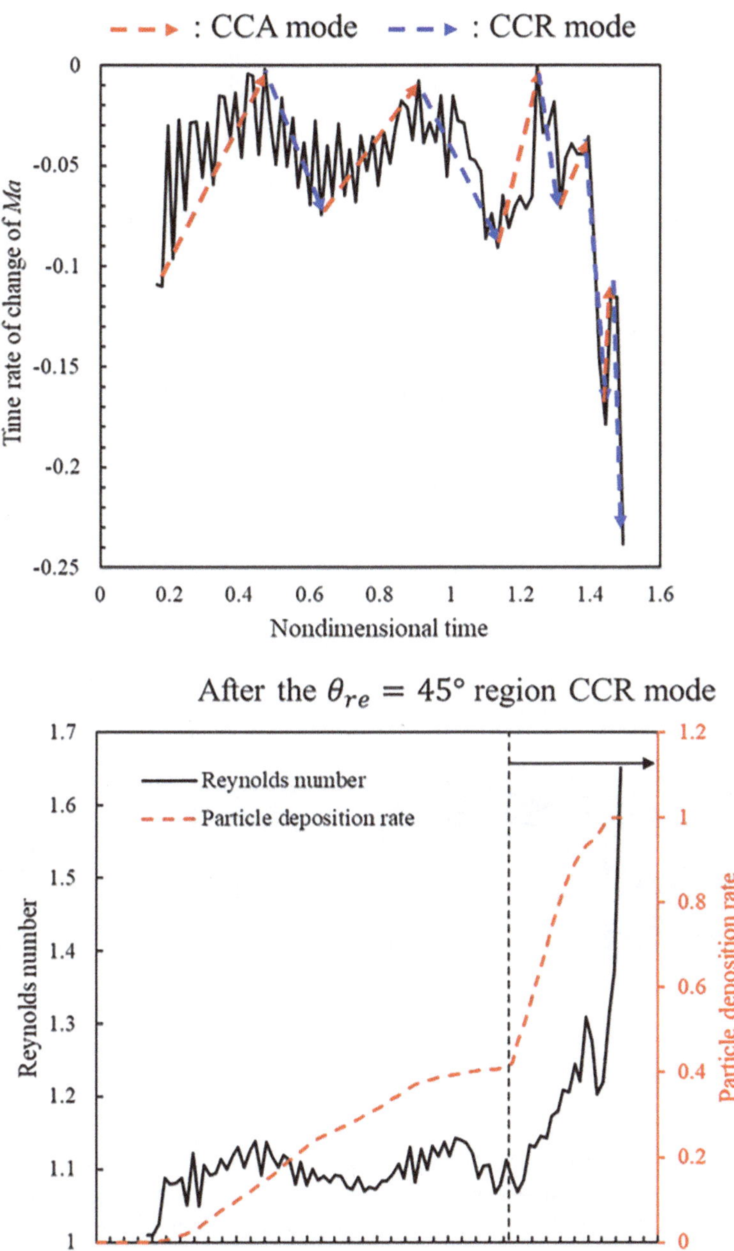


**Fig. 10.9** (**a**) Temporal variation of the Marangoni number (*Ma*). (**b**) Time profiles of the Reynolds number (*Re*) and particle deposition rate. The black dashed line marks the moment when the contact line reaches the region with a receding angle of 45°

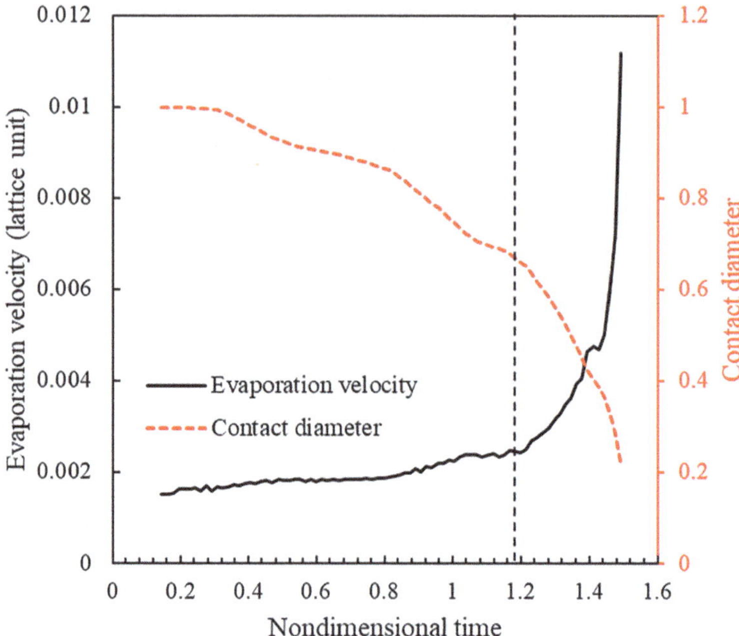


**Fig. 10.10** Time evolution of parameters contributing to *Re*. A marked rise in evaporation velocity is observed after crossing the region with a 45° receding angle, indicated by the black dashed line

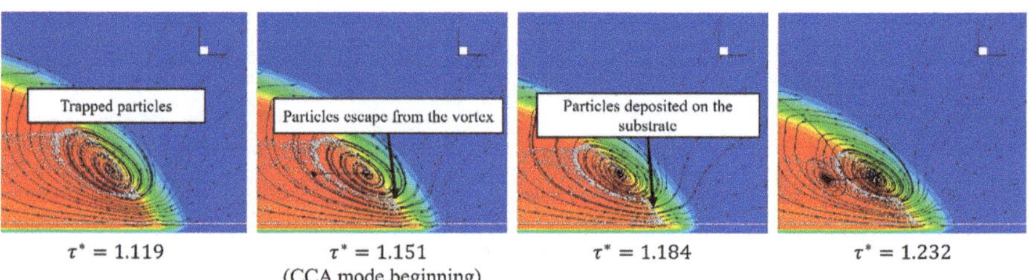


**Fig. 10.11** Flow field and particle distribution inside the droplet within the 45° receding angle region. Particles are visualized in white

However, upon entering CCR mode, strong evaporation at the contact line allows these particles to break free from the vortices. They are then carried outward and deposited at the contact line, leading to a sharp increase in deposition.

In conclusion, the patterned substrate induces the formation of multiple rings, leading to a more uniform particle distribution. This behavior is illustrated schematically in Fig. 10.12. The internal droplet flow varies depending on whether the contact line is pinned or advancing. When fixed, capillary flow dominates toward the contact line, enhancing particle deposition. In contrast, a moving contact line induces Marangoni-driven vortices that trap particles and suppress deposition. On a

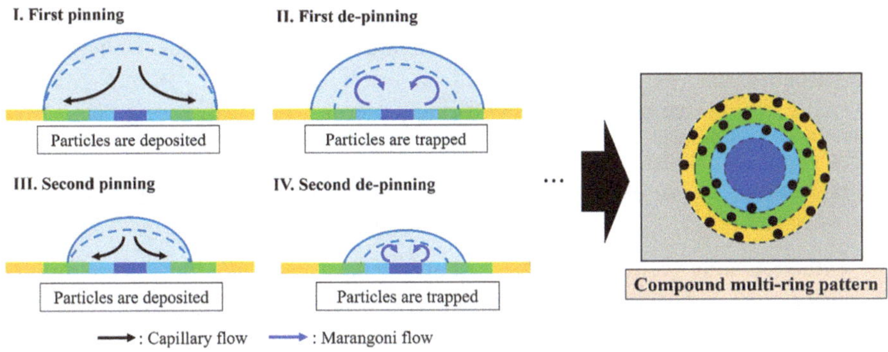


**Fig. 10.12** Mechanism of particle deposition on a patterned substrate

patterned surface, this alternating promotion and suppression of deposition results in successive ring formation. Consequently, particles accumulate at the boundaries where the contact line becomes pinned, as seen in Figs. 10.4 and 10.7.

**Exercise 10.3 Multi-ring Formation Mechanism**

(a) Explain why the Marangoni number decreases as the droplet moves inward across patterned regions.
(b) Compare the flow regimes where capillary flow dominates versus where Marangoni flow dominates.
(c) Predict deposition outcomes if Marangoni convection remained strong throughout the entire evaporation.

**Solution**

(a) As the droplet moves inward across patterned regions, the lateral temperature gradients along the interface are reduced because the droplet becomes smaller and the heating/cooling pattern becomes more uniform over its footprint. The characteristic length $L$ also decreases with time. Meanwhile, viscous and diffusive time scales do not decrease as rapidly.

Since $Ma \propto (\partial\sigma/\partial T)\Delta TL/(\mu\alpha)$, the combined reduction of $\Delta T$ and $L$ leads to a decrease of $Ma$. Therefore, Marangoni convection is strongest in the early, larger-radius stages and weakens as the droplet shrinks and moves inward.

(b) Marangoni-dominated regime (high Ma):

- Strong surface tension gradients drive vigorous recirculating vortices inside the droplet.
- Flow can reverse direction relative to simple outward capillary flow, transporting particles both outward and inward.
- Particle deposition can become more uniform or even biased toward the center, with coffee-ring patterns suppressed.

Capillary-dominated regime (low Ma, significant Pe):

- Evaporation-induced capillary flow from the center to the edge dominates.
- Particles are advected primarily toward the contact line, where they accumulate.
- Classical coffee-ring-like or multi-ring patterns form, with strong peaks near pinned regions.

(c) If Marangoni convection remained strong (high Ma) during the entire evaporation:

- Recirculating vortices would persist even in late stages, continuously remixing particles and preventing them from being irreversibly trapped at the contact line.
- Radial outward flux would be partly balanced by Marangoni-driven inward motion along parts of the interface.
- Deposition would likely be much more uniform, with weaker rings or even a central "bump" depending on the specific flow topology.

In other words, the coffee-ring and multi-ring features would be suppressed, and the patterned substrate would primarily modify the internal vortex structure rather than create distinct ring peaks.

**Exercise 10.4 Pattern Optimization and Region-Width Effects**
(a) Explain why Cases 34, 16, and 42 yield the highest uniformities in Fig. 10.6.
(b) Summarize the key design principles suggested by the optimized cases.

**Solution**
(a) Figure 10.6 shows that Cases 34, 16, and 42 achieve the highest uniformity index $U$. These cases share two key features:

- Balanced wettability sequence: They combine sufficiently hydrophilic outer regions (to control the outer ring) with moderately hydrophobic inner regions (to avoid central over-accumulation). This balance prevents both a dominant outer ring and a strong central peak.
- Appropriate region widths: The widths of the patterned zones are chosen so that no single region monopolizes the pinning time. Each region contributes to deposition without creating excessively sharp peaks.

As a result, particles are distributed over several zones with comparable density, producing a smoother radial profile and higher uniformity.

(b) From the optimized cases, two main design principles emerge:

- Outer control, inner smoothing
  - Use an outer hydrophilic zone to capture and weaken the classical coffee-ring at the perimeter.
  - Follow with inner zones whose wettability and width are tuned to redistribute particles and smooth out density peaks.

- Distributed pinning, not single dominant pinning

- Ensure multiple pinning sites with comparable importance rather than one overwhelmingly strong pinning radius.
- This distributes deposition across several rings of moderate intensity rather than concentrating it into a single sharp ring.

## 10.4 Conclusions

This chapter presented a detailed numerical investigation into particle deposition behavior in evaporating sessile droplets using a radial-band-patterned substrate. A pseudopotential lattice Boltzmann framework was developed, incorporating thermal phase-change dynamics and nanoparticle motion governed by hydrodynamic, capillary, Brownian, van der Waals, and drag interactions. GPU-based parallelization was employed to efficiently simulate large droplet systems and explore a wide range of pattern configurations.

The results confirmed that radial patterns of receding contact angles modulate the dynamics of the contact line through repeated pinning and depinning, which in turn shape the internal flow structure and affect deposition outcomes. Compared to uniform substrates, patterned substrates generated multiple ring formations and significantly improved particle distribution uniformity—up to 6.57 times higher. Among 51 configurations, the highest uniformity was achieved when the outer pattern widths were carefully tuned to balance deposition timing and internal flow evolution. Flow-field analysis revealed that as the contact line progresses inward, Marangoni convection weakens while capillary-driven flow intensifies, especially near regions with a 45° receding angle. This transition releases particles from vortex traps and enhances deposition along the pinned boundary. The competition between capillary and Marangoni flows was further quantified using time-resolved Marangoni and Reynolds numbers.

These findings provide physical insight into how substrate wettability patterns regulate deposition mechanisms in drying droplets. The proposed approach offers a practical route to suppress coffee-ring effects and tailor particle distributions, with relevance to precision applications such as inkjet printing, bio-patterning, and functional coatings.

**Summary**

1. **Motivation and conceptual shift from Chap. 9**

   - Chapter 9 controlled particle adhesion directly via interparticle forces (DLVO + magnetic field).
   - This chapter instead manipulates internal droplet flow by engineering the contact line dynamics through radial wettability patterns.

- By spatially varying the receding contact angle, the droplet alternates between constant contact radius (CCR) and constant contact angle (CCA) modes, producing multiple flow regimes.

2. **Patterned wettability modeling**
   - Substrates contain concentric radial zones with stepwise receding angles (e.g., 15° → 30° → 45° → 75°).
   - Contact line remains pinned in regions of low receding angle, then depins when the local apparent angle drops below the next region's receding angle.
   - This enforces periodic CCR–CCA transitions and generates flow-field oscillations aligned with substrate patterns.

3. **Deposition behavior on uniform wettability substrates**
   - Low receding angle (15°) → long pinning → classical coffee-ring.
   - Mid receding angle (45°) → mixed deposition.
   - High receding angle (75°) → strong coffee-eye central peak.
   - A single fixed receding angle cannot achieve uniform distribution.
   - Motivates the need for spatially varying wettability.

4. **Equally spaced patterned substrates**
   - Radial receding-angle patterns produce multiple deposition rings instead of a single dominant ring.
   - Particle density exhibits several peaks aligned with CCR–CCA transition radii.
   - Uniformity index shows patterned substrates outperform uniform ones by large margins.
   - Highest uniformity occurs in substrates with multiple but not excessive pinning–depinning events.

5. **Optimization of region widths (51 substrate designs)**
   - 51 designs varying the widths of each receding-angle zone.
   - Best-performing cases (16, 34, 42) achieve uniformity up to 6.57× higher than uniform-15° substrates.
   - All high-uniformity cases share:
     - Wide outer region (where most early deposition occurs).
     - Increasing region widths toward the center, maintaining balanced pinning despite shrinking droplet size.
   - Shows that outermost region geometry strongly influences final deposition.

6. **Flow-field evolution: Marangoni vs. capillary convection**
   - Marangoni convection
     - Marangoni number (*Ma*) oscillates with each CCR–CCA transition.
     - Ma decreases as droplet shrinks due to reduced surface temperature gradients and characteristic length.
     - Strong in early stages but weakens inward.
   - Capillary convection
     - Reynolds number (*Re*) rises sharply when the droplet reaches the 45° region.
     - Increase in *Re* corresponds with increased evaporation flux near the contact line.
     - Capillary flow eventually dominates, transporting accumulated particles outward and forming a ring.
7. **Mechanistic understanding of multi-ring formation**
   - During CCR: strong capillary flow pushes particles outward → ring deposition.
   - During CCA: strong Marangoni vortices trap particles → suppressed deposition.
   - Patterned wettability creates alternating promotion and suppression cycles, producing multiple rings.
   - This alternating flow field is the key mechanism behind enhanced uniformity.

## References

1. H. Ding, and P. D. Spelt, "Wetting condition in diffuse interface simulations of contact line motion," Physical Review E—Statistical, Nonlinear, and Soft Matter Physics **75**, 046708 (2007).
2. N.-A. Goy, N. Bruni, A. Girot, J.-P. Delville, and U. Delabre, "Thermal marangoni trapping driven by laser absorption in evaporating droplets for particle deposition," Soft Matter **18**, 7949-7958 (2022).
3. M. E. Shanahan, "Simple theory of "stick-slip" wetting hysteresis," Langmuir **11**, 1041-1043 (1995).
4. Y. O. Popov, "Evaporative deposition patterns: Spatial dimensions of the deposit," Physical Review E—Statistical, Nonlinear, and Soft Matter Physics **71**, 036313 (2005).
5. R. Bhardwaj, X. Fang, and D. Attinger, "Pattern formation during the evaporation of a colloidal nanoliter drop: A numerical and experimental study," New Journal of Physics **11**, 075020 (2009).

Zeitfracht Medien GmbH
Ferdinand-Jühlke-Straße 7
99095 Erfurt, Deutschland
produktsicherheit@kolibri360.de